For Pamela, Lawrence and Howard;
and for Mary, Courtenay and Christopher.

Contents

America

Advanced

Geography

Ian Price and Paul Guinness

Hodder & Stoughton

A MEMBER OF THE HODDER HEADLINE GROUP

The authors and publishers would like to thank the following for permission to reproduce copyright material in this book:

American City and Country, Figure 12.3.9; *The Atlanta Journal*, Figure 10.3.6; *Canada Focus*, Figure 8.1.4; *Canada Year Book, 1994*, Figures 5.1.7, 10.1.5 and 12.1.4; *CQ Researcher* Figure 3.1.13; *The Daily Mail*, Figure 12.6.3; *The Economist*, Figures 3.1.11, 3.2.13, 4.3.5, 6.3.6, 6.4.3, 8.1.5, 9.4.2, 10.2.1, 10.2.4, 11.4.8 and 13.2.8; *Environment* Figure 5.1.4; *The Guardian*, Figure 6.4.2; *Harpers Magazine*, Figure 5.3.5; *New Scientist*, Figures 5.6.5, 5.7.5 and 5.7.6; *The New York Times Magazine*, Figure 5.5.5; *US News and World Report*, Figure 12.6.1.

The authors and publisher would like to thank the following for permission to reproduce copyright photographs in this book:

AeroGraphics Corp., Figure 4.1.8; J Allan Cash, Figures 5.3.3 and 9.3.2; Annapolis Royal Tidal Plant, Figure 5.7.9; Communications Partnership, Figure 11.3.2; Corbis, Figures 1.2.4, 2.2.1 and 5.5.4; Environmental Images, Figure 7.3.1; Gamma, Figures 9.2.9, 9.4.3, 9.4.4, 9.4.6 and 10.5.4; Hutchison Library, Figures 5.5.3, 7.1.6 and 9.1.4; Life File, Figures 1.2.6 and 4.5.6; Magnum, Figures 5.4.9 and 9.2.8; MARTA, Figure 10.3.2; Mary Evans Picture Library, Figure 4.3.1; Per Briehagen, Figure 6.2.8; Phern, Kemp & Associates, Figure 4.3.7; Pictor, Figure 5.7.3; Pictures Colour Library, Figures 4.1.9, 4.5.5, 6.2.2, 6.3.1, 11.3.3, 11.4.3 and 12.5.11; Planet Earth, Figure 12.5.16; Popperfoto, Figure 2.2.2; Shell International, Figure 5.4.3; St Lawrence Seaway Authority, Figures 10.6.2 and 10.6.3; Science Photo Library, Figures 1.2.2, 7.2.3, 9.3.3 and 12.3.3; Spectrum Colour Library, Figure 11.4.5; Telegraph Colour Library, Figures 6.3.5, 10.3.1 and 12.4.3; Tony Stone, Figures 11.1.2, 11.3.4 and 12.5.5; Trip, Figures 3.4.4, 4.3.4, 5.2.1, 5.4.6, 5.7.4, 6.2.4, 10.4.1, 10.4.4, 12.3.8, 12.5.3, 12.5.6 and 12.6.5; Zefa, Figures 6.2.6 and 7.1.2.

All other photos belong to the authors.

Inside artwork by Philip Ford Design and Illustration and 1–11 Live Art.

Every effort has been made to contact the holders of copyright material, but if any have been overlooked, the publishers will be pleased to make the necessary alterations at the first opportunity.

Brian Price BA, MSc, PhD is a Senior Lecturer
in American Studies at the Crewe and Alsager Faculty,
Manchester Metropolitan University.

Paul Guinness MSc is Head of Geography at
King's College School, Wimbledon.

British Library Cataloguing in Publication Data

Price, Brian
   North America: an advanced geography
   1. North America
   I. Title   II. Guinness, Paul
   970

ISBN 0 340 62109 5

First published 1997
Impression number   10 9 8 7 6 5 4 3 2 1
Year   2002 2001 2000 1999 1998 1997

Typeset by Wearset, Boldon, Tyne and Wear.
Printed in Great Britain for Hodder & Stoughton Educational, a division of Hodder Headline Plc, 338 Euston Road, London NW1 3BH by Redwood Books, Trowbridge, Wiltshire.

Preface

The focus of this book is on place, space and environment set firmly in the context of 'real world' issues and problems. It has been designed to embrace and exemplify the latest developments in the subject within the geographical area of the USA and Canada by examining a wide range of topics and relevant case studies, many of which accentuate the interaction between people and the environment.

The authors are British geographers who have long held a special interest in North America, due initially to a combination of personal travel and participation in the 'Advanced Geography of the United States' MSc course at Birkbeck College, University of London. Both of the authors have previously contributed to the literature on the subject, and both remain active teachers about North America.

The themes – presented in an attractive format of text, maps, photographs, statistical data, integrated questions and other resources – have been carefully selected to meet the demands of A-level and similar syllabuses which have recently been revised. The book will therefore be of particular value in sixth-form and college geography courses and wherever geography plays a role in American/North American Studies. Moreover, its contents will usefully complement investigations of other advanced regions such as Western Europe and Japan, and serve as a contrast to parallel studies of the developing world.

The continent is introduced by presenting its principal physical characteristics as the natural arena in which the human drama is set, and to illustrate how nature offers North Americans both opportunities and constraints. Then follows an historical section which provides a chronological basis for understanding the human developments that shaped the evolution and destinies of Canada and the United States as distinct – and distinctive – nations. A series of themes and topics exemplified by carefully selected case studies then forms the main body of the book, while reviews of socio-economic disparities within the two countries and of their relationships with the wider world serve as a conclusion.

The physical, historical and contemporary links between the USA and Canada are intense. This will be one of the underlying themes of this book. The strength of the relationship was aptly summarised by President John F Kennedy in a speech to the Canadian Parliament in 1962 when he said 'Geography has made us neighbours. History has made us friends. Economics has made us partners. And necessity has made us allies.'

Teachers and students will find much of interest and value in the book's maps and other illustrations but are advised to have access to an atlas with good coverage of North America. Other useful aids include relevant news, current affairs and documentary programmes broadcast on radio and TV. Better still, readers should themselves visit Canada and the United States if at all possible. Indeed, if this book helps to provide travellers with insights into, and an informed understanding of the lands and peoples of these fascinating countries, then a wider objective will have been achieved.

1 The Physical Background

Country	Area in millions of km²
Russia	17.1
Canada	10.0
China	9.6
USA	9.4
Brazil	8.5
Australia	7.7

Figure 1.1.1
Countries with the largest land areas

1.1 Boundaries

Canada and the USA are the second and fourth largest countries in the world in terms of land area (Figure 1.1.1). Canada crosses six time zones. Its greatest east-west expanse is 5500 km from Cape Spear, Newfoundland to the Yukon-Alaska border. The maximum north–south distance is 4600 km from the southern border with the US to the northern tip of Ellesmere Island where Cape Columbia is only 768 km from the North Pole. Canada (Figure 1.1.2) is more than 40 times bigger than the United Kingdom, and is the only nation in the world bounded by three oceans, the Atlantic, Arctic and Pacific. Not surprisingly it has the world's longest coastline at almost 244 000 km.

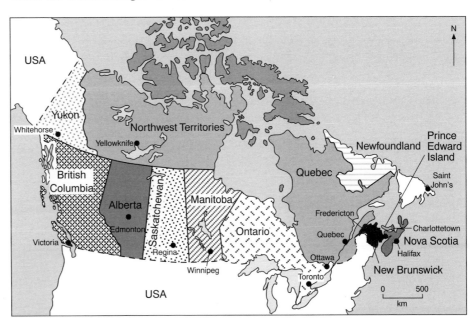

Figure 1.1.2
Canada's provinces and territories with their capitals

The main United States–Canada border is at latitude 49°N. To the east it follows the Great Lakes and, for a relatively short distance, the St Lawrence River. The border then runs south of the St Lawrence, reaching the Atlantic Ocean at about 45°N.

The 48 contiguous states of the USA stretch from the 49th parallel to below 25°N at the Florida Keys. The two non-contiguous states are Alaska and Hawaii. Nearly one-fifth the size of the rest of the United States, Alaska is the largest state in the Union. At its extremity, Alaska extends beyond 70°N. Hawaii consists of a group of eight major islands and numerous islets, located just south of the Tropic of Cancer. Honolulu, the capital of Hawaii, is 3220 km from San Francisco, the nearest city on the US mainland.

To the east, America is separated from Europe by the Atlantic Ocean which is only 2900 km wide between Ireland and Newfoundland, but between Spain and Florida it widens out to double this distance. To the west the continent is separated from Asia by the Bering Strait which is only 58 km wide. Here the two continents have the appearance of being linked together by the Aleutian Islands. These islands stretch in a long festoon westward from Alaska to the Kamchatka Peninsula in Asiatic Russia.

Apart from its 50 states, the US has a federal district (the District of Columbia) in which the capital city, Washington, is located. Like Canada's provinces the US states have considerable decision-making powers. They also vary a great deal in size from Alaska (1 518 776 km²) to Rhode Island (3144 km²). In general, the largest states are in the west. To make regional patterns clearer, the 50 states are often grouped into nine official divisions (Figure 1.1.3). Many of the data in this book refer to these divisions.

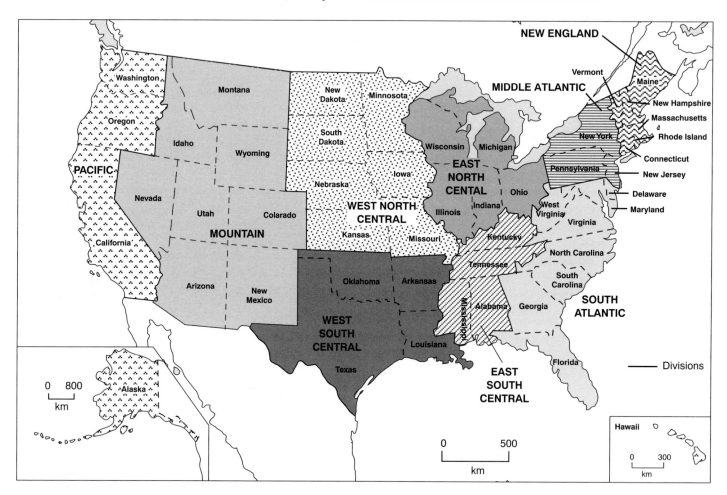

Figure 1.1.3
States and divisions of the US

The United States has possession of various island territories in the Caribbean Sea and the Pacific Ocean. Some of them, such as the Virgin Islands and Guam have a great degree of self-government. Puerto Rico is a commonwealth associated with the United States that has been given wide powers of self-rule by the US Congress. American Samoa, Guam, Puerto Rico, and the Virgin Islands each send to Congress a representative who may vote in committees but not in plenary sessions.

Canada is a federation of ten provinces and two territories. Ottawa, located on the boundary between Ontario and Quebec, is the capital city. Under Canada's system of government the provinces make many major decisions for themselves, particularly in education, local government, direct taxation, property and civil rights. In the north the huge Yukon and Northwest Territories have very small populations and are directly ruled by the federal government in Ottawa. They may become provinces at some time in the future with the agreement of the federal government and the existing provinces.

1.2

1. The Laurentian shield
2. The Hudson Bay lowlands
3. The Appalachian highlands
   (a) The southern section
   (b) New England section
   (c) Maritimes–Newfoundland extension
4. The Atlantic and Gulf coastal plains
5. The Interior Plains
   (a) The Mississippi–Great Lakes section
   (b) Interior low plateaux
   (c) The Great Plains
6. The Ozark–Ouachita highlands
7. The Rocky Mountains
8. The high plateaux
   (a) The Colorado plateau
   (b) The Great Basin
   (c) The plateaux of the Columbia and Snake
   (d) The plateaux of British Columbia and the Northwest
9. The Pacific coastlands

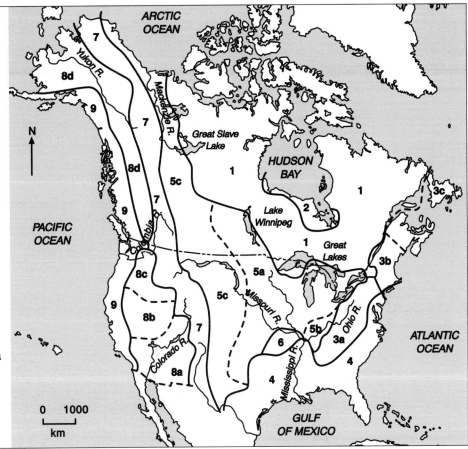

Figure 1.2.1
Physiographic regions of North America

Figure 1.2.2
The Laurentian Shield

The Laurentian Shield

The Shield (Figure 1.2.1) is a huge horseshoe-shaped area of ancient rocks which forms the core of the continent. Forming more than half of Canada, it stretches 1900 km from the Arctic coast to the Great Lakes and 2900 km from Labrador almost to the mouth of the Mackenzie River. The Great Lakes lie along the southern boundary of the region while the St Lawrence River skirts its south eastern margin. Another series of large lakes – Winnipeg, Athabasca, Great Slave, and Great Bear – marks the western boundary of the Shield. The region is almost entirely within Canada but there are two extensions into the United States. These are the Adirondack Mountains of northern New York State and the Superior Upland in northern Wisconsin and Michigan.

From above, the Shield has the appearance of a plateau interrupted by rounded hills, numerous lakes and swamps (Figure 1.2.2). In the east the average height is 500 m but much of the region, which generally slopes towards the shores of Hudson Bay, is less than 300 m above sea level. Strong relief occurs only where uplift has been greatest or where the Shield terminates in an abrupt rim. It rises sharply to over 700 m above the St Lawrence Lowlands and attains heights of over 1600 m in the Adirondacks and in the Tomgat Mountains of Labrador. The Shield rises spectacularly to over 3000 m in Ellesmere Island.

Geological evidence in the ancient Precambrian igneous and metamorphic rocks, indicates that parts of the Laurentian Shield were once extremely rugged and mountainous. However, hundreds of millions of years of erosion and weathering have reduced the surface to its present elevation. The Shield is one of the great stable blocks of the earth's crust. In particular, it has acted as a buffer zone against which subsequent fold mountains have been formed.

The surface pattern has been greatly complicated by Pleistocene glaciation, which will be examined later in this chapter. The ancient rocks contain valuable mineral deposits including silver, gold, iron, nickel and copper which are of importance to the Canadian economy. However, the Shield is almost entirely without agricultural potential.

The Hudson Bay lowlands

This relatively small physical region is sometimes considered along with the Laurentian Shield but its almost horizontal strata are of more recent formation. In structure and relief, these lowlands are very similar to the Interior Plains.

The Appalachian highlands

The eastern side of North America is dominated by the Appalachians, the remnants of an ancient fold mountain range of Palaeozoic age, which have been much reduced in height since their formation.

The region trends from north north east to south south west, from Newfoundland to central Alabama. In general the eastern areas of the Appalachians are more rugged than the western; Mt Mitchell (2039 m) in the Blue Ridge Mountains is the highest peak.

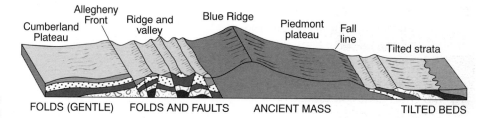

Figure 1.2.3
Simplified block section of the Appalachians

The Appalachian highlands can be divided into four belts (Figure 1.2.3) lying longitudinally and running parallel to each other. However, not all of the belts extend the entire length of the system. From east to west these belts are:

- the Piedmont Plateau stretching from Alabama to New York, with a relatively small outlier in New England. This is a gently sloping dissected plateau of severely eroded rocks. It rises gradually from an altitude of 150 m at the Fall Line, its junction with the coastal plains, to between 350 m and 450 m where it merges with the wooded mountains. The plateau is 200 km across at its widest point in North Carolina;
- the inner western margin of the Piedmont is formed by the Blue Ridge, a bold forested ridge, linear in the north but to the south becoming the tangled upland mass of the Great Smoky Mountains. This belt extends into New England as the Green and Hoosac Mountains. Resistant rocks such as granite form a barrier that is virtually uninterrupted in the Great Smoky Mountains and the Blue Ridge, rising in a number of places to over 1800 m. However, the highlands subside northward, and in northern Virginia and in Pennsylvania the belt becomes low and discontinuous;
- in the Ridge and Valley province, folded and faulted Palaeozoic sediments have been etched by erosion into a series of parallel crests rising above incised river valleys. This belt is between 40 km and 125 km wide and through it run the north-flowing Shenandoah and the south-flowing Tennessee rivers with their tributaries;
- farther west, above the high steep scarp face of the Allegheny Front, lie the Appalachian Plateaux. The almost horizontal Carboniferous strata lie at a surface level of about 600 m on the Cumberland Plateau in Tennessee and

Kentucky, and rise to over 1000 m in West Virginia. The plateaux dip gently westwards to Lake Erie and the lower valleys of the Ohio and Tennessee rivers.

The Atlantic and Gulf coastal plains

The Atlantic coastal plain extends eastwards from the Piedmont to the Atlantic Ocean. Rivers that reach the Fall Line tumble down from the Piedmont to the lower coastal plains in a series of falls and rapids. This physical region ranges from a narrow strip of land in New England to a broad belt that covers much of North and South Carolina, Georgia and Florida. Numerous rivers, including the Hudson, Potomac and Savannah cross the plain and flow into the Atlantic Ocean. Bays, such as Chesapeake and Delaware cut deeply into the plain in some areas. These two large bays are broad river valleys drowned at the end of the Ice Age.

The Gulf coastal plain borders the Gulf of Mexico from Florida to southern Texas. It is crossed by many rivers flowing south into the Gulf. By far the largest is the Mississippi, which originates in the interior plains to the north.

The interior plains

This region occupies a huge expanse of land, stretching from the Appalachian highlands to the Rocky Mountains. The character of the plains is far from uniform throughout. In general they consist of undulating lowland in which there is one single break – the Ozark Mountains. The plains are drained by the huge Mississippi river system. The Mississippi which is approximately 4000 km long, attains its maximum width of nearly 2 km near Cairo (Figure 1.2.4). West of the Mississippi the plains slope gradually upward to the foothills of the Rockies. The western part of the region, known as the Great Plains rises from 610 m at the Missouri Escarpment to 1500 m where the Front Range of the Rockies rises abruptly (Figure 1.2.5). Understandably, this is by far the most important agricultural region in the country.

Figure 1.2.4
The Mississippi at Cairo, Illinois

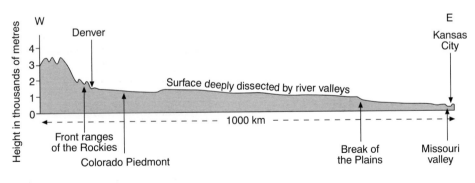

Figure 1.2.5
Cross-section of the Great Plains along latitude 40° North

The Ozark – Ouachita highlands

These highlands, rising respectively to 620 m and 810 m, form a scenic landscape in southern Missouri, north west Arkansas and eastern Oklahoma. Rivers have cut deep gorges through the rugged terrain. Much of the region has poor soil, but fertile land lies along the river valleys.

The Rocky Mountains

This is the major mountain system in North America, extending more than 4800 km from central New Mexico to north west Alaska. Mt Elbert (4399 m) in Colorado is the highest peak. The Rockies are located between the Great Plains on the east and a series of broad basins and plateaux on the west. The mountains form the continental divide, separating rivers draining to the Pacific Ocean from those draining to the Arctic and Gulf. Glaciers and snowfields, which today cover the northern ranges and the high peaks of the south, were at one time more extensive. Topographically the Rockies are usually divided into five sections: the Southern Rockies, Middle Rockies, Northern Rockies (all in the USA), the Rocky Mountain system of Canada, and the Brooks Range in Alaska. The Wyoming Basin, the system's principal topographic break, is sometimes considered a sixth sector. The Southern Rockies are the system's highest section and include many peaks over 4000 m.

Mining is important throughout the entire system. The Rockies are also a year-round recreational attraction as well as being a major barrier to overland transcontinental travel.

The high plateaux

The high plateaux lie to the west of the Rockies. They form a series of plateau steps, some of them almost as high as the Rockies and almost all above 900 m. The picturesque landscape is the result of volcanic activity, crustal faulting, and intense fluvial action. The plateaux of British Columbia and the Northwest have been intensely affected by glaciation. The far north of the region is occupied by the Yukon basin. The plateaux of the Columbia and Snake and the Colorado plateau take their names from the large rivers which cross them. The former are partly covered by gently folded basaltic lava sheets. The latter is characterised by horizontal sedimentary strata gashed by mile-deep canyons, the best known example of which is the Grand Canyon (Figure 1.2.6). The Great Basin is an extensive arid area of inland drainage between the Wasatch Mountains and the Sierra Nevada. Desert basins of recent sedimentation are separated by up-faulted ridges carved by arid erosion into spectacular landforms. The region includes the Great Salt Lake which occupies over 5000 km².

Figure 1.2.6
The Grand Canyon

The Pacific coastlands

In California, the Sierra Nevada and the Coast Ranges enclose the Central Valley of the Sacramento and San Joaquin rivers. This pattern is continued to the north, with the Willamette Valley–Puget Sound lowland separating the Cascades and Coast Ranges. The Cascades continue as the Coast Mountains in British Columbia. The system culminates in the Alaska Range with Mt McKinley (6178 m) forming the highest point in North America. The valleys of the Columbia, Fraser and Skeena rivers are the only low-level crossings in the entire system. In Canada the central trough has been drowned, with the extension of the Coast Ranges of the USA becoming the offshore island chain further north.

1.3 Glaciation in North America

Quaternary ice sheet

At its maximum, ice covered over 70 per cent of North America. It is likely that ice streams and ice caps, like those existing today in the Arctic, expanded, coalesced and then merged with an independent centre of ice development in the Ungava and Labrador plateaux of mainland Canada. This formed a proto-Laurentide ice sheet that developed and flowed south westward into central interior Canada, eventually becoming thickest over Hudson Bay. At about the same time, ice accumulations in the western mountains were creating the Cordilleran or Rocky Mountains ice sheet. The latter covered the northern intermontane zone and also flowed east to merge with the Laurentide ice to form a gigantic integrated ice sheet.

At its greatest extent the area covered by ice in North America was bigger than that covering Antarctica then and now. It stretched from the Pacific to the Atlantic and from the Mid-West of the USA to Ellesmere Island where, at times, it was joined to the Greenland ice sheet. In the south the ice formed a series of lobes, reaching almost to the junction of the Ohio and Mississippi rivers. In the east it reached as far south as Long Island.

The Laurentide ice sheet in general resembled the ice currently covering east Antarctica, which completely overwhelms the underlying topography. However, in the higher altitude Ellesmere Island–Baffin Island area it was more like that of the West Antarctic sheet, which largely but not completely buries an island archipelago and rough, mountainous terrain, so there were many projecting islands of bedrock (nunataks) and mountain ranges rising above the surrounding ice.

From the degree of depression caused in central Canada it has been estimated that the ice sheet reached a thickness of over 4500 m. Since the end of the Ice Age the land has been readjusting and rising. This has been at a rate of about 40 cm a century in the north of the Superior Basin and along the St Lawrence. When glacial rebound is complete, Hudson Bay will cease to exist as a significant body of seawater. It is largely a temporary relic of the depression of central Canada.

Some areas that seemingly should have been glaciated were not. Interior Alaska is an example. Its modestly elevated Yukon Plateau lay west of the north-west corner of the Laurentide sheet, south of a large ice mass on the Brooks Range and north of an extensive glacier complex in the Alaska Range. It was almost fully enclosed by large areas of ice on three sides, but it contained hardly any glacier ice as the area was simply too dry.

Landscape erosion

The dominant action of the ice over most of the area it covered was erosive. North of the Great Lakes and Lake Winnipeg there was scouring but very little deposition. Soil was removed, relief smoothed and a new drainage pattern created, dominated by a maze of lakes and swamps.

It is difficult to determine the magnitude of ice sheet erosion. Some estimates for the Laurentide ice sheet say that an average of only 10 m of soil and rock were removed from the land. This surprisingly low figure, based on the extent of drift deposits, may be partly due to the low relief and hard rocks of the Shield. Recently some investigators have proposed that the amount of fine glacial silt carried away by glacier-fed rivers and deposited in the oceans far exceeds the volume of drift deposited on land. On this basis they estimate that a rock layer averaging 120 m thick has been removed from the central area covered by the Laurentide ice sheet.

Glacial widening and deepening was considerable in many valleys in the western cordillera. The Kicking Horse Valley near Field and the upper Athabasca near Jasper are often quoted as classical glacial troughs, while Mount Assiniboine is an obvious pyramidal peak also featuring well developed cirques. The floor of part of Yosemite Valley in California is known from geophysical studies to have been overdeepened at least 450 m by glaciers, whereas in the highest parts of the Sierra Nevada mountains, great cirques with their confining walls over 500 m high were formed along with associated arêtes and pyramidal peaks. Good examples of hanging valleys, rock steps and ribbon lakes are in evidence in the region.

Glacial lakes

For reasons related to the nature and structure of the underlying bedrock and/or to preglacial topography, ice sheets can locally excavate closed basins of great size and depth. The large lakes that border the Shield, including the Great Lakes, the Great Bear Lake and the Great Slave Lake, owe part at least of their existence and much of the detail of their form to glaciation. The Great Slave Lake exceeds 600 m in depth and the Great Bear Lake, 400 m. Many other lakes have since been drained, among them the huge Lake Agassiz (Figure 1.3.1).

The Prairies rise in three steps towards the Rockies. Much of the first step was once the floor of this ancient glacial lake. When the ice-sheet retreated to the north, the lake was formed by the damming of meltwater between uplands to the south, the hills to the west, the Shield to the east and the ice-front to the north. At its maximum, it extended for over 250 000 km². Today, Lakes Winnipeg, Winnipegosis, Manitoba, Moose Lake and the Lake of the Woods are all that is left of this once more extensive water body. The remaining area, consisting of fertile lacustrine deposits, mainly in the form of clays, was in places subsequently covered with alluvium.

Because ice accumulation was so much greater in the north the inter-montane basins became completely covered. Farther south this was not so but glacial retreat did have similar effects when meltwater filled the inter montane basins. The Great Basin of Nevada and Utah became the bed of numerous lakes. The largest was Lake Bonneville of which the Great Salt Lake is a relatively small relic. The shoreline of the latter lies approximately 300 m below that of the former.

The form of glacial lakes varied as the ice-front retreated. Huge areas of water overflowed, creating channels, some of which now form important routeways. One is the Mohawk Gap which links the Great Lakes to the Hudson River, another carries a ship canal south from Chicago to the Mississippi. To the south the Great Lakes are a remnant of the Laurentian Lakes system, the bed of which now provides flat and fertile farmland in Manitoba, Minnesota and Illinois.

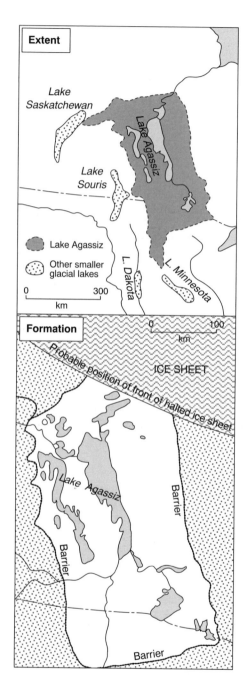

Figure 1.3.1
Lake Agassiz

Glacial deposits

In total 2.5 million km² were covered in drift which varied in thickness from 50 cm to over 40 m. Areas of till, due to direct glacial deposition, are generally extremely fertile. However, in the east, the Maritime provinces of Canada and the New England states were left a poorer legacy. The deposition of rocks and boulders over considerable areas has greatly impeded agricultural development. Fortunately the floors of ancient glacial lakes, the largest of which extended along the Connecticut valley, have produced corridors of good agricultural land. These contrast with terminal moraines, today represented by the islands of Nantucket and Martha's Vineyard. Considerable esker and kame deposits dot the landscape between the terminal moraines. Drumlins are also a prominent feature, particularly in south-east New Hampshire. Drumlins are also prominent further west. There are over 20 000 of them in a huge 'field' to the north of Lake Ontario. They are also numerous along the Great Lake shorelands of Wisconsin, Michigan and western New York. The shallow waters around Newfoundland and New England, the famous 'banks' also result from glacial deposition.

South of the Missouri, extensive areas were covered with outwash sands and gravels and smaller areas with fertile wind-blown loess. Because there were four main glacial periods separated by clear interglacials (Figure 1.3.2), glacial and interglacial deposits are often interleaved. Wide belts of loess were formed during interglacials, particularly in Iowa which is sometimes called the 'loess state'.

Name of Pleistocene	Years ago
WISCONSIN	9000– 67 000
Sangamon	
ILLINOIAN	128 000–180 000
Yarmouth	
KANSAN	230 000–300 000
Aftonian	
NEBRASKAN	330 000–470 000

Glacials are in upper case, inter glacials in lower case

Figure 1.3.2
Subdivisions of the Pleistocene in North America

Present-day glaciation

It is in the western Cordillera that the classic features of highland glaciation can be found. The Barnard Glacier of Alaska is a spectacular compound valley glacier made up of more than a dozen separate ice streams, illustrated by the number of parallel medial moraine strips on its surface. The Malaspina Glacier in Alaska's southeast region, at 70 km across is arguably the finest example of a piedmont glacier in the continent. Alaskans have clearly witnessed the power of ice in recent times. In 1986 the Hubbard Glacier in the St Elias Mountains hit the headlines when it surged forwards at an average rate of about 10 m a day, blocking the entrance to Russell Fjord. In 1983 the Variegated Glacier also surged to reach a speed of 65 m a day.

1.4 Climate

Air masses

The dominance of particular source regions, the characteristics of the air produced and the preferred paths of air-streams go far to account for the seasonal and spatial variations in climate in North America.

To the east of the Rockies two air masses are dominant. Polar, or arctic continental air moves south 'en masse' from its source region in northern Canada or over frozen Arctic waters. In winter such air brings very cold, sometimes bitter weather to the northern and, to a lesser extent, the central interior of the continent. At times the danger of frost can penetrate to the Gulf coast. In summer, polar continental air brings clear but cool conditions. The tropical maritime air mass originates in the Gulf of Mexico and in the Atlantic Ocean to the southeast of Florida. In winter it brings mild spells and rain or fog to the cold interior. However in summer this air mass is characterised by high temperatures and considerable humidity over much of eastern North America. It is moisture, transported by these southerly or southeasterly air streams, which accounts for most of the precipitation of the interior.

The boundary between these two major air masses changes during the course of the year. In summer it runs south of the Great Lakes, bringing alternating conditions to the Midwest. Conditions are cool with relatively low humidity when the northerly air stream dominates, but warm and humid when tropical maritime air prevails. In winter the boundary shifts south towards the Gulf coast so that the interior is dominated by cold air; interspersed by occasional warm breaks brought by the southerly air stream. However fog and dangerous thaws can occur. In spring and autumn the boundary moves between the two extremes. Depressions form along this zone of conflict, moving across southern Canada and the United States from southwest or west to the east.

Tropical continental air forms in a source region straddling the US–Mexican border. This air is always dry and in summer extremely hot.

The West is influenced by both polar maritime and tropical maritime air masses. The former affects the western coastline from Alaska to San Francisco, its depth of penetration varying with both latitude and season. On the central Californian coast polar maritime air arrives only during the winter when the major belts of wind and pressure have been displaced to the south. This is a moist air mass yielding abundant precipitation when it ascends the west-facing slopes of the Cordillera. In contrast the tropical maritime air masses coming from the southwest are relatively stable and consequently not great rain-bringers.

The north Atlantic coastline is affected by polar maritime air which originates off southern Greenland, extending on occasion as far south as Chesapeake Bay. This air mass brings lower temperatures and moderate precipitation in summer and cold and drizzle in winter.

Temperature

Winter

The influence of low temperature is very clear in the north. This largely explains why 75 per cent of Canada's population live within 160 km of the US border. Over two-thirds of Canada has a mean January temperature below −20°C (Figure 1.4.1). The far north is affected by permafrost. Here the ground is permanently frozen to a depth of about 300 m. In summer the top metre or so thaws out resulting in a marshy, waterlogged landscape.

The intense cold results partly from the spreading south of polar air masses over the vast open plains and partly from vigorous local cooling of the land surface. The interior provides a wide open passage from the Arctic due to the absence of an east–west barrier. It is possible to go from the extreme north to the Gulf without rising much above 300 m. Thus cities such as St Louis and Kansas City, in the same latitude as Lisbon and Athens have prolonged spells of below-freezing temperatures in mid-winter.

Hudson Bay is largely frozen in winter, causing polar ice to reach far into the continent. In contrast, temperature increases significantly near the Great Lakes. The latter cause a slight northward swing in the isotherms in October by their retention of summer heat. This effect becomes more pronounced until January, after which it diminishes, and in March the Lakes cease to provide warmth. The warmest part of Canada in winter, excluding the west coast, is the Lake peninsula of Ontario, which lies farthest south and is most favoured by the influence of the Lakes.

The isotherms swing farthest south in the Mississippi valley and trend northwards as they approach the east coast. However, here the cold Labrador Current ensures an ice-bound St Lawrence during the winter. Farther south where the Gulf Stream operates, the prevailing offshore winds minimise its warming influence.

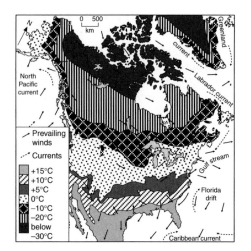

Figure 1.4.1
North America: average January temperatures

In the west the isotherms diverge from their general east–west pattern to run almost parallel to the coastline. Here the warm waters of the oceans are as effective a source of heat as direct insolation. The North Pacific Current is particularly effective in keeping the south coast of Alaska and ports such as Vancouver and Seattle ice-free in winter. However temperature declines rapidly inland when mountain ranges shut off the warm ocean conditions. While coastal temperatures in British Columbia compare to the British Isles in winter, the Fraser Valley, in the interior, experiences conditions closer to central Europe. Many different micro-climates are found in the western Cordillera but the high altitudes are intensely cold, windswept and often snow-covered. To the east of the Rockies the sudden onset of warm, dry Chinook winds can result in spectacular temperature increases. The most extreme change in temperature took place in January 1962 in Pincher Creek, Alberta when temperature increased from −19°C to 22°C in one hour!

The warmest winters in the continent are in southern Florida and the lower Colorado valley. Key West, off the south coast of Florida, is the only weather station in North America exempt from frost.

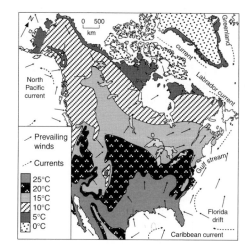

Figure 1.4.2
North America: average July temperatures

Summer

The difference in temperature between south and north is much less in summer than in winter, partly because the decrease in the sun's altitude with increasing latitude is compensated by the greater length of the day (Figure 1.4.2). In July the isotherms bend sharply poleward over the warm land. Denoting sea level temperatures, they run considerably farther north on the west than in the east since the lower incidence of cloud offers less obstruction to insolation. The highest temperatures are achieved in the lowlands of Arizona, and in the Colorado and Mojave deserts where figures comparable to the Sahara are recorded.

Ocean currents also have a significant influence in summer. On the east coast the cold Labrador Current moves south from Baffin Bay to Newfoundland. Its arctic waters and masses of ice are responsible for the very cool summers of the Labrador coast. Cold water creeps further south along the US coast but here its influence is not prominent, since it is narrow in extent and is overshadowed by the Gulf Stream. In summer the east coast of the USA benefits from hot and moist onshore winds. When these qualities are particularly pronounced they form 'heat waves', an unpleasant element in the climate of the eastern states.

The Gulf of Mexico, the source of part of the Gulf Stream, is always hot and the air over it charged with moisture, which explains in part the heavy rainfall of the south east. In winter the warm damp air of the Gulf is conducive to low pressures and favours the spread of 'cold waves' from the interior to its sub-tropical shore.

Precipitation

The highest annual precipitation totals (Figure 1.4.3) are near the Pacific coast where the warm westerly winds meet the mountains. Their ascent, together with frequent cyclonic activity results in considerable condensation, massive low cloud and high levels of precipitation. The cloud and precipitation provides welcome moisture for the coniferous forests especially in the neighbourhood of the US-Canadian border where the annual mean exceeds 2500 mm.

The intermontane plateaux are very dry, the double mountain barrier of the west robbing the westerlies of much of their vapour. The Rocky Mountains to the east also intensify the aridity. As in the west generally, the precipitation on the plateaux is heaviest in the north. The interior of Washington and Oregon and the west of Idaho have 250–500 mm, but Nevada and the Colorado basin receive less than 250 mm. Here the westerlies are less strong and the enclosing

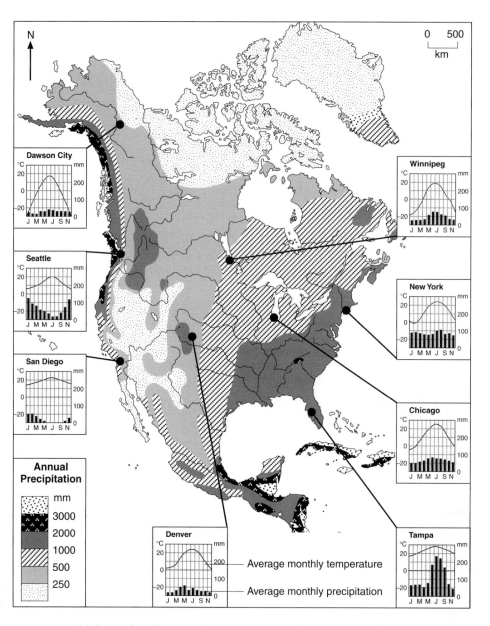

Figure 1.4.3
North America: average annual precipitation

mountains higher. Also the usual tendency to aridity in the sub-tropics asserts itself. The driest region is the lower Colorado basin which also receives a mean sunshine total of over 3500 hours, the highest in the continent.

Moving east, precipitation increases on the Rockies but surprisingly little for the altitude. A small area in British Columbia has over 1250 mm, but in the USA most of the range has less than 600 mm.

East of the Rockies precipitation again decreases on the Great Plains but is still greater than on the arid intermontane plateaux. Most of Canada east of the Rockies has below 110 days with precipitation, except the St Lawrence basin which ranges between 150 and 170 days. Most areas between the Rockies and 100°W in both countries have means between 250 and 500 mm, but actual totals are very variable from year to year.

From the 100th meridian precipitation increases towards the east and south. The southern Appalachians, which are within range of the moisture-bearing winds from both the Gulf and the Atlantic, receive over 1750 mm, the highest total for a large area in the USA excluding the Pacific coast. The Gulf coast exceeds 1500 mm between New Orleans and Mobile where heavy rainfall is associated with the hurricanes of late summer.

1.5 Natural Vegetation

Due to great physical and climatic contrasts within the continent, North America exhibits enormous variety in its natural vegetation (Figure 1.5.1). However the huge landscape impact of human activity in both the United States and Canada means that a map of natural vegetation must be viewed more as an historical document than a description of the contemporary landscape. The period of European settlement has witnessed forest clearance on a massive scale, the ploughing up of huge areas of grassland, the irrigation of considerable areas of dry land, and the spread of urbanisation.

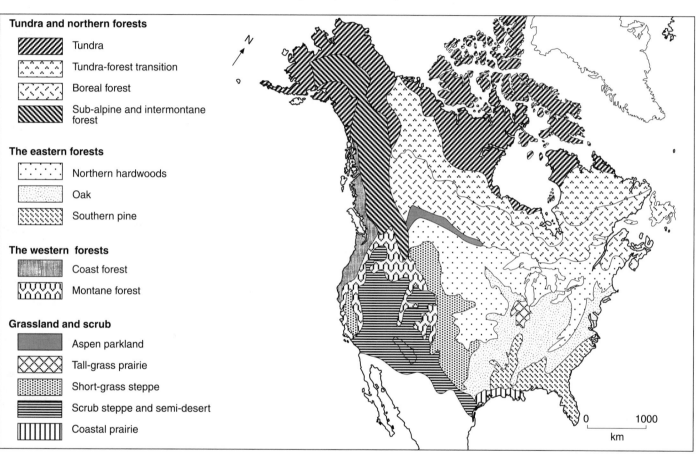

Tundra and northern forests

- Tundra
- Tundra-forest transition
- Boreal forest
- Sub-alpine and intermontane forest

The eastern forests

- Northern hardwoods
- Oak
- Southern pine

The western forests

- Coast forest
- Montane forest

Grassland and scrub

- Aspen parkland
- Tall-grass prairie
- Short-grass steppe
- Scrub steppe and semi-desert
- Coastal prairie

0   1000
km

Figure 1.5.1
North America: natural vegetation

Tundra covers a large area of northern Canada and Alaska where the intense physiological drought of the winter rather than the low temperature is the limiting factor for tree development. This is an area of permafrost supporting low forms of vegetation such as mosses, lichens, sedges and sporadic grasses along with a few miniature willows, birches and other trees tolerant of a harsh environment. The tree line generally follows the course of the 10°C July isotherm.

The Boreal forest region comprises the greater part of the forested area of Canada. It forms a continuous belt from Newfoundland and the coast of Labrador westward to the Rocky Mountains and northwest to Alaska. Among the dominant species are spruce, tamarack, balsam fir and various pines. Although the forests are primarily coniferous, there is a general admixture of deciduous trees such as poplar and white birch; these are important in the central and south-central regions, particularly along the edge of the prairie. Towards the north the close forest gives way to an open lichen-woodland which gradually merges into tundra.

In the north eastern United States and continuing into southern Canada is an area of hardwoods such as beech, maple and oak. Except for relatively minor areas of red alder, oak and aspen in the west, the bulk of the continent's hardwood is located in this region. Although the region underwent major exploitation because of the quality of its timber and relative accessibility, each hardwood inventory made by the US Forest Service since 1952 has shown an increase in volume due to improvements in forest management. Conifers can be found on poorer soils, on high ground, and in the northern parts of this zone.

After an oak–pine transition belt, conifers dominate the south east, forming the southern pine forests. This is potentially the most important timber resource region in North America where a warm and moist climate allows pines to mature in about 35 years.

In the west, forest cover significantly reflects contrasts in rainfall. On the wet coastal slopes of the Pacific ranges, fine stands of Douglas fir give way northwards to sitka spruce and southwards to the redwood forests of California. The redwood is the world's tallest tree at over 90 m high in some cases. In California too is the sequoia which grows on the western slopes of the Sierra Nevada. On the drier and often higher slopes of the Cascades, Sierra Nevada and Rocky Mountains are considerable areas of lodgepole and ponderosa pine. The treeline increases in altitude from north to south and in the latter part of the region fir and spruce can be found at altitudes of 3500 m.

The grasslands of the interior lowlands originally took two forms, the tall-grass prairie and the short grass steppe. The former dominated the wetter eastern part of the region, gradually giving way to the latter as precipitation declined to the west. Little of the original vegetation now exists on what is now one of the world's great agricultural regions.

The greater aridity of the intermontane plateaux results in degeneration to semi-desert scrub, dominated by saltbush, sage and creosote. True deserts are of limited extent, the largest being in north west Utah. In the 'deserts' of southern California and southern Arizona are found a variety of large cacti along with other drought-tolerant plants.

2 The Historical Perspective

The First North Americans

About 20 000 BC, Mongoloid mammoth hunters crossed the Bering Strait between north east Asia and Alaska, via a land bridge caused by falling sea level during the Ice Age. These were the first people to inhabit the Americas. Their descendants are the Indians and Inuit of today (Figure 2.1.1).

Figure 2.1.1
Indians in North America prior to European incursion

By AD 1 the area now covered by the USA and Canada had a population of about 300 000 people. This increased to 500 000 by AD 1000. The first Europeans to reach North America were the Norse people who landed on the Labrador coast at about this time. They were few in number and it is thought that their community was abandoned after just a few generations.

Between AD 1000 and AD 1500 the population of North America doubled (Figure 2.1.2). However from about 1500 onwards the continent was influenced by a new and powerful source: sustained migration, first from a few and then from many different countries, mainly European.

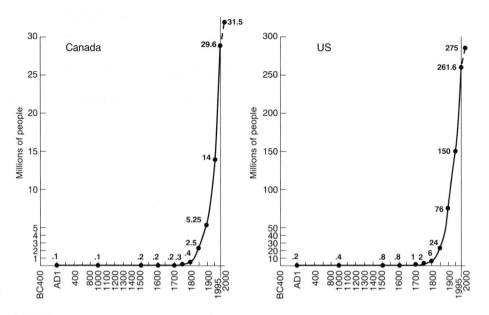

Figure 2.1.2
Population growth in Canada and the US

2.2 European Colonists

The Spanish in the South and West

Although the Spanish laid claim to a considerable part of what is now the USA they were not successful in settling it, and their interest was half-hearted once they decided gold would not be found there.

In 1492 Christopher Columbus landed at San Salvador in the Bahamas, heralding the beginning of the modern age of settlement in the Americas. Soon other explorers sailed westward to the New World; among them was Amerigo Vespucci whose name was used to describe the new lands. Within twenty years of Columbus' landing, much of the Caribbean area had been settled by Spaniards, who then turned to the exploration of the eastern and southern coasts of the mainland. Juan Ponce de Leon named Florida in 1513, and from then to the 1550s a series of expeditions landed and ventured into the interior, failing to find gold and succeeding only in leaving a trail of death through deliberate slaughter and from the European diseases to which the Indians were not immune.

In 1559 the Spaniards attempted to colonise Florida, but this and later attempts were unsuccessful until mission stations were planted after 1580. By 1650 the Franciscans controlled 26 000 Indians, but after that date there was a decline in recruitment from Spain and missions were abandoned in the face of an anticipated English takeover from the north.

In the west the Spanish occupation began with attempts to discover mineral wealth, but these soon petered out in failure. When no riches were found, the Spaniards developed the northern frontier area of New Spain (later to become Mexico) as a buffer zone for the main part of the country and reacted to threats (real or imagined) from outside. A movement to occupy the upper Rio Grande Valley led to the establishment of Socorro (1598) and Santa Fe (1610) and the setting up of farming amongst the local Indians. Missionaries were encouraged to move to Texas to try and convert the Indians. In 1718 the villa of San Antonio was founded and acted as a place of refuge during Indian attacks. However, when the Louisiana Territory became Spanish after 1763 little advantage was taken of this area and even Texas was neglected.

In the mid-eighteenth century Russians moved their fur-hunting activities down the north west coast of the continent causing the Spaniards to occupy California after 1767. A land expedition under Portola reached San Francisco and reputedly discovered the Bay for the first time. Missions were set up all along the coast including San Diego (1769) and San Francisco (1776).

The French in the North

While Spain led the initial phase of exploration, France and England were not far behind. Jacques Cartier was sent across the Atlantic by Francis 1 in 1534, exploring the Gulf of St Lawrence and penetrating up that river in a series of voyages to 1541 when an attempt to plant a colony proved unsuccessful. He misunderstood the Algonquin word 'kanata', meaning 'village' for the name of the whole country, which became 'Canada'. Cartier returned to France with news of the fishing grounds around Newfoundland and this stimulated a wave of seasonal expeditions from western Europe.

Contact with Indians brought access to beaver furs which were to provide a major trading item for the next 300 years. In 1603 the lower St Lawrence region was declared a French royal province, and a group led by Champlain was given monopoly rights in the fur trade for a short period. He and others explored deeply into the interior of the continent and eventually reached the Mississippi River. Champlain also established colonies at Port Royal in Acadie (1605) and Quebec (1608). The French, however, formed trading alliances with the Montagnais, Algonquin and Huron Indians, and became embroiled in the running feuds between these tribes living north of the St Lawrence and the Iroquois Federation to the south.

The fur trade remained the basis for the French presence in the area during the seventeenth century, and traders extended through the Great Lakes area to the Mississippi in the west and south to the Ohio River. The strategic portage points became the sites of forts to protect French trading interests and formed the initial settlements which grew into modern Detroit and Chicago.

Numbers of permanent settlers grew slowly until Louis XIV made New France a royal province in 1661. This move led to formal possession of the interior lands and extension down the Mississippi with the founding of the port of New Orleans (1717) as the base for the new colony of Louisiana, which extended across the entire interior of the present USA from the Appalachians to the Rockies. The move also encouraged further settlement, mainly along the St Lawrence from Montreal to the sea.

The British southern tidewater plantations

Although it had been visited by a number of explorers, the central section of the eastern coastline was unsettled before 1600. Sir Walter Raleigh had attempted to set up trading colonies on Roanoke Island in the 1580s but they were not successful and it was clear that the only way ahead would be to set up permanent, self-sufficient settlements. The Jamestown (Virginia) settlement was established in 1607 and began what became a great tide of immigration from Europe. The early travellers to the New World came in small overcrowded ships and during their six to twelve-week voyage they lived on meagre rations. Many of them died of disease and some were lost at sea.

The southern tidewater area consisted of low lying and swampy coasts and inlets extending over 100 km inland, especially around Chesapeake and Delaware bays. The soils in the valley floors were fertile, but those on the hills were thinner and acidic. The plentiful annual rainfall of 1000–1500 mm was matched by long warm summers. Indian burnings had removed much of the forest so the land was easy to cultivate. Game and fish were plentiful. And yet many of the first settlers nearly starved to death due to disease and the planting of wheat in soils with such a low nutrient content. Fortunately the Indians demonstrated that the burning of vegetation increased soil fertility and they also provided seeds of their own crop, corn (maize). John Rolfe grew tobacco experimentally in 1612 and realised that he had found the staple crop which would provide cash and trade with England. Virginia became a Crown Colony in 1624, although life was still far from certain; while 3500 immigrants arrived between 1618 and 1624, the total population grew by only 1000 due to disease and the start of the Indian attacks after 1622.

In 1630 the area north of the Potomac was granted to Lord Baltimore by Charles I as a place where Roman Catholics could live without prejudice. The first settlers of the Maryland colony in 1634 bought their land from Indians and immediately built a fortress on a high point. Both colonies (Virginia and Maryland) continued to expand by first occupying the land along the valley floors as far as the Fall Line, and then spreading over the interfluves between. Settlers came in good numbers, Puritans driven out of England by the harsh policies of Charles I, and later Anglicans driven out by Cromwell (1649–60). Policies were again reversed after the restoration of Charles II in 1660.

The abundance of land and shortage of labour was a major factor of economic life in the colonies at this stage; poor people were brought over to work for 4–7 years for the cost of their passage and 100 acres of land at the end. There was a constant need for such indentured labour although it was seldom of high quality.

Farther south there was more development in the Carolinas after a grant by the Crown of lands from Virginia to the 31st parallel in 1663. However development was slower in these southern areas, since tobacco did not do so well in the swampy coastal area or the barren sandy soils behind. A fur trade was established by the 1680s and Savannah became a centre for this and the collection of naval stores (e.g. pitch from pine trees). Rice was introduced in the 1690s.

The British in New England

Figure 2.2.1
Map of New England and surrounding area
(1614)

The British colonies in New England were quite different in purpose and form from those in the southern tidewater lands (Figure 2.2.1). The first settlers landed from the Mayflower in 1620 and their success encouraged others to follow. The settlement of Boston, founded by the Massachusetts Bay Company, was the greatest draw and 24 500 settlers arrived between 1630 and 1640. Many of the Puritans discovered that Boston was ruled in a rather restrictive manner and moved to Narragansett Bay, where Providence was founded in 1636. They also moved westwards along the coast to the better soils of the Connecticut valley (Hartford 1636) or northwards into New Hampshire (Exeter 1638).

Posts were also established inland to attract the fur trade.

Settlement was related to the distinctive New England system of land allocation. The predominantly Puritan groups wished to see a commonwealth peopled by men and women of like faith; they abandoned the individual basis of land ownership and gave the colonial legislature control over settlement. This method had the advantages of planned migration to new frontier towns, and communal defence against Indian attack, but some townships were slow to grow because they were regarded as unattractive.

In the later part of the seventeenth century the conquest of the interior led to more frequent clashes with the Indians and a major war in 1675–7, after which the Indian power in New England was largely broken.

Competition in the Middle Atlantic area

The lands between the Hudson River and Delaware Bay were occupied slightly later than the lands immediately to north and south, and had a more chequered start. Henry Hudson sailed up the river in 1609 and Dutch fur traders became the first permanent settlers. Fort Orange on the site of Albany was set up in 1624, and Fort Amsterdam on the southern tip of Manhattan Island in 1626.

Meanwhile Swedes had set up a colony on the Delaware River and built Fort Christina in 1638 and Fort Gothenburg in 1644. New Englanders also attempted to settle there, but the Dutch removed both groups by 1655. Squabbles between traders in the Hudson valley eventually led Charles II of England to order the capture of New Amsterdam and the Dutch lands. This was carried out in 1664, and there was a rapid phase of British settlement in the area. New Amsterdam was renamed New York after being granted to the King's brother, the Duke of York.

In 1681 William Penn, a Quaker, received the Delaware lands from the Crown as a payment of debt. He arrived in 1682 and selected the site for his main centre at Philadelphia. His colony was marked by liberty of religious expression and the granting of cheap lands. It became the main focus of settlement from England, Wales and Germany in the late seventeenth century.

The predominant subsistence farming outside the Chesapeake tobacco area was beginning to develop greater specialisations near the coast. At the same time the British West Indies had been occupied and developed for sugar plantations; their needs for food, horses and lumber could be supplied from the American colonies.

The colonies expand

After the initial phase of settlement and consolidation came the process of expansion into the interior. The years 1700–75 saw a major incursion westwards of the British colonies so that conflict with Indians, French and Spaniards became a major feature of the period. The tensions of this initial era of expansion also led to the revolution which established the United States of America as an independent nation.

The north eastern lands were the poorest, but were occupied from the coastal lowlands as the growth of population led to overcrowding, while others wished to escape the restrictive form of township organisation. Speculative wealthy merchants bought up the best lands ahead of settlement and then cashed in by attracting settlers to their newly acquired lands. This process resulted in rapid settlement in the valleys leading north to the hill country, but it also changed the character of New England. Settlements became more scattered and exposed as townships were created more rapidly than could be filled with people. Many of the new settlers came from the older parts of New England, but they also included new Ulster Protestant immigrants, settling especially around Worcester in central Massachusetts in 1718. The group settling west of the

Connecticut river spilled over into the Hudson Valley, causing tenant wars and leading to the establishment of the New York–Massachusetts boundary 20 miles east of the Hudson River.

The southern piedmont had been settled from the 1650s onwards as expeditions moved out from the Virginia forts. There was also a major move westwards in the early eighteenth century by farmers into the rolling hills of the piedmont area, with its rich, residual red soils, as many smaller coastal plantations were crowded out. Immigrant Germans also joined this movement and by 1750 the entire Virginia and Carolina piedmont was settled.

The major development of this period however took place through Pennsylvania, where the tolerant Quakers opened their lands cheaply to anyone who would come. They also maintained good relations with the Indians, and William Penn bought large areas of land from them in the Susquehanna valley. This was a contrast to the areas north and south; in New York for instance, settlement in the Hudson Valley was discouraged first by fur traders and then by Indian troubles, and the colony placed many difficulties in the way of land acquisition.

The early eighteenth century saw the influx of two major groups of settlers, mainly to Pennsylvania. The Palatine Germans arrived first in New York in 1708, but soon moved to land along the Hudson and Schoharie rivers. Various problems caused them to move to Pennsylvania and from 1710 groups emigrated from Germany via England direct to Pennsylvania and on to the Susquehanna valley. In 1726 there were 100 000 squatters in addition to those who had bought land. These became known as the Pennsylvania 'Dutch' by mistake, but soon dominated the fertile lands around Lancaster. Settlers moved into the lands of the Great Valley to the south east, entering Virginia and the Carolinas by the 1750s. Some moved back east through gaps in the Blue Ridge to fill in unsettled piedmont lands.

The second major migration was by the Ulster Protestants from the northern part of Ireland. They left in large numbers when their leases came up for renewal in 1717–18, and again in the depression of 1740–1. Most went to Pennsylvania to seek isolated areas where they could squat. As settlement expanded, the British colonists came into conflict with the French or Spanish, or were plagued by running skirmishes with the Indians whom they sought to dispossess. Through much of the eighteenth century the main battles occurred where the French or British persuaded the Indians to fight with them against the other, culminating in the Seven Years War of 1756–63.

However, the settlers continued to move west into the Ohio lowlands, the Kentucky Bluegrass area and the upper Tennessee valleys. The main migration continued to thrust forward from Virginia and Pennsylvania, where the major routeways began. The area around Fort Pitt (later Pittsburgh) was settled rapidly, especially after the redrawing of the Indian Boundary Line in 1768.

The period between 1763 and 1776 saw a worsening of relations between Britain and the American colonies. There were two major economic reasons for this rift which united north and south in rebellion. First, the British mercantile policy viewed the colonies as providers of raw materials and markets for manufactured products, and trade was restricted to British ships. This irked the New England and Middle Atlantic merchants who wished to trade more widely. Secondly, the restrictions on movement beyond the Appalachians annoyed the southern frontiersmen and the speculators, and the Quebec Act of 1774, which extended the Quebec frontier to the Ohio River, was a source of irritation to those Americans pushing forward along the Ohio. Various taxes instituted to help pay Britain's heavy military bills in the Seven Years War and in maintaining protection against Indian attacks also resulted in anti-British feelings. The combination of grievances brought together those who signed the Declaration of Independence in 1776 (Figure 2.2.2).

Figure 2.2.2
Signing the Declaration of Independence

The War of Independence which followed was fought on several fronts, and the British, armed with better gifts, set the Indians to attack the frontier areas beyond the Appalachians and also northern New England. Gradually the colonists came to terms with the Indians and reduced the Loyalists and British military forces to ineffectiveness, but only after much bloodshed. In 1783 a treaty was signed between Britain and the USA giving the new nation all the lands to the Mississippi and opening the way for a new phase of expansion.

2.3 From Independence to the Twentieth Century

The Acquisition of new lands

No.	State	Date of foundation
1	New Hampshire	1629
2	New York	1665
3	Massachusetts	1620
4	Rhode Island	1638
5	Connecticut	1633
6	Pennsylvania	1680
7	New Jersey	1665
8	Delaware	1664
9	Maryland	1632
10	Virginia	1607
11	North Carolina	1628
12	South Carolina	1663
13	Georgia	1727

Figure 2.3.1
The thirteen original United States

The 13 original colonies (Figure 2.3.1) which formed the new nation could scarcely have visualised the extension of their lands to the Pacific coast within less than 70 years of independence. They began by owning the land from the Atlantic coast to the Mississippi River (apart from Florida), although much was still occupied by Indians. Many of the 13 colonies claimed areas west of the Appalachians, since they had been set up with sea-to-sea charters. By 1790 all except Georgia (which followed in 1802) ceded their western lands. This was the first large area of 'public domain', or land owned by the federal government.

The public domain was enlarged in 1803 by the Louisiana Purchase (Figure 2.3.2). France, occupied in Europe with the Napoleonic Wars, sold this huge area for $15 million. All or parts of 13 new states were created from this land later in the century.

Figure 2.3.2
Territorial expansion

Florida was the next area to be purchased – from Spain in 1819. Spain also occupied Texas, which became part of the new Republic of Mexico when it achieved independence in 1821. A small number of Americans tried to halt the Mexican army at the Alamo in San Antonio, but were all killed. However, the Mexicans were subsequently defeated by a Texan force under Sam Houston in 1836. Texas was an independent nation from 1837 to 1845, but in the latter year it was accepted into the USA as a new state.

The lands on the west coast south of latitude 42°N also belonged to Spain and then Mexico. However, American settlers had been moving into California, and the USA offered to buy these lands. But in 1846 war erupted in which the American forces took over California with scarcely a shot fired. In 1893 the Gadsden Purchase added another strip of land to Arizona and New Mexico.

Britain claimed the west coast lands north of 42° but American settlers moved into the area over the Oregon Trail and in 1846 a treaty between the two countries set the boundary at the 49th parallel.

The final major addition to the US was Alaska which was purchased from Russia in 1867. Later still, war with Spain in 1898 brought the dependencies of Puerto Rico (now a 'commonwealth') and Guam. Hawaii became an American possession in 1898 and along with Alaska gained statehood in 1959.

The Federation of Canada

The story of Canadian expansion is really one of accepting what was left after claims by the USA following the War of Independence, the War of 1812 and the 1846 agreement. The Constitutional Act of 1791 had divided Quebec into Upper Canada (present-day Ontario), predominantly British and Protestant; and Lower Canada (present-day Quebec), predominantly French and Catholic. Each new province had its own legislature and institutions. After 1815 thousands of immigrants came to Canada from Scotland and Ireland. Movements for political reform arose and the two Canadas were made one province by the 1841 Act of Union and responsible government was achieved in 1849. Federation of all the Canadian provinces was achieved with the British North America Act in 1867. The four original provinces were Ontario, Quebec, Nova Scotia, and New Brunswick (Figure 2.3.3).

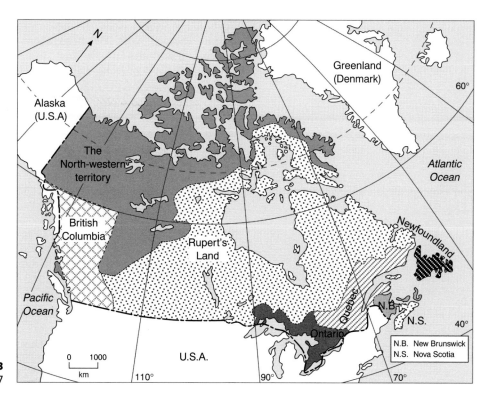

Figure 2.3.3
The Canadian provinces in 1867

The Removal of the Indians

The Indian 'problem' began during colonial days, when it was soon discovered that each new area had to be won for settlement by conquest and clearance. Although a policy of purchasing land from the Indians was pursued from time to time throughout this period, they were often paid with small sums or worthless trinkets and could not understand that they had given over possession of the land for all time.

Even when an agreement between the federal government and the Indians looked like being kept by the Indians, explorers, trappers, miners or farmers would violate it. As soon as the Indians retaliated, however, the army came back with sledge-hammer blows to humiliate or even exterminate whole nations. It is significant that the Bureau of Indian Affairs was under the War Department until late in the nineteenth century, when any threat the Indians had once posed was removed.

The first Indian problems after independence came when there was an attempt to remove them from the Ohio country. Some Indian peoples ceded their lands at conferences in 1783 and 1785, but others, encouraged by the British, attacked the early settlers.

War broke out in 1789 and the Indians were defeated at the battle of Fallen Timbers in 1794. In the South speculators and traders attempted to move the Creek Indians by treaty, but the latter allied themselves to the Spaniards in Florida and went to war in 1785. Other attempts to move west into Kentucky came up against Indians as well, but the worst conflicts were in the War of 1812 against Britain.

The process of piecemeal removal of the Cherokee in the South began with a series of treaties between 1816 and 1821, and started the idea of providing land farther west for the Indians. Persuasion eventually caused many members of the south east nations (Creeks, Cherokees, Choctaws, Chickasaws, together with the Seminoles), to move out of lands in Georgia, Alabama, Mississippi and Florida to a large area prepared for them west of the Mississippi River. The new lands took up much of present Oklahoma and Kansas, but the moved peoples found it difficult to come to terms with the new environment and with the hostile Plains Indians who resented their intrusion. The Shawnee and Delaware nations from the north east joined the five southern nations, but they kept largely to the wooded eastern section of Indian Territory. Thus the policy of 'one large reserve' was implemented on land that was regarded as unfit for white settlement, and many hoped that the end of Indian troubles had been reached.

Then the push to settle lands to the north west in the 1840s led to an ousting of tribes from Iowa, but resistance from the Sioux, until a second large reservation was created in the Black Hills of Dakota. There were also troubles as mining developed in the mountainous west after 1850. A pattern had been established in the California Gold Rush of 1849, where local legislation removed all protection from the Indians and the miners were able simply to turn them off the land or shoot them.

The last major Indian resistance took place in the area occupied by the strong, mobile Plains Indians when the transcontinental railroads extended their lines across their lands, dividing up the 'one large reserve'. A new policy was implemented during the phase of demand for the previously unwanted plains lands: the Indians were 'concentrated' into reservations instead of having free access to the plains. Meanwhile the Plains Indians' main staff of life, the buffalo, was virtually exterminated between 1867 and 1883. After 1868 there was a further stage in the restriction of Indian lands as the 'small reservation' policy was extended to all new areas being settled. Many Indians did not accept this at once, since it meant further retreat and it led to more wars with inevitable results. The last successful Indian uprising was the Battle of the Little Big Horn in 1876. The final debacle at Wounded Knee in Dakota in 1890 was a massacre of several hundred unarmed Indians.

Land policy

Among the earliest problems of the new American republic were the questions of how to dispose of the public domain lands, and how the new lands should be governed. The federal government wished to make land available for sale, and it was decided that survey before sale would be the best approach. A law of 1785 set the basis; government surveyors would establish east–west 'base lines' and north–south 'principal meridians' for each state or groups of states. There would then be divisions into townships of 36 square miles (each square mile of 640 acres [260 ha] would be one section) but the lands would be sold to individuals. Today the results of this process of rectilinear land survey and sale dominate the landscapes of all but the thirteen original states which have a much less regimented appearance.

Pressure of opinion for smaller and cheaper holdings led to gradual reductions of both the minimum area and price per acre (Figure 2.3.4). However, in many places the first pioneers entered an area before the surveyors, and thus before the land sales. But by 1820 these squatters, who understandably often

Figure 2.3.4
The changing bases for the disposal of public domain lands

Date	Minimum area to be purchased (acres)*	Minimum price per acre (dollars)	Minimum farm price (dollars)	Credit
1796	640	2	1280	50%, 1 year
1800	320	2	640	50%, 25% in 2 years
1804	160	2	320	25% in 4 years
1820	80	1.25	100	None
1832	40	1.25	50	None
1841	Squatters Rights (Pre-emption Act – could buy 160 acres at minimum price)			
1854	Land unsold for 30 years available at 12½c per acre			
1862	Homestead Act (160 acres for head of family; to be lived on and cultivated for 5 years)			

*1 acre = 0.405 hectares

took the best land, had gained recognition, and the Pre-emption Act of 1841 allowed anyone settled on unsurveyed land to buy 160 acres at the minimum price.

Thus the progression of events was in the direction of encouraging more and more people to settle land in the west. In 1862 the Homestead Act was passed, allowing any head of family over 21 years of age to have 160 acres of public domain lands for a small payment; this would be his if he lived on it, and cultivated the lands, over five years.

The new lands north west of the Ohio River were soon divided into states under the following process:

- a Territory would be established in newly settled areas under the control of a Governor and Secretary with three judges, all appointed by Congress in Washington;
- when the adult population passed 5000 the Territory could elect its own legislature, sharing its power with a Council of five including the Governor and members of Congress;
- once the total population reached 60 000, this legislature could frame a constitution and apply for admission to the Union as a new State on equal terms with the older states.

As new territories petitioned for statehood the issue of slavery came to the fore. The new territories north west of the Ohio River prohibited slavery, but the southern states allowed it. For many years the balance of slave and non-slave states in Congress was maintained, with one of each being admitted at a time, but in the 1850s there was increasing difficulty in doing this as more western states were opposed to slavery. Following the 1860 Presidential election of Abraham Lincoln, believed by the South to favour the abolition of slavery, the southern states left the Union to become the Confederate States of America, which precipitated the Civil War (Figure 2.3.5).

Figure 2.3.5
The 11 Confederate States

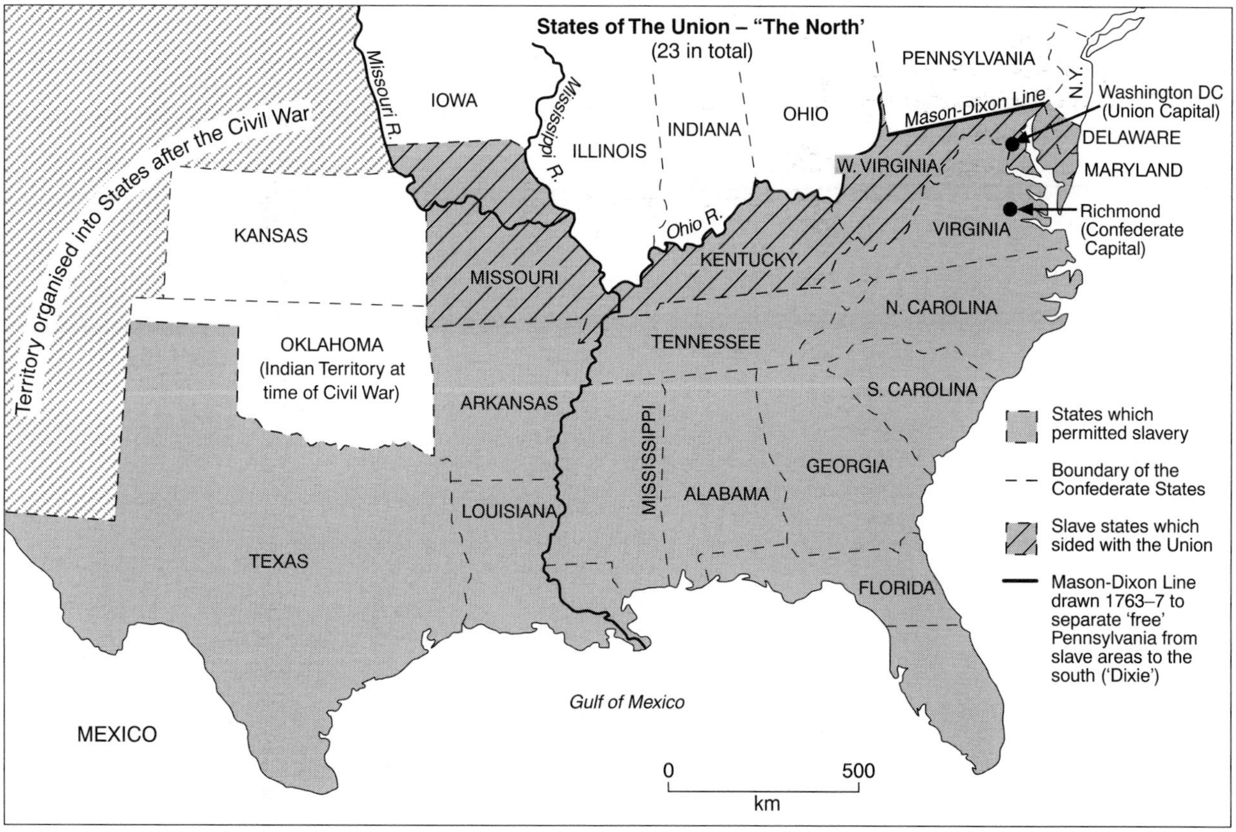

The Civil War and after

On 12 April 1861 Confederate guns opened fire on Fort Sumter in Charleston harbour. While the populations of both sides entered the war with high hopes for an early victory, the odds heavily favoured the North, whose 23 states with a population of 22 million were set against 11 states inhabited by 9 million people. The North also benefited from a huge industrial superiority, providing it with abundant facilities for manufacturing arms and ammunition, clothing, and other supplies. Similarly the network of railways in the North contributed to Union military prospects.

On 1 January 1863 President Lincoln issued an Emancipation Proclamation, freeing the slaves in the Confederate States. This became an objective of the war in addition to the declared aim of saving the Union. After a number of bloody encounters the three-day battle of Gettysburg, Pennsylvania in July 1863 proved to be the turning point. After crippling losses General Robert E Lee's army retreated to the Potomac River. Defeat followed defeat and on 9 April 1865 at Appomattox, Virginia, Lee had no option but to surrender.

Within a week of Lee's surrender Lincoln was assassinated and succeeded in office by Andrew Johnson. That December Congress ratified the Thirteenth Amendment to the US Constitution, abolishing slavery. This and general heavy-handed treatment caused deep resentment in the South. As time passed, it became more and more obvious that the problems of the South would not be solved by harsh laws and continuing rancour against former rebels. In May 1872, Congress passed a general Amnesty Act, and five years later northern troops stationed in the South were withdrawn. After 12 years of false reconstruction, real efforts to rebuild the South began.

Coming of age

In the 50 years following the Civil War the United States was transformed from a rural and agricultural to an urban and industrial society, in which developments in transport and communications proved to be a vital catalyst. The spread of the frontier westward from the Mississippi relied heavily on the railway. As the network developed, people, farming and industry followed. Track length grew from 14 500 km in 1850 to 386 000 km in 1910. Huge land grants were given to the railway companies by the federal government to encourage track construction. In 1869 the first transcontinental railway terminating at San Francisco was completed. By 1883 there were four routes to the west coast. New settlements grew up along these routes. Another year of great importance was 1861 when the first transcontinental telegraph service linking New York to San Francisco opened.

The granting of patents mushroomed, a factor of great significance in a rapidly expanding economy. In the years before 1860, 36 000 were granted; in the next 30 years, 440 000 were issued. In the first quarter of the twentieth century, the number reached nearly one million.

The initial growth of almost all American cities was related to commercial functions. It was not until the last 25 years of the nineteenth century that manufacturing industry increased in scale to become the major employer in urban areas. Between 1850 and 1900 the number of settlements with more than 5000 people grew from 147 to 905 (Figure 2.3.6).

By 1900 the value of manufactured products was twice that of all farm products and the US had become the world's chief industrial power. This was due to a combination of technological development and changing concepts of management. The latter encouraged the concentration of large manufacturing units; by 1905 40 per cent of national manufacturing capital was controlled by 300 large companies.

	Number of places 5000+	Total population (millions)	Urban %
1800	21	5.3	5.2
1810	28	7.2	6.3
1820	35	9.6	6.2
1830	56	12.9	7.8
1840	85	17.0	9.8
1850	147	23.2	13.9
1860	229	31.4	17.9
1870	354	38.6	22.9
1880	472	50.2	24.9
1890	694	62.9	31.5
1900	905	75.9	35.9

Figure 2.3.6
The growth of urban population in the USA in the nineteenth century

Locational specialisation was part of this trend. Chicago, Cleveland, Pittsburgh and Birmingham became iron and steel centres; Cincinnati dominated machine tool manufacture; Philadelphia had copper and nonferrous metal manufacture together with chemicals. New York dominated the manufacture of clothing; Boston was still the centre for textile and shoe manufacture, although textiles were beginning to move south; and Winston–Salem dominated cigarette making. Industrially the north eastern and midwest urban centres were outstanding at this time, and even as late as 1910 cities west of the Mississippi were still largely commercial in function, lacking any significant industrial development.

The commercial functions of cities also changed and developed. City centre department stores and specialist shops in large cities encouraged shoppers with delivery and credit facilities after the 1870s, and extended this to most medium-sized cities by the 1920s. Chain stores such as F W Woolworth (1879) extended the benefits of mass production in low prices. Thus the central shopping areas of cities developed, and the growth of financial services resulted in the second major aspect of central business district concentration and employment.

The United States boasted the fastest growing cities in the world at this stage. By 1890 New York's population reached one and a half million while Chicago and Pittsburgh each had over a million people. By 1920 there was a system of inter-related cities of different sizes providing a range of services in each region and across the nation. Within the major cities, each spreading outwards in a star-shaped pattern reflecting the development of residential suburbs along street railway lines, could be distinguished a central business district, a number of manufacturing zones close to railway lines, and residential areas which were becoming increasingly segregated. Segregation was at its most extreme for black people. In 1910 New York had 92 000 blacks, of whom 60 500 lived in Manhattan, and 49 000 of these were concentrated in a 23-block area of Harlem.

Canada after federation

The new federation gained the extensive lands of the Hudson's Bay Company in 1869, and the Red River Settlement became the province of Manitoba in 1870. British Columbia was admitted to the federation in 1871, followed by Prince Edward Island in 1873. Alberta and Saskatchewan joined in 1905 and Newfoundland became the tenth province in 1949. Earlier, in 1880 the Arctic Archipelago passed under the sovereignty of Canada, while in 1903 the British Columbia–Alaska boundary was settled. In 1920 the districts of the North West Territories were given their present boundaries.

In Canada the frontier had moved forward at a slower pace compared with its southern neighbour. Only 106 km of railway track existed in the whole country in 1850. Nevertheless, the first transcontinental railway was completed in 1885. The cold climate of the northlands ensured that agriculture, industry and settlement largely hugged the border with the USA. Between 1891 and 1914 more than three million people emigrated to Canada, mostly from Europe.

The early twentieth century

Many further changes have occurred since the start of the twentieth century, especially in increased productivity, transport and manufacturing, and urban–rural links, but the nineteenth century laid the basis for the development of modern geographical conditions that prevailed by 1920:

- patterns of movement and land-use were on a much larger scale and specialisation on a regional, and even a national, scale had become established;

QUESTIONS

1 Estimate the proportion of the present-day USA occupied by the original 13 states.

2 Describe and account for the expansion of the USA beyond the original 13 states.

3 Identify the major stages in the formation of present-day Canada.

4 Explain the trends illustrated by the data presented in Figure 2.3.4.

5 With reference to Figure 2.3.6 comment on the process and pace of urbanisation in the USA in the nineteenth century.

- economic, rather than environmental control had become strong in determining the location and concentration of human activities;
- distance was scarcely a barrier as transcontinental railroads enabled transport to take place to virtually anywhere in the continent. Intra-urban transport had led to major changes in the internal geography of cities;
- innovation and speculation had resulted in rapid developments which took the North American economy ahead of the rest of the world;
- reliance on animal and human muscle power had been replaced by machinery based on energy from fossil fuels in transport, factories and farms;
- per capita productivity had been raised greatly by this use of machines and by increases in the size of production units;
- the restrictive mercantilistic economic system of colonial times had given way to the free market system. The freedom of trading across internal boundaries made a vast market available and encouraged the growth of production.

By the early 1920s geographical space in North America was organised into a hierarchy of settlements with their manufacturing and service activities linked by an extensive rail network. Manufacturing industry was concentrated in large units, especially in the north east and Midwest of the USA, together recognised as the 'manufacturing belt', and farming had differentiated into a series of specialist regions.

3

Population Studies

3.1 Continental Overview

Population growth

At the dawning of the twentieth century the populations of the USA and Canada stood at 76 million and 5.4 million respectively (Figures 3.1.1 and 3.1.2). By 1990 the populations had reached 249.9 million and 26.2 million, a continental increase of almost 195 million. The 1995 estimates of population were 262.7 million in the USA and 29.6 million in Canada.

Figure 3.1.1
Population growth in the USA, 1790–1990

| Census date | Resident population | | | |
| | Number | Per km^2 of land area | Increase over preceding census | |
			Number	Per cent
CONTERMINOUS US[1]				
1790 (2 August)	3 929 214	1.7	(X)	(X)
1800 (4 August)	5 308 483	2.4	1 379 269	35.1
1810 (6 August)	7 239 881	1.7	1 931 398	36.4
1820 (7 August)	9 638 453	2.1	2 398 572	33.1
1830 (1 June)	12 866 020	2.9	3 227 567	33.5
1840 (1 June)	17 069 453	3.8	4 203 433	32.7
1850 (1 June)	23 191 876	3.0	6 122 423	35.9
1860 (1 June)	31 443 321	4.1	8 251 445	35.6
1870 (1 June)	39 818 449	5.2	8 375 128	26.6
1880 (1 June)	50 155 783	6.5	10 337 334	26.0
1890 (1 June)	62 947 714	8.2	12 791 931	25.5
1900 (1 June)	75 994 575	9.9	13 046 861	20.7
1910 (15 April)	91 972 266	12.0	15 977 691	21.0
1920 (1 January)	105 710 620	13.7	13 738 354	14.9
1930 (1 April)	122 775 046	15.9	17 064 426	16.1
1940 (1 April)	131 669 275	17.1	8 894 229	7.2
1950 (1 April)	150 697 361	19.6	19 028 086	14.5
1960 (1 April)	178 464 236	23.2	27 766 875	18.4
UNITED STATES[2]				
1950 (1 April)	151 325 798	16.4	19 161 229	14.5
1960 (1 April)	179 323 175	19.5	27 997 377	18.5
1970 (1 April)	203 302 031	22.2	23 978 856	13.4
1980 (1 April)	226 542 199	24.7	23 240 168	11.4
1990 (1 April)	248 709 873	27.1	22 167 674	9.8

X Not applicable.   [1]Excludes Alaska and Hawaii.   [2]Includes Alaska and Hawaii.

Period	Total population growth 1000	Births 1000	Deaths 1000	Natural increase 1000	Ratio of natural increase to total growth %	Immi-gration 1000	Emi-gration 1000	Net migration 1000	Ratio of net migration to total growth %	Population at the end of the Census period 1000
1851–1861	793	1281	670	611	77.0	352	170	182	23.0	3230
1861–1871	459	1370	760	610	132.9	260	411	−151	−32.9	3689
1871–1881	636	1480	790	690	108.5	350	404	−54	−8.5	4325
1881–1891	508	1524	870	654	128.7	680	826	−146	−28.7	4833
1891–1901	538	1548	880	668	124.2	250	380	−130	−24.2	5371
1901–1911	1836	1925	900	1025	55.8	1550	739	811	44.2	7207
1911–1921	1581	2340	1070	1270	80.3	1400	1089	311	19.7	8788
1921–1931	1589	2415	1055	1360	85.6	1200	971	229	14.4	10 377
1931–1941	1130	2294	1072	1222	108.1	149	241	−92	−8.1	11 507
1941–1951[2]	2141	3186	1214	1972	92.1	548	379	169	7.9	13 648
1951–1956	2072	2106	633	1473	71.1	783	184	599	28.9	16 081
1956–1961	2157	2362	687	1675	77.7	760	278	482	22.3	18 238
1961–1966	1777	2249	731	1518	85.4	539	280	259	14.6	20 015
1966–1971	1553	1856	766	1090	70.2	890	427	463	29.8	21 568
1971–1976	1425	1756	822	934	65.5	841	350	491	34.5	22 993
1976–1981	1350	1820	842	978	72.4	588	216	372	27.6	24 343
1981–1986	1011	1873	885	988	97.7	500	477	23	2.3	25 354
1986–1991	1754	1930	945	985	56.2	874	105	769	43.8	27 108

Figure 3.1.2
Population growth in Canada[1] 1851–1991

[1]Includes Newfoundland since 1951.
[2]Data on components of growth shown for 1941–51 were obtained by excluding data for Newfoundland.

In the 1920s and 1930s the birth rate in both countries was at a low level compared with previous decades. In general, people now wanted smaller families, not least because of the great economic uncertainty of the time. However, after the end of the Second World War the birth rate rose sharply (Figure 3.1.3). This period of high fertility, lasting almost 20 years, is generally referred to as the 'post-war baby-boom'.

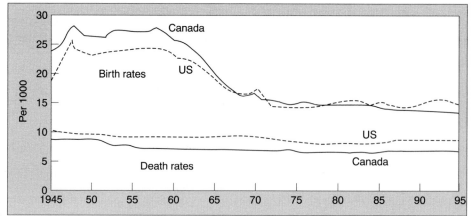

Figure 3.1.3
Birth rates in Canada and the USA

Since the mid-1960s the birth rate in the USA and Canada has fallen to a lower level, mainly because:

- children were perceived as increasingly expensive to bring up and many couples opted to maintain or attempt to improve their standard of living instead;
- more women put their careers ahead of starting or extending a family;
- contraception became more widely used and more reliable.

Canada's population is growing comparatively quickly and, at 1.5 per cent in 1990, the rate of population growth was actually the highest in the developed world. Growth rates in the USA and Europe were 1.1 and 0.4 per cent respectively in the same year.

The two factors that affect the changing size of a population are natural change and net migration. Figure 3.1.2 shows the relative importance of both variables in Canada since 1851. Since Confederation, natural increase has accounted for about 80 per cent of Canada's population growth. However, in the 1986–1991 period the contribution of net migration was at one of its highest ever levels. Immigration has also been a major factor in the growth of the US population. Figure 3.1.4 shows its contribution in recent years.

Figure 3.1.4
Components of US population change, 1982 to 1995

YEAR	Population at start of period (1000)	Net increases		Natural increases		Net migration (1000)
		Total (1000)	Per-cent	Births (1000)	Deaths (1000)	
		TOTAL (1 January–31 December)				
1982	230 645	2157	0.9	3681	1975	595
1983	232 803	2066	0.9	3639	2019	592
1984	234 868	2070	0.9	3669	2039	589
1985	236 938	2171	0.9	3761	2086	649
1986	239 109	2158	0.9	3757	2105	661
1987	241 267	2195	0.9	3809	2123	666
1988	243 462	2243	0.9	3910	2168	662
1989	245 705	2438	1.0	4041	2150	712
1990	248 143	2549	1.0	4148	2152	594
1991	250 692	2976	1.2	4111	2165	1030
1992	253 667	2888	1.1	4087	2166	967
1993	256 516	2651	1.0	4039	2268	880
1994	259 141	2485	1.0	3979	2286	792
1995	261 626	2397	0.9	3961	2329	762

Canada's total fertility rate (the number of children likely to be borne by one woman in her lifetime) has recently, after decades of decline, shown a modest increase. Canada's rate is now greater than that of Europe, but slightly lower than that of the US. At 1.8 in 1990 it was at its highest level for 14 years. Fertility has also increased recently in the US, in 1989 and 1990 the total fertility rate topped 2.0, the first time since 1971.

Mortality rates in both countries also reflect their economically developed status. Canada's crude death rate was 6.9 per 1000 in 1993–4 while its infant mortality rate at 6.3 per 1000 live births was almost the lowest in the world. In the same year the respective figures in the US were 8.8 and 8.4.

Population density and distribution

North America has a low population density compared with most other parts of the world. The US has an average of 27 persons per km², while Canada has only two per km². In both countries population is highly concentrated in some areas while large expanses of land elsewhere are very sparsely settled (Figure 3.1.5).

Sparsely populated areas

Very few people live in the cold, dry and mountainous regions. The influence of low temperature is very clear in the north and largely explains why 75 per cent of Canada's population lives within 160 km of the main border with the USA. Life is extremely difficult in the permafrost environment of the northlands and, apart from the native Inuit and Indians, the few people living there are mainly involved in the exploitation of raw materials and in maintaining defence installations.

The mountain ranges of the west form imposing landscapes but few people live in this rugged environment since economic opportunities are scarce.

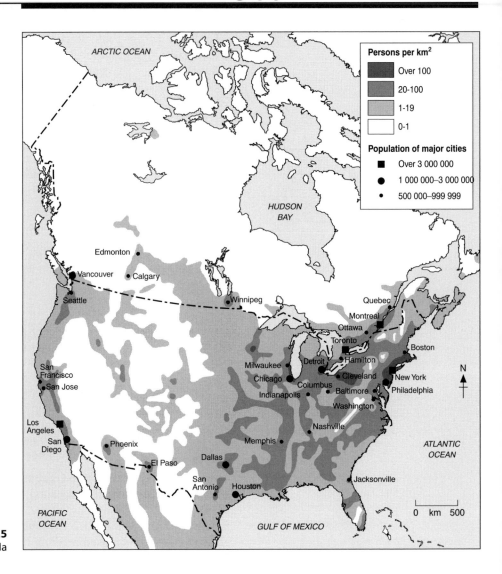

Figure 3.1.5
Density of population in the US and Canada

Mountain settlements are related mainly to mining, tourism and agriculture and are small in size. For example, Cheyenne (altitude 1850 m), the largest settlement in Wyoming, has a population of only 50 000. The Appalachian Mountains in the east are both lower in height and less extensive in area; nevertheless, the most isolated parts of the Appalachians are very sparsely peopled.

Although much of the south western USA is desert or semi-desert, no country in the world has been more successful in watering its drylands. Expensive irrigation schemes have opened up many parts of the region to farming, settlement and industry. Cities such as Phoenix, Tucson and Las Vegas, standing out like oases in the desert, are clear evidence of the huge level of investment. Yet large parts of the South West still remain empty and may well do so in the future as the water supply problem has now reached crisis point.

Densely populated areas

In the US the greatest concentration of population is in the North East, the first region of substantial European settlement. By the end of the nineteenth century it had become the greatest manufacturing region in the world and by the 1960s the highly urbanised area between Boston and Washington had reached the level of a megalopolis. Although other parts of the country are growing at a faster rate the intense concentration of job opportunities in the North East will ensure that it remains the most densely populated part of the continent in the foreseeable future.

The coastline and major lowland valleys are as attractive to settlement now as they have been in the past. More than half of all Americans live in the counties adjacent to the Atlantic and Pacific Oceans, the Gulf of Mexico and the Great Lakes. This coastal population increased by 39 per cent between 1960 and 1994, a rate of growth slightly less than the USA as a whole. The fastest coastal growth has been along the Pacific and Gulf of Mexico coasts, where the population has approximately doubled since 1960.

Soil fertility is another influence on population distribution, but the high level of mechanisation in modern agriculture means that this factor was more important in the past than it is today. The location of other natural resources is also an attraction to settlement, but again such influence has lessened. Most North Americans today live in urban areas and are employed in the service or manufacturing sectors of the economy. The availability of employment in these sectors is the most important influence on the distribution and density of population today.

Canada

The main concentration of population in Canada is in the southern parts of Ontario and Quebec. This region has a combination of physical and human advantages greater than that found in most other parts of the country. Two of Canada's three 'million-size' cities, Toronto and Montreal, are located here. The third ranking city, Vancouver, is situated on the Pacific coast. A large area of low to moderate population density is found in the cereal farming region of the Prairie Provinces. In terms of political boundaries population density ranges from a high of 22.9 per km^2 on tiny Prince Edward Island to less than 0.1 per km^2 in the Northwest Territories. In comparison, in the USA, population density varies from 410 per km^2 in New Jersey to 0.4 per km^2 in Alaska.

Regional changes

The westward movement of population has continued unabated in the twentieth century in both the USA and Canada, while the former has also experienced rapid growth in many southern states. Regional changes have been so great that the organisation of economic and political power is in its greatest state of flux since the 1860s. In both countries the position of the traditional industrial heartland has been eroded as the new growth regions have expanded.

The mean centre of population in the USA has moved steadily to the west. The mean centre of population is that point at which an imaginary flat, weightless, and rigid map of the USA would balance if weights of identical value were placed on it so that each weight represented the location of one person on the date of the census. In 1980 the mean centre crossed the Mississippi River for

Division	Population [1000]		Percentage Population Change			
	1960	1995	1940–60	1960–80	1980–90	1990–5
New England	10 546	13 312	25.0	17.1	6.9	0.8
Middle Atlantic	34 269	38 153	24.4	7.4	2.2	1.5
East North Central	36 340	43 456	36.5	14.7	0.8	3.4
West North Central	15 425	18 348	14.1	11.4	2.8	3.9
South Atlantic	26 066	46 995	46.2	39.3	17.9	7.9
East South Central	12 073	16 066	12.0	21.5	3.5	5.8
West South Central	17 008	28 828	30.2	39.6	12.5	8.0
Mountain	6897	15 645	66.2	64.8	20.1	14.5
Pacific	21 352	41 951	119.4	48.9	23.0	7.2
Total	179 977	262 755	36.7	25.9	9.8	5.6

Figure 3.1.6
USA: Population growth by division, 1940 to 1995

the first time and for the first time in US history more Americans lived in the South and West than in the Northeast and Midwest. In 1990 the mean centre was in Crawford County, Missouri.

Figure 3.1.6 shows the extent of US divisional population changes in recent decades. In the three time periods illustrated, the two western divisions of Mountain and Pacific grew at by far the fastest rates. Rapid growth has also been experienced by the South Atlantic, recently taking over from the East North Central as the most populous division. Of the two remaining divisions in the South, the West South Central ranked fourth in population growth during the overall period while growth was markedly more restricted in the East South Central. The latter has clearly suffered from a lack of dynamic urban centres and a relatively low rate of job creation.

Figure 3.1.7
Canada: Population growth by division

	Population [1000]		Percentage Population Change		
	1961	1995	1941–61	1961–81	1981–91
Newfoundland	458	575	51.0	24.0	0.8
Prince Edward Island	105	136	10.0	17.1	6.5
Nova Scotia	737	938	27.5	15.0	6.1
New Brunswick	598	760	30.7	16.5	4.2
Quebec	5259	7334	57.8	22.4	6.2
Ontario	6236	11 100	64.6	38.3	14.5
Manitoba	922	1138	26.4	11.3	6.5
Saskatchewan	925	1016	3.2	4.7	2.9
Alberta	1332	2747	67.3	68.0	13.6
British Columbia	1629	3766	99.2	68.4	17.4
Yukon territory	15	30	192.0	58.9	16.7
NW Territories	23	66	91.7	99.0	20.4
Total	18 238	29 606	58.5	33.5	10.8

Figure 3.1.8
USA: Population by state in 1995, and projected changes 1990–2000

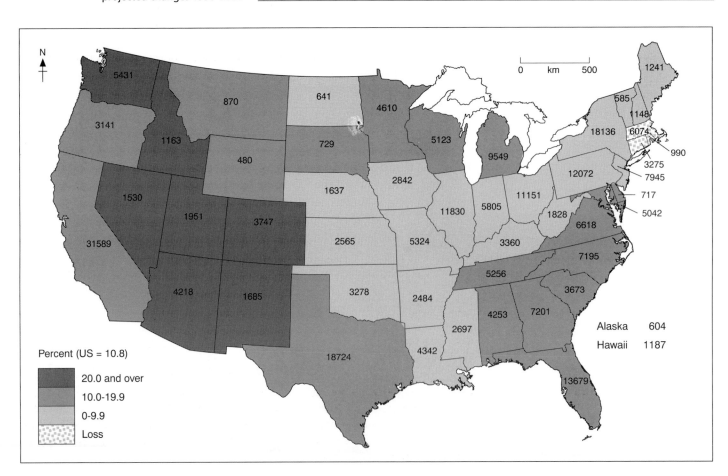

While the benefits of population and general economic growth are well documented, communities in the west are becoming increasingly aware of the costs. This issue has moved rapidly to the top of the political agenda in some areas and the scene is set for more local and state legislation to limit further development.

Some individual states have actually recorded population losses. Between 1980 and 1993 reduced numbers were recorded in Iowa (2.91 to 2.81 million), North Dakota (0.65 to 0.64 million) and West Virginia (1.95 to 1.82 million). Figure 3.1.8 shows 1995 population by state and the projected change in state populations 1990–2000.

Canada

The distribution of Canada's population (Figure 3.1.7) is markedly shifting westwards, particularly to Ontario from more easterly provinces, and to British Columbia and Alberta. In 1991 the two western provinces accounted for some 21 per cent of Canada's total population, up from 16.2 per cent in 1961. However, slightly more than 62 per cent of the population still reside in Ontario and Quebec. Although the Yukon and Northwest Territories have recorded high rates of growth these were registered on low base levels of population.

Immigration

Immigration has arguably been the most dominant trend in the demographic history of North America. In the USA almost 60 million people have entered the country since 1820. However during this time both the rate of entry and the origin of immigrants have changed considerably. The highest recorded rate for any decade was 10.4 per 1000 during 1901–10, when 8.75 million newcomers arrived (Figure 3.1.9), although other decades in the nineteenth century were not far behind in proportional terms. The flood of immigrants continued until the outbreak of hostilities in Europe in 1914, whereupon it sharply abated from 12.3 per 1000 in 1914 to 3.2 in 1915. It has rarely risen above the latter figure since, apart from a few exceptional years in the early 1920s and in very recent years. The main reason for this was a growing concern about the numbers and origin of migrants.

Figure 3.1.9
Immigrants to the USA per 1000 population

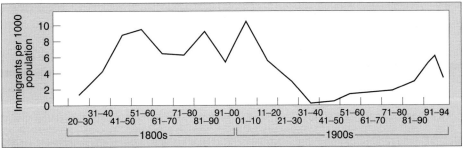

National origins quotas

In 1924 a system of 'national origins quotas' was introduced which operated with only slight modification until 1965. The new legislation was designed to greatly reduce immigration and in particular to stem the influx of eastern and southern Europeans who poured into the USA at the turn of the century. Anti-Chinese restrictions had already been in force for many years. The system aimed to preserve the ethnic balance which existed in the country at the 1920 census, offering the largest quotas of entry permits to British, Irish and German immigrants (70 per cent in total), as people from these countries made up 70 per cent of the existing population.

The racist overtones of this system, resulting in considerable international and internal opposition, led to its abolition in 1965. The 1965 Act, which became fully effective in July 1968, set an annual limit of 120 000 immigrants from the Western Hemisphere (the Americas), and 170 000 from the Eastern Hemisphere. People from every country within each hemisphere now had an equal chance of acceptance into the US. However, total annual immigration exceeded this level because relatives of US citizens could be admitted without numerical limitation. The Immigration Act of 1990 revised the numerical limits and preference system regulating legal immigration, raising quotas of legal immigrants by 40 per cent.

As the intervening obstacles to immigration into the US were lowered for potential migrants from a number of world regions, so the ethnic composition of new arrivals changed significantly (Figure 3.1.10). Europe, the previous major source region, has been overtaken in the last three decades by the rest of the Americas and by Asia, a trend that is likely to continue in the future.

Figure 3.1.10
Sources of immigration into the US

	Immigrants (1000)			
	1820–1979[a]	1971–80[b]	1981–90[b]	1991–94[b]
Europe	36 267	801	706	600
Asia	3038	1634	2817	1366
The Americas	9248	1929	3581	2405
Africa	142	92	192	118
Total (including others)	49 124	4493	7338	4510

[a]by country of last permanent residence   [b]by country of birth.

Recent trends

In 1991 the immigration rate rose to 7.2 per 1000, up from an average of 3.1 in the previous decade, with over 1.8 million people entering the country that year. The respective figures for the next three years were 3.8 (974 000), 3.5 (904 000), and 3.1 (804 000). Such an increase, coinciding with a period of intense economic recession, reopened the immigration debate in the USA in a big way (Figure 3.1.11). Some Americans argue that recent immigrants are taking scarce jobs that should be theirs, while others voice concern about racial tension and the impact on the welfare system. A poll by Newsweek magazine in mid-1993 recorded a distinct hardening in attitudes to immigration with 60 per cent of Americans seeing current high levels of immigration as bad.

Immigration into the USA is very spatially selective. In 1994, 58 per cent of all immigration was to just four states – California, New York, Florida and Texas. The main reasons for such concentration are:

- the location of existing immigrant communities;
- the availability of employment in the four most populous states;
- the land border with Mexico and Florida's proximity to the Caribbean.

Illegal immigration

Because of its very nature it is difficult for the authorities to be precise about the level of illegal entry. According to The Urban Institute's 'best estimate', the illegal population peaked in 1986 at between 3 and 5 million. The illegal population dropped following passage of the 1986 Immigration Reform and Control Act, which made it harder for illegal aliens to find work and allowed more than 3 million illegals already in the US to obtain legal status. Since 1989, however, the number of illegal immigrants has risen sharply, reaching 2.5 to 4 million in 1992. Of these, one million are thought to live in southern California. A large number of potential illegals are caught trying to enter the USA each year (Figure 3.1.12).

Figure 3.1.11
They're coming

They're coming

WASHINGTON, DC, AND LOS ANGELES

A debate about America's immigration policy, quietly under way all last year, has come into the open. Facing the prospect of yet more boatloads of shivering Chinese crossing the Pacific, both the administration and some influential congressmen are considering new legislation. In California, the state that feels most hurt by illegal immigration, various ideas to stem the flow are being considered. Meanwhile, in the wake of the arrests of alleged terrorists in New York last month, most of them foreign nationals, public opinion has become more anti-immigrant than for many years. A recent Gallup poll found 65% of Americans in favour of tighter controls. The young were slightly more tolerant; the South, which until the past 20 years saw very little immigration, was, as ever, the most suspicious of newcomers.

The context for the new nativism is the sheer volume of immigration in the 1980s. In absolute numbers, the 1980s probably saw more immigrants (legal and illegal) than any other decade in America's history. This wave of immigration had been building up since the late 1960s, but it followed a period when there was not much immigration to America. As a proportion of population, more immigrants were admitted to America in the 1850s, when it was a sparsely populated rural society, than in the 1960s, when it was an industrial giant. And they came from different places. In the years after 1945 it was almost impossible for Asians to get in unless they married a GI. Now Asians make up the fastest-growing minority in America.

In California, with 1.3m undocumented immigrants, some estimates put the annual cost to the state at $5 billion, of which $1.2 billion goes to the prisons, where illegal immigrants are about 15% of inmates.

Meanwhile, the age-old debate about immigration rumbles on. Are immigrants good for the economy—at all times and in all numbers—or not? Those who argue that immigrants, and especially illegal ones, depress wages and drive natives out of work are getting more of a hearing than they used to.

But they are not having things all their own way. A carefully documented new book by Thomas Muller for the Twentieth Century Fund makes the traditional case: that immigrants are, as they have ever been, a source of vitality and entrepreneurship for the economy.

Mr Muller tackles head-on the common recent claim that immigrants have hurt the prospects of blacks in the inner cities, which is something that many blacks believe. Historically, he points out, the influence of immigrants on blacks has been benign, since immigrants have displaced blacks from the most menial jobs and allowed them to move into better-paid employment. Granted, it may be the case that immigrants compete with the poorest blacks for jobs; but anyone who seriously thinks that the problem of the black urban underclass has been caused by immigration, or that stopping immigration would solve it, is living in a dream world.

The Economist
24 July 1993

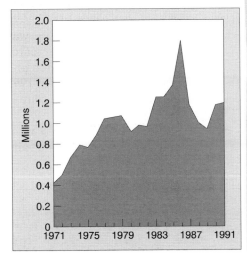

Figure 3.1.12
Number of people caught entering the USA illegally each year

In 1991 more than one million undocumented migrants were apprehended coming from Mexico. 'The general rule of thumb is that for every one caught, three or four get across,' says Ernesto Rodriguez, a migration expert at the University of Houston. Immigration, particularly illegal immigration, has become such a controversial issue that a certain mythology has been built up. The federal government publication, *CQ Researcher* takes a look at some of these myths (Figure 3.1.13).

Figure 3.1.13
Common myths about illegal immigrants

Common Myths About Illegal Immigration

The 1980s witnessed the largest wave of arrivals in the United States since the turn of the century. In recent years, some Americans have concluded that too many newcomers are being admitted, and too many of the "wrong kind" at that. Though it is difficult to separate the impact of illegal immigration from recent arrivals who are here legally, recent research indicates several myths have arisen about both groups:

Myth 1: Mexicans make up nearly all illegal immigrants. Mexicans make up no more than 55 percent of the illegals in the United States. The balance come from all points on the globe, and most of them enter the country legally but overstay their visas.

Myth 2: People who cross the border illegally stay permanently. More than 3 million people illegally cross the U.S.– Mexican border each year. The common assumption is that they all stay but, in reality, more than nine out of 10 don't. Experts say only 200,000 to 300,000 become permanent inhabitants each year.

Myth 3: Illegal immigrants take jobs from native-born workers and legal immigrants. Illegal immigrants do take unskilled jobs from Americans, mostly minorities and legal immigrants in border states such as California and Texas. Illegals make up as much as 6 percent of the U.S. labor force and undoubtedly push wages — and hence consumer prices — downward.

At the same time, immigrants, both legal and illegal, start businesses, consume goods and services and, in so doing, help the U.S. economy grow, benefiting all workers. According to the 1990 report of the President's Council of Economic Advisers, "numerous studies suggest that the long-run benefits of immigration greatly exceed any short-run costs. The unskilled jobs taken by immigrants in years past often complemented the skilled jobs typically filled by the native-born population, increasing employment and income for the population as a whole."

Myth 4: Most illegal immigrants work in agriculture, picking fruits and vegetables. In fact, researchers estimate only 9 percent of undocumented workers are pickers. Significantly larger numbers work in the service sector (hotels, restaurants, etc.), in industry and as domestic helpers. Some experts say the 1986 Immigration Reform and Control Act (IRCA), which enabled 1.2 million temporary farm workers to become U.S. citizens, has encouraged more illegals to seek employment in urban areas.

Myth 5: Illegal immigrants depend heavily on welfare and other social services, placing a heavy burden on U.S. taxpayers. The legal status of immigrants — whether they are refugees, legal immigrants or undocumented — has a major effect on their use of welfare and all public services. Ironically, refugees, who garner the most public sympathy, tend to rely most heavily on welfare and Medicaid. Illegal immigrants are barred from most forms of government assistance and tend to avoid all contact with public officials.

Myth 6: Recent arrivals are being assimilated more slowly than previous waves of newcomers. Today's immigrants are often thought of as uneducated, unwilling to learn English and reluctant to become fully assimilated. Recent research findings, however, tell a different story. According to 1980 census data, of the immigrants who entered the United States from 1975 to 1980, almost 24 percent had completed four or more years of college. Only 16 percent of the U.S.-born population had an equivalent level of education. And contrary to popular beliefs, most second- and third-generation Hispanics are doing fine, about on a par with the record of Italian-Americans. Of the 40 finalists in the 1991 national high school science competition sponsored by Westinghouse Corp., 18 were foreign-born children of foreign-born parents. In Boston, 13 of the 17 public high school valedictorians in the class of 1989 were foreign-born.

Myth 7: U.S. immigration policy is becoming more restrictive. Anxiety about the economy and illegal immigration has created a false perception that the United States is closing the door on immigrants. In fact, the 1990 Immigration Act expanded immigration by 40 percent above previous levels, already the most generous in the world.

CQ Researcher 1992

Canadian immigration

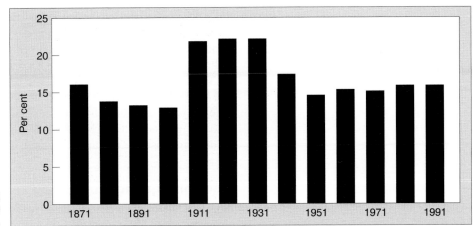

Figure 3.1.14
Immigrant population as a percentage of Canada's total population

When the first Canadian census after Confederation was taken in 1871, about 16 per cent of the population were recorded as immigrants (Figure 3.1.14). This proportion fell steadily until the turn of the century, as emigration was consistently higher than immigration. Many immigrants were en route to the USA where more employment opportunities were available. Those born in Canada also formed part of this movement southwards in search of work. The proportion of immigrants was significantly higher, about 22 per cent, between 1911 and 1931, owing to an immigration boom fuelled partly by a government campaign to attract settlers to the farmlands of western Canada.

The 1930s was the only decade this century when emigration outpaced immigration. After 1945 immigration increased again as employment opportunities expanded for a diversified range of skilled workers. Although not as contentious an issue as in the USA, there is an increasing feeling in the country that a high level of entry cannot continue indefinitely.

For most of its history Canada had a discriminatory immigration policy. However, in 1962 substantial changes made Canada accessible to all, regardless of ethnic origin. In 1967, a points system was initiated to select independent immigrants on the basis of merit, with further amendments made in 1976. Partly as a result of these changes an increasing proportion of immigrants are non-European, including large numbers from Asia, particularly Hong Kong.

New legislation, which came into effect in 1993, continues to give priority to the reunification of families, offers protection to refugees who seek Canada's help and improves Canada's capacity to select immigrants having particular economic skills or qualifications. It also includes measures to protect Canadian society from those who would abuse its fundamental laws.

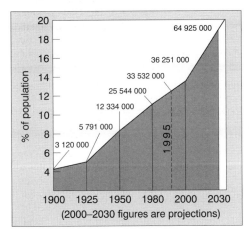

Figure 3.1.15
Percentage of US population over 65 years of age

Ageing populations

Twentieth century trends

The populations of both the USA and Canada are ageing at a demographically rapid rate. Four per cent of the USA population was 65 years of age and older in 1900 (Figure 3.1.15). By 1995 this had risen to 12.8 per cent and by 2030 it is likely that one in five Americans will be senior citizens. The trend in Canada has been very similar, with 12 per cent of the population aged 65 or more in 1995.

Demographic ageing is partly due to increased life expectancy, but more especially to a drop in fertility since the baby-boom. With fewer births and the block of baby-boomers heading towards middle age, the average age of the population has steadily pushed upwards. As a consequence of the decline in mortality, life expectancy rates have risen. In the USA, life expectancy at birth has increased from 47 years in 1900 to 76 years in 1993. For those reaching 65, average life expectancy for men is projected to rise from a current 81 years to 85 in 2040; the respective figures for women are 85 and 88. Recent data for Canada show that in 1992 men could expect to live to 74.6 and women to almost 81 years. The fastest growing segment of the population in both countries is the so-called 'oldest-old': those who are 85 years or more.

The age structure of both countries varies considerably by ethnic group. In the USA 14.5 per cent of the non-hispanic white population are 65 years or over. The respective figures for non-hispanic blacks and hispanics are 8.4 per cent and 5.5 per cent. Clearly, without a high level of immigration from 'younger' ethnic groups, demographic ageing in North America would have progressed further.

The consequences of ageing

Some of the major consequences of this most significant of demographic trends are:

- a steady fall in the ratio of the working to the non-working population. In the USA the projected change will be from 5:1 in 1980 to 3:1 in 2030;
- pressure on social security and private pension schemes as more people spend a greater part of their life in retirement;
- increasing demand for expensive medical care;
- greater gender imbalance. In the USA there are 131 women for every 100 men at age 65. At age 85 there are 224 women for every 100 men;
- the rising political power of the elderly.

Providing for the elderly is going to be an even more significant issue in the future. Very few countries are generous in looking after their elderly at present. Poverty amongst the elderly is a considerable problem but technological advance might provide a solution by improving living standards for everyone. If not, other less popular solutions, such as increasing taxation will have to be examined.

QUESTIONS

1 Explain the changing pattern of fertility in North America in the 20th Century.

2 Analyse the components of Canadian population growth shown in Figure 3.1.2.

3 Describe and account for the variations in population density in North America (Figure 3.1.5).

4 To what extent has the distribution of population changed in the US and Canada since the middle of the century?

5 Account for the variations in the rate of immigration into the US illustrated by Figure 3.1.9.

6 Identify the different reasons why some Americans are in favour of a high rate of immigration while others want to see immigration considerably reduced.

7 Suggest why so many myths abound about immigration.

8 Account for the ageing of the US and Canadian populations.

9 What are the alternative ways of coping with such a rate of demographic ageing?

3.2 Minority Groups in the United States

Still a melting pot?

The term 'The Melting Pot' originated from a play of that name, written by Israel Zangwill, which opened in Washington in 1908. The speaker is David, a young composer: 'America is God's Crucible, the great Melting-Pot where all the races of Europe are melting and reforming.... Germans and Frenchmen,

Irishmen and Englishmen, Jews and Russians – into the Crucible with you all! God is making the American!' The imagery comes from steel making which was the high technology industry then. The analogy was that the USA would fuse together a diversity of cultures and create a typically American way of life.

Many observers are, however, cynical about the validity of the melting-pot theory of majority–minority relations. This has sometimes been expressed by the formula $A + B + C = A$, where A, B and C represent different ethnic groups, and A is the dominant one. Over time the other groups gradually conform to the attitudes, values and lifestyle of the dominant group, while A will change only marginally.

The melting-pot theory may be applicable to white immigrants but beyond that it has obvious limitations. Clearly, many people from minority groups do not regard full Americanisation as either necessary or particularly desirable, while many white Americans practise direct or indirect discrimination against minority groups which do not meet with their approval. The former is certainly true of many American Indians who seek to maintain, as far as possible, their traditional way of life, of a significant number of blacks, and more recently of Hispanics who want to preserve their language. In parts of the Sunbelt the USA is rapidly becoming a bi-lingual society.

The deprivation of minority groups

Figure 3.2.1 shows the populations of the main minority groupings in the USA and their demographic characteristics. Blacks form by far the largest group after whites but Hispanics are increasing more rapidly in number. It must be remembered, however, that as persons of Hispanic origin may be of any race they will also be counted under another grouping as well. The Asian population of the US has experienced the most rapid increase in recent years.

Figure 3.2.1
USA: Components of population change, by racial origin

	Population 1 Jan 1995 (million)	Birth rate	Death rate	Net migration rate
White	217.4	14.1	9.2	2.2
Black	32.9	20.3	8.7	2.7
American Indian, Eskimo, Aleut	2.2	17.9	5.0	1.6
Asian, Pacific islander	9.1	17.8	3.0	20.8
Hispanic origin	26.5	25.4	3.8	12.5

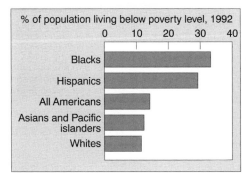

% of population living below poverty level, 1992

Figure 3.2.2
USA: Poverty rates

All minority groups have levels of poverty above those of whites (Figure 3.2.2), while the rates for blacks and Hispanics are over twice the national average. Such a dichotomy occurs in a society where the gap between rich and poor is steadily widening. One way in which minority groups are seeking to improve their socio-economic status is through increased political representation and more minorities than ever before have won congressional office in recent years, thanks in large part to the creation of electoral districts with majorities of minority voters. However, this system of electoral district design is being challenged in a number of southern states. Civil rights lawyers contend that requiring states to draw majority–minority districts in their legislative maps is an appropriate remedy for past voting discrimination. In November 1992, Alabama, Florida, North Carolina, South Carolina and Virginia elected the first blacks to the House of Representatives this century, while Illinois elected a black woman to the Senate (Figure 3.2.3).

A number of recent studies have highlighted the poor environmental conditions in which many minority people live. The general contention is that low income and minority communities bear greater health and environmental burdens while the more affluent and whites receive the bulk of the benefits of a high level of economic and social activity.

Black Americans

The black population of North America began when 20 young people were taken in slavery from Africa to Virginia in 1619. Today almost 33 million blacks live in the USA. Theirs is a history of almost four centuries of struggle against discrimination:

- in practice, slavery was not abolished until 1865. At this time most blacks worked on plantations in the South where conditions were harsh and families were often split up. A small number were employed in domestic service in the north eastern states;
- after 1865 many blacks became sharecroppers. They rented small farms and paid the landowners a share of their crop, but incomes from such activity were very low. The sharecropping period lasted a hundred years and during that time many farmers struggled to buy their land. Black land ownership reached a peak of 6.2 million hectares in 1910;
- from 1910 many blacks left farming for work in the cities, partly because their small farms could not compete with the large mechanised farms of the big white landowners. They migrated in large numbers to the rapidly growing industrial cities of the north east and midwest, in one of the great migration streams of US history;
- overt discrimination and the violation of civil rights reached a climax in 1965, in Selma, Alabama, when 600 civil rights marchers were attacked with tear gas, bullwhips and club-wielding police on horseback. The majority of Americans were outraged when they saw such wanton violence on television news. Since then legislation such as the Voting Rights Act and the Civil Rights Act has been passed to end discrimination. However, covert racism is much more difficult to counter and by virtually every socio-economic measure black Americans are in a position well below the national average.

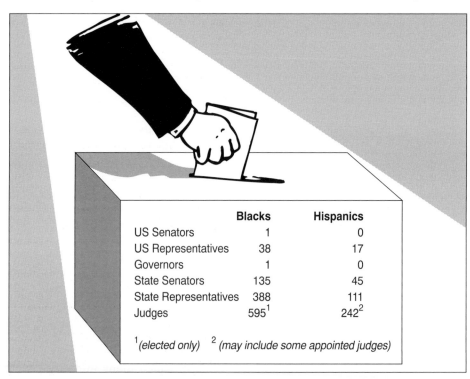

	Blacks	Hispanics
US Senators	1	0
US Representatives	38	17
Governors	1	0
State Senators	135	45
State Representatives	388	111
Judges	595[1]	242[2]

[1] (elected only)   [2] (may include some appointed judges)

Figure 3.2.3
USA: Black and Hispanic officials by category, 1993

Figure 3.2.4
USA: Distribution of the black population (%)

Region	1960	1970	1980	1990
North east	16.0	19.2	18.3	18.7
Midwest	18.3	20.3	20.2	19.1
South	60.0	53.0	53.0	52.8
West	5.7	7.5	8.5	9.4

Figure 3.2.4 shows the changing distribution of the black population from 1960. The most significant trends have been the relative decline in the South and the steady increase in the West. The pattern of movement is of course much more complicated than the final data might suggest; one such movement that should be mentioned is the considerable number of blacks who have returned south as manufacturing jobs were lost in the North East. The fact that the East South Central was the only southern division to show a decrease in its share of the nation's black population since 1970 is evidence that the era of high out-migration is over.

The states with the largest black populations (1995 projected figures) are New York (3.2 million), California (2.5 million), Texas (2.3 million), Florida (2.1 million), Georgia (2.0 million) and Illinois (1.8 million). Together these six states account for almost 42 per cent of the country's black population and only Georgia is in the traditional 'South'. However, a distinctly different rank order emerges if the black proportion of each state's total population is considered. Under this criterion the southern states clearly dominate, led by Mississippi, South Carolina, Louisiana, Georgia and Alabama. An important contrast also exists in the intra-regional distribution of blacks. In the South, while migration to the towns and cities of the region has been considerable, many blacks still remain on the land. However, in the other regions the black population is extremely concentrated in inner cities. Although there has been a greater degree of black suburbanisation in recent years it still operates on a relatively small scale and tends to maintain segregation. In the 75 largest metropolitan areas as defined by the 1990 census, blacks formed over 25 per cent of the population in the following: Washington-Baltimore, Atlanta, Norfolk-Virginia Beach-Newport News, New Orleans, Memphis, Richmond-Petersburg, Birmingham, Baton Rouge, and Charleston-North Charleston.

The influence of the black population on the geography of the USA has been substantially greater than in either Canada or Mexico. In Mexico slavery was abolished much earlier and the black population dispersed, inter-married and gradually lost its separateness. In Canada blacks form only a very small minority and for the most part have been absorbed into the wider community. It is the relative separateness and isolation of the blacks in the USA that sets the country apart from its neighbours and which has been the cause of so much internal conflict within the nation.

The root cause of black deprivation is high unemployment and low incomes. In 1993, the unemployment rate for blacks was 9.8 per cent, almost twice that for whites. In the same year, median weekly earnings for full-time workers were $478 for whites and $370 for blacks, a significantly wider gap than ten years earlier. While only 7.2 per cent of white families had incomes below $10 000, in 1992, 26.3 per cent of black families had incomes below this level (Figure 3.2.5). Not surprisingly, such an economic gap has clear social consequences. The

Figure 3.2.5
USA: White and black family income distribution compared

Family income in previous year in constant (1992) dollars	Percent distribution			
	White		Black	
	1980	1993	1980	1993
Total families	100.0	100.0	100.0	100.0
Less than $5000	1.8	2.7	6.9	11.3
$5000 to $9999	4.2	4.5	14.5	15.0
$10 000 to $14 999	6.2	6.6	13.0	11.8
$15 000 to $24 999	15.7	15.2	21.9	18.8
$25 000 to $34 999	16.1	15.3	13.9	13.0
$35 000 to $49 999	23.4	20.0	15.5	14.0
$50 000 or more	32.5	35.8	14.3	16.1

Figure 3.2.6
USA: Infant mortality rates by division, 1993

Division	White	Black
New England	6.0	13.1
Middle Atlantic	6.5	16.8
East North Central	7.4	18.7
West North Central	7.2	17.1
South Atlantic	6.9	16.4
East South Central	7.5	15.7
West South Central	7.0	14.9
Mountain	6.9	18.0
Pacific	6.3	15.8
USA	6.8	16.5

infant mortality rate is frequently cited as the best single indicator of social progress (Figure 3.2.6). In the USA as a whole the rate for black children is more than twice that for whites. The gap is also considerable in terms of life expectancy, as Figure 3.2.7 indicates. Another social indicator which has been the subject of much debate is the prison population. In 1991 47.5 per cent of all state prison inmates were black, up from 46.9 per cent in 1986.

The most devastating social change to hit the black community in recent years has been the abandonment of the family unit. A black child today has only a 1-in-5 chance of growing up with two parents until the age of sixteen. This is a staggering transformation in the space of less than half a century, for in 1950 black and white marriage patterns were very similar. The reasons for the low marriage rate among black parents have been keenly debated but academics and officials are closely agreed about the consequences.

Marriage is therefore one area where there appears to be a difference in values and attitudes between black and white. Another area is in the pattern of voting where blacks are much more likely to vote Democrat.

Figure 3.2.7
USA: life expectancy in years, 1994

	White Male	White Female	Black Male	Black Female
At birth	73.1	79.5	64.6	73.7
At 40	35.9	41.0	30.2	36.9
At 65	15.4	19.0	13.4	17.1

Hispanics

The word 'Hispanic' appeared as a census term only in 1980. It applies to people of white, black, Indian and, frequently, thoroughly mixed ancestry who hail from countries that sometimes seem to have little in common except language and historical traditions. For most Hispanics, the Spanish language is the most important symbol of their common cultural heritage. Although many emigrate from rural areas, in the USA they are largely resident in urban areas.

In mainstream US society, misconceptions about Hispanics tend to overshadow the reality. Many Americans feel Hispanics are prone to violence and gang warfare. Youth gangs are a problem in some areas but police analyses show that 'barrio' crime rates are no worse than in poor black and poor white areas. Illegal immigrants in particular seem to be less the perpetrators than the victims of crimes, which, for obvious reasons, they are often reluctant to report.

People of Latin descent are increasingly embracing the term 'Hispanic' as a way of enhancing their individual rights and collective identity. The socio-political system has created clear incentives for Hispanics to group themselves together. In the political arena, it has obviously helped them gain 'safe seats' in Congress, through aggressive redistricting in recent years. It also pressurises society to allocate more jobs and entitlements to an obvious minority group campaigning for equality. Hispanics lagged behind non-Hispanics in various economic and social indicators in 1991, including education levels. But Hispanics as a group had higher household incomes, lower unemployment and fewer female-headed households than blacks did that year (Figure 3.2.8).

Some commentators see a conflict between black and Hispanic civil rights campaigns. In Los Angeles County, for example, blacks, who make up 10 per cent of the population, hold 30 per cent of the public sector jobs. Hispanics, who constitute 33 per cent of the population, hold only 18 per cent of the jobs. 'Blacks think we want to take jobs away from them, so they're fighting us tooth and nail,' says Raul Nunez, president of the Los Angeles County Chicano Employees Association. 'They are doing the same thing to us that whites did to

Figure 3.2.8
USA: How Hispanics compared, 1991

	Non-Hispanics	Blacks	Hispanics	Mexicans	Puerto Ricans	Cubans	Central and South Americans
Population (in millions)	227.4	30.9	21.4†	13.4	2.4	1.1	3.0
Median household income	$30 513	$18 676	$22 330	$22 439	$16 169	$25 900	$23 568
Per cent households with income of $50 000 or more	25.4	11.9	13.4	11.6	11.9	19.8	15.7
Per cent female-headed households	11.4	47.8	19.1	15.6	33.7	15.3	21.5
Per cent urban households	72.8	N/A	91.8	90.5	95.2	95.7	97.0
Per cent households owning or buying home	65.8	42.4	39.0	43.5	23.4	47.3	22.2
Per cent completed high school	80.5	66.7	51.3	43.6	58.0	61.0	60.4
Per cent with four or more years of college	22.3	11.5	9.7	6.2	10.1	18.5	15.1
Per cent unemployed	6.9	12.4	10.0	10.7	10.3	6.4	10.3
Per cent of individuals below poverty level	12.1	31.9	28.1	28.1	40.6	16.9	25.4

† Note: This figure does not include the 1.6 million people in the category 'other Hispanic.'

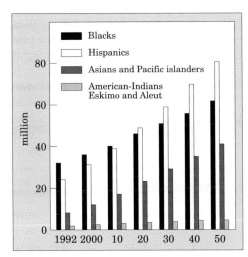

Figure 3.2.9
America's minority populations

them.' However, many civil rights leaders play down any idea of conflict, contending that it is largely a product of media hype.

What is undeniable is the fact that the Hispanic population of the US is growing at a rapid rate. By 2010 it will have overtaken the black population as the main minority group (Figure 3.2.9).

American Indians

Native Americans, or 'Indians', form a minority more unique and distinctive than any other in the USA. Their segregation is more intense and their socioeconomic standing lower than all other ethnic groups. However, in the 1970s Indian groups began to assert themselves more effectively in the political arena. An early result of such action was the formation of the American Indian Policy Review Commission. The commission's 1977 report painted a bleak picture of reservation life in the USA, recording the lowest standards in the nation for health, education, housing and employment. American Indians want to share the benefits of modern society but are acutely fearful of what is frequently referred to as 'cultural genocide'. The root of Indian poverty lies in the poor land quality of their reservations. As Figure 3.2.10 shows the geography of the American Indian today is very different from how it was before European colonisation.

The 1990 census showed a significant jump in the population of Indians from 1.4 million in 1980 to almost 1.9 million. Some of the increase, experts say, reflects non-demographic factors: people are more willing than they once were to identify themselves as American Indian on census forms. New-found racial pride and cultural awareness may have been a factor here. J S Passel of the Washington-based Urban Institute also speculates that there 'may be a certain romanticism on the part of the people answering the census – a tendency to romanticise their heritage'. Many additional 'switches' might have occurred if a

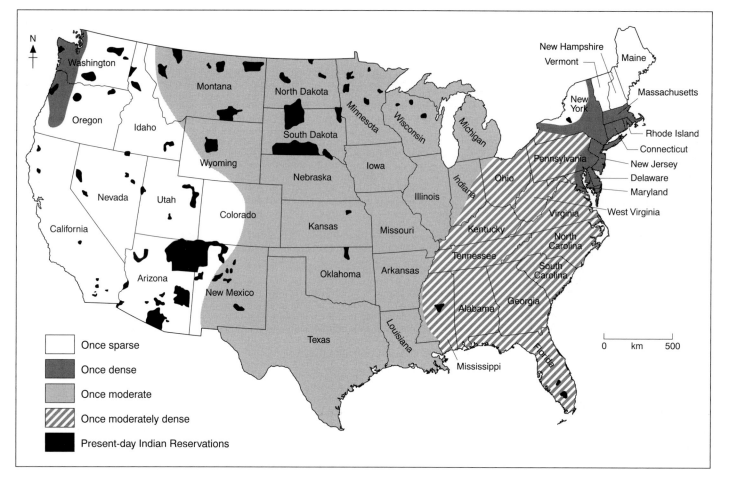

Figure 3.2.10
The changing population distribution of the Indians in the USA

hit movie sympathetic to Indians had been released earlier, Passel believes. ' "Dances with Wolves" came out at the end of 1990,' he says, 'if it had come out before the census [April 1990] we'd have seen a lot more American Indians'.

Indians live throughout the nation (Figure 3.2.11) and comprise small but significant minorities in a number of states. The greatest absolute concentration is in a line of four states from Oklahoma to California, together accounting for over 43 per cent of all Indians. Indians also make up a sizeable presence in several major metropolitan areas. The Cherokee and Navajo are by far the largest tribes (Figure 3.2.12).

Of the almost one billion hectares that Native Americans roamed 500 years ago, less than 19 million remain in Indian hands today. The Indians' lost heritage helps explain their opposition to the 500th anniversary celebrations in 1992 of Christopher Columbus's first voyage to the New World. The attitude of the Indians is that Columbus was not a brave discoverer but the bringer of slavery, disease and genocide.

After a prolonged period of relative neglect by the federal government, Indian activism began to emerge on a nationwide scale in the mid-1960s. As a result there is now a much wider awareness of the Indians' crushing economic, social and health problems. The death rate from diabetes, tuberculosis and sudden infant death syndrome is substantially above the US average.

Poverty and social isolation make many Indians vulnerable to chronic ailments like depression, alcohol and drug abuse. According to the 1980 census, 28 per cent of Indians lived in poverty. A quarter of all reservation houses lacked complete plumbing, while 16 per cent lacked electric lighting. Indians under the age of 19 are nearly three times as likely to commit suicide than youths of other races. As many as one in three Indian deaths is caused by alcohol abuse.

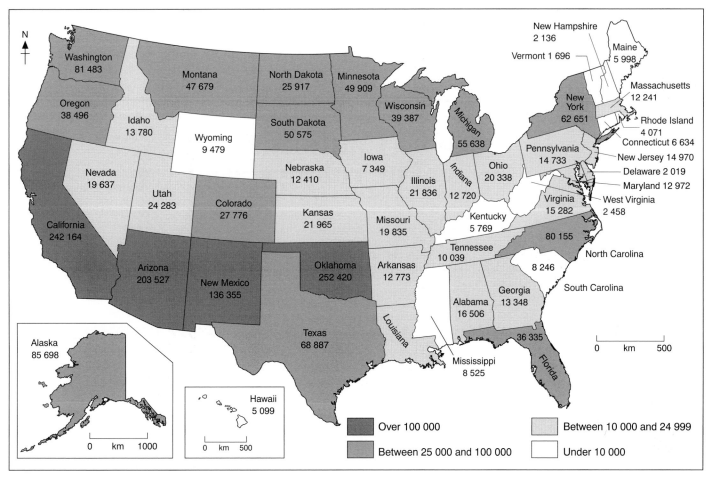

Figure 3.2.11
USA: Indian population by state, 1990

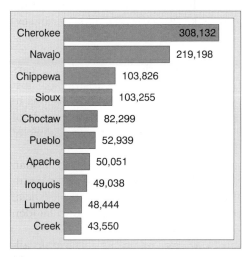

Figure 3.2.12
The ten largest American Indian tribes in 1990

However, some degree of progress has been made. The Indian Health Service, created by Congress in 1921, has helped bring communicable diseases like tuberculosis, pneumonia and gastrointestinal ailments under control. As a result the life expectancy of Indians has risen dramatically over the past few decades and infant mortality has been brought down to the US average.

Certain Indian demands, particularly the return of millions of hectares of their ancestral lands, inevitably provoke intense opposition when they clash with the interests of other groups. In February 1972 Maine's Passamaquoddy and Penobscot tribes sued the state, demanding 5 million hectares and $25 billion in damages. The disputed land area, covering more than half the state's land area, embraced some 350 000 non-Indian residents and the city of Bangor.

The two tribes acknowledged that the land had been relinquished more than 150 years earlier in treaties signed with local authorities. However, the Indians claimed the treaties were invalid because they were never approved by the federal government, as required by a 1790 law. Under a 1980 agreement, approved by Congress and the Maine Legislature, the tribes accepted a $27 million trust fund and 120 000 hectares. In return they surrendered all further land claims. Another longstanding Indian grievance centres on the desecration of natural sites regarded as sacred. At South Dakota's Mount Rushmore National Monument, for example, Indians say the four gigantic heads of US presidents carved into the Black Hills defile a spiritual landmark.

Recent cuts in funding for economic development on reservations have caused resentment. More and more tribes have looked to the provision of casinos as a source of money and jobs on reservations, where unemployment rates above 40 per cent are not uncommon.

Asian Americans

Between 1984 and 1994 the Asian American population doubled to 7.5 million. Census Bureau projections and immigration patterns suggest there will be three times as many Asian Americans by the year 2020. Ethnic diversity characterises the Asian community, encompassing large numbers of Chinese, Korean, Japanese, Filipino, Vietnamese, Cambodian and Laotian Americans.

California
The ruling class

SAN FRANCISCO

For the moment, whites run California. White governor; two white senators; white mayors in the two largest cities. But their hegemony may be ending. Their successors will not be Latinos, who are set to surpass them in numbers by about 2020, but Asians, who are surpassing them in brains.

On several University of California campuses, freshman classes last year enrolled more Asian than white students. At UCLA, 38% of the freshmen were Asian; 30.5% were white. At Berkeley, so many Asians came in last year that they tipped the undergraduate total to 36.6% Asian, 34.5% white. Mutterings about quotas are beginning to be heard. Yet in the 1990 census Asians comprised only 9.5% of California's population.

Meanwhile Latinos, who now constitute 26% of all Californians and who have accounted for half the state's population growth in the past decade, lag at the bottom of the academic charts. A survey of all nine University of California campuses found that they accounted for a mere 13% of all new students last year,

while Asians made up 36%. Latinos are also at the bottom for the percentage of college graduates among Californians over 25: a mere 7%. Asians, yet again, were top with 34%; whites made up 25%.

The discrepancy starts in high school. In 1992, 52% of Asian students in California's high schools were taking the relatively demanding courses that prepare a student for college. Among white students, only 34% were taking these courses; among blacks, 27%; and among Latinos, last again, 21%. In the 1990 census, Latinos also made up 62% of the Californians between the ages of 16 and 19 who had not graduated from high school.

The state's business leaders are alarmed that Latinos, who are about to become the largest segment of the population, show the least interest in higher education. Many colleges are now going out of their way to encourage Latino pupils at high school, even offering them summer courses on campus to motivate them to go there, or giving them grants to finish their degrees. But the reasons for their bad

performance are still puzzling.

Sanford Dornbusch, a professor of sociology and education at Stanford, blames the "tracking" system used in elementary grades, by which teachers classify children early on into "low" or "college-bound" streams. He has found that the process demoralises Latino and black children, who give up trying. Others say there is too much pressure on Latino boys to leave school to take (badly paid) jobs. Poverty, in general, takes much of the blame. In 1989 (when the last such study was done), Latino children accounted for more than half of the children aged 17 or younger who were living below the official poverty level in California.

In the same study, Asian children made up 11% of those living in poverty. That is higher than their percentage in the population. It suggests, at the least, that many Asian families are struggling too. But cultural and social factors seem to make a difference: no matter how poor they are, Asian families prize education, and believe that their children should be trained to their full potential. Although it may be politically incorrect to say so, race matters.

The Economist
17 Sept 1994

Figure 3.2.13
The ruling class

Among the issues facing this diverse group are the development of political cohesion and documenting discriminatory voting patterns. In the past, Asian-Americans outside of Hawaii have been unsuccessful in persuading officials to create majority-Asian districts. In San Francisco, where Asian-Americans make up 30 per cent of the city's population, only one Asian-American has served on the Board of Supervisors. In New York City, where the Asian-American population exceeds half a million, Asian-Americans have never been elected to legislative office this century. One major obstacle is language – in New York City, for example, 44 per cent of Asians do not speak English well. However young Asians are rapidly assimilating into American society, many through high academic achievement (Figure 3.2.13).

QUESTIONS

1 Compare the demographic characteristics of minority groups shown in Figure 3.2.1.

2 To what extent and why is there a correlation between race and residential environment?

3 (a) Why is the infant mortality rate so much higher for blacks than whites (Figure 3.2.6)?
   (b) Suggest reasons for the considerable divisional variation in black infant mortality.

4 To what extent do the socio-economic characteristics of the main groups that make up the Hispanic population vary (Figure 3.2.8)?

5 Compare the population projections for the main minority groups in the USA (Figure 3.2.9).

6 Describe and explain the changing population distribution of the Indians illustrated by Figure 3.2.10.

7 Outline the economic, social and political problems facing American Indians.

8 Discuss the contention that Asian-Americans have assimilated more rapidly into mainstream American society than any other minority group in the US.

3.3 The Aboriginal Peoples of Canada

Canada's aboriginal population includes Inuit, Metis, Status Indians and non-Status Indians. Together they make up about 2 per cent of the nation's population. Metis people are of mixed Indian and non-Indian blood.

The Inuit

The word Inuit means 'people'. The Inuit of Canada were once called Eskimos, an American Indian word meaning 'eaters of raw meat'. Most of Canada's 33 400 Inuit live in the Northwest Territories, northern Quebec or Labrador (Figure 3.3.1). Inuit communities are sited mainly on the coast, reflecting the historic use of coastal waters for food and transport. Sites were also determined by the location of trading posts, giving Inuit hunters a local centre to sell their furs and skins. Today the Inuit depend much less on hunting, trapping and fishing. Work in tourism, arts and crafts, oil exploration and municipal services has steadily become more important. Many Inuit now work for their own cooperatives rather than outside companies. However, while the Inuit have achieved much, their standard of living is still well below the Canadian norm.

Figure 3.3.1
Inuit communities in Canada

Figure 3.3.2
Canada's Indians fight for their rights

Canada's Indians Fight for Their Rights
Newly designated Inuit territory of Nunavut

Indians in Canada share many of the grievances and hopes voiced by Indians in the United States. This is hardly surprising, considering that many tribes have members in both countries. But the legal status of Canadian Indians differs in fundamental ways from that of U.S. Indians.

Unlike tribes in the United States, Canadian tribes are not treated as sovereign nations. The Indian Act of 1876 made Canada's Indians wards of the federal government. It also made their reservations, known in Canada as reserves, subject to both provincial and federal law.

While Indians in the United States can use their reservation land as collateral for loans, Canadian Indians cannot. The ban, Indian leaders say, acts as a daunting obstacle to economic advancement. "We are sitting on millions of dollars of real estate, and we can't use it to access financial institutions," says Darrell Boissoneau, chief of the Ojibwa tribe on Canada's Garden River First Nation Reserve, near northern Michigan.

Still, Canadian Indians are making headway in their battle to win self-government — including the right to make their own laws and tax themselves. Their cause received an unexpected boost in June 1990 when the sole Indian member of the Manitoba legislature, Elijah Harper, a Cree, effectively killed a series of constitutional amendments known as the Meech Lake Accord, arguing that it failed to address the concerns of native peoples.

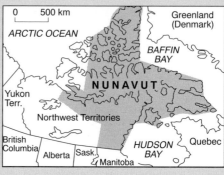

A development of more tangible benefit to Canada's aboriginal peoples came May 4, when residents of the Northwest Territories approved the western boundary for the vast, new territory of Nunavut ("Our Land" in the Inukitut language). The vote followed a Dec. 16 announcement by the Ottawa government that it had reached agreement with leaders of the Inuits, or Eskimos, on a land-claim settlement including the creation of the new territory in 1999. The accord calls for the existing Northwest Territories to be split into two sections — Nunavut, embracing 772,000 square miles [1.2 million km²] in the north and east, and an unnamed western sector. Eskimo voters are expected to vote on ratification of the land claim agreement in November.

The Inuit are to come into outright ownership of 135,000 square miles [217 000 km²] within Nunavut — a subarctic territory slightly larger than New York, Pennsylvania and Ohio. They also will receive the equivalent of about $1 billion U.S. dollars over a 14-year period, as well as hunting and fishing rights in the rest of the sparsely populated Nunavut region. Parliament is expected to vote on legislation endorsing the new territory and land-claim settlement in early 1993.

However, things have not gone so smoothly for Canadian Indians when it comes to hunting and fishing rights on lands with majority non-Indian populations. In May 1991, for instance, Ontario Premier Bob Rae created an uproar when he announced that native Canadians would be allowed to hunt and fish for food, or for ceremonial purposes, in contravention of the province's wildlife laws. Many white Ontarians vehemently objected, claiming that Indians were depleting wildlife resources.

"It is not 1492," said Matthew Murphy, a spokesman for the Ontario Federation of Anglers and Hunters. "You cannot hunt and fish with abandon."

CQ Researcher,
8 May 1992

The language of the Inuit is Inuktitut and in an effort to protect this language and culture the Inuit Broadcasting Corporation was established in 1951, transmitting up to six hours of Inuktitut programmes a week. Inuktitut newspapers and magazines have also been established and Inuit schools now teach more and more in Inuktitut rather than in English. Television Northern Canada (TVNC) began broadcasting in January 1992 to 100 000 people, over half of whom are of aboriginal ancestry. By May 1993 TVNC was broadcasting over 100 hours per week of aboriginal language, culture, information and educational programming. There are broadcasts in 15 aboriginal languages as well as in English and French.

The Inuit now have more control over local government than in the past through settlement councils and, like the Indians to the south, they have made a number of land claims. Some agreements have already been reached, the most important of which will be the establishment in 1999 of the new territory of Nunavut, covering one-fifth of Canada (Figure 3.3.2). John Amagoalik, policy advisor for the Tungavik Federation of Nunavut hailed the agreement as his people's entrance into the Confederation. Nunavut will be run by a public government similar to that of the Northwest Territories and Yukon. Inuit leaders have begun education and training programmes designed to prepare their people for the positions that will be created as their government develops.

The Indians

Non-Status Indians are native by birth but not registered as members of specific bands, with most living away from Indian reserves. Canada's total registered Status Indian population was 533 461 in December 1992, belonging to 604 bands (Figure 3.3.3). About 295 000 live on Indian reserves, 217 800 outside reserves and 20 600 on Crown-leased land.

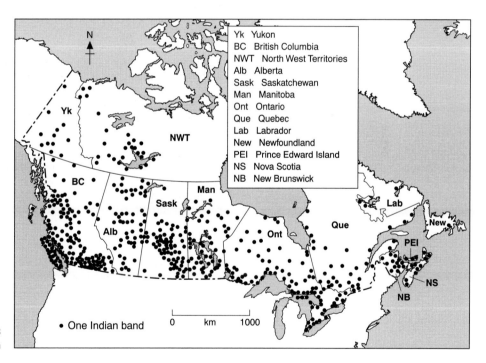

Figure 3.3.3 Indian bands in Canada

While the standard of living of Indians has improved over the years it is still significantly below the Canadian average:

- life expectancy is ten years less;
- violent death is three times more likely;
- university entrance is half the national average;
- unemployment is more than twice the average;
- less than 50 per cent of Indian homes are properly serviced compared with over 95 per cent nationally.

Indian politicians claim that such deprivation is a result of a history of discrimination against aboriginal peoples. In particular they argue that:

- much of their land was taken away from them without compensation;
- many reserves have very poor quality land;
- Canadian law unfairly restricts the ways in which reserve land can be used;
- government investment in Indian projects has been too low;
- firms are less likely to employ Indians than other Canadians.

The Indians have become more and more aware of the need to state their case clearly (Figure 3.3.4). To facilitate this process bands have grouped together to form associations. The most influential is the Assembly of First Nations. Canada's First Nations speak 50 languages, belonging to 11 distinct linguistic families. Cree, spoken by almost 94 000 Canadians, is the most common aboriginal language.

Figure 3.3.4
Taking control: Micmacs bring Indian views to education

Taking control: Micmacs bring Indian views to education

The sign on St. Anne's Roman Catholic School in the remote Micmac community of Conne River has been taken down. A new sign – Sent Anneway Kegnamo'-gwom – is to be affixed to the brick and clapboard school.

The 600 Micmacs who live on Newfoundland's only Indian reserve are assuming control of the education of the 220 young people living in Conne River. The Micmacs are the first native group in Atlantic Canada to be granted responsibility by the federal Department of Indian Affairs and Northern Development for both elementary and high schools.

"Our kids are treated as if they were Newfoundlanders, English people. We are saying, no, they are going to learn the same facts but from an Indian point of view," Jerry Wetzel, director of planning and policy for the Conne River Band Council, said.

He cited an example of such bias in a history text that refers to the English explorer John Guy, who visited Newfoundland in the early 1600s. Mr Wetzel said the book notes that the explorer and his fellow sailors, as they came ashore, "saw savages on the beach."

"Those were not savages. Those were our grandparents," he said. "That is the kind of discrimination we intend to stop."

Toronto Globe and Mail,
8 August 1986

Since 1986, the Canadian government has pursued community self-government negotiations with individual aboriginal communities. As of March 1993, two native communities in Canada, the Cree-Naskapi of Quebec and the Sechelt Band of British Columbia, became self-governing in the sense that they have their own local governments which are legal and political entities of municipal status, accountable to an aboriginal electorate. Negotiations were in progress on 15 other community government projects involving 45 First Nations. In addition, negotiations concerning comprehensive land claims were in progress with four groups: the Nisga'a First Nation in British Columbia, the Council for Yukon Indians, and the Gwich'in and Sahtu First Nations, both in the Northwest Territories.

In September 1990 the government launched a Native Agenda focussing on four main areas: settling land claims, improving social and economic conditions on reserves, addressing the needs of native peoples in contemporary society, and building a new relationship between Canada and its native peoples.

Canada enacted a land claim settlement in February 1995 that gives native Indians in the Yukon control of an area bigger than Switzerland. The deal gives the Indians ownership of 41 440 km² of land bordering Alaska and cash compensation of C$ 243.7 million paid over 15 years.

The land claims process

Native groups, in a process of complex negotiations with the federal government, surrender claims to large tracts of territory they traditionally utilised in return for retention of ownership of smaller land areas. At the outset bands have to legitimise claims over territory by proving historic and effective use and occupancy of land. Subsequently they select land to which they wish to retain ownership from within this territory. Legitimacy of a claim is communicated through mapping of areas of traditional use and occupancy. Cultural landscape features, including harvest areas, camp-sites, settlement sites, traditional route-ways and indigenous toponymy are mapped in a process that often involves cartographic representation of information traditionally stored mentally and transmitted orally. Until recently conventional cartographic techniques have been utilised to depict land-claim information. However there is an increasing interest in the application of digital cartography and Geographical Information Systems to land claims.

QUESTIONS

1 Account for the location of Inuit communities in Canada.

2 Why is the creation of Nunavut regarded as so important by the Inuit (Figure 3.3.2)?

3 Why do most Indians claim that their present socio-economic and political status is a result of a history of discrimination against them?

4 Why do Indian bands like the Micmacs (Figure 3.3.4) want to control their children's education?

5 Suggest why some Canadians are against the government settling aboriginal land claims.

6 How can Geographical Information Systems aid the land claim process?

7 Why are Canada's urban Indians often said to be worse off than their counterparts living on reserves (Figure 3.3.5)?

METRO'S INVISIBLE NATIVES

For many of the estimated 40 000 to 70 000 Indians in Metro Toronto, life is a choice between booze and the mean streets, or invisibility and absorption into the white man's world.

Natives live in pockets throughout Metro: Danforth and Pape, the rooming house district around Parliament and Dundas, a non-profit housing project in Scarborough, the West End. Some live in squalor; some live in comfortable homes on comfortable streets.

However, an inordinately high proportion of native Indians is homeless. According to Clarence Southwind, who heads an organisation called Street Workers Anishnawbe Toronto which helps the homeless, between 1000 and 2000 natives live on the streets.

20% of native Canadians in Toronto are estimated to be unemployed. That is certainly a far better ratio than exists on some reserves, but it is about four times the average jobless rate in the city. In one of the richest cities in Canada, Indians have the highest unemployment rate and the lowest level of skills.

The Toronto Star,
18 April 1990

Figure 3.3.5
Metro's invisible natives

3.4 Quebec and the French Canadians

Quebec is Canada's largest province (Figure 3.4.1). Over $1\frac{1}{2}$ million km² in area, it is more than twice the size of Texas and nearly seven times bigger than Britain. Its population (7.3 million in 1995) is, within Canada, second only to that of Ontario ; its capital, Quebec City (695 000), is the oldest European settlement in Canada; and its metropolis, Montreal (3.3 million), is Canada's second largest city and its busiest port.

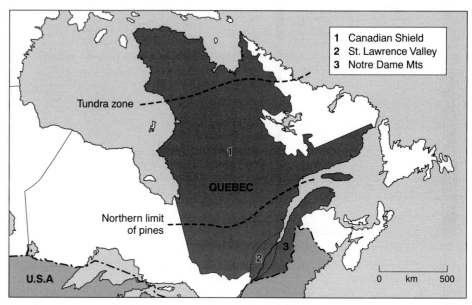

Figure 3.4.1
The Province of Quebec: natural regions

Almost all of Quebec's population live in the St Lawrence Valley and adjoining lowland areas penetrating the Laurentian and Notre Dame mountains (Figure 3.4.2). In fact, southern Quebec contains a quarter of Canada's entire population and comprises the eastern section of the nation's economic core or heartland – the western section being southern Ontario.

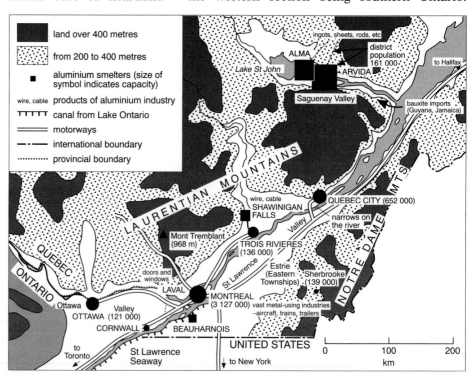

Figure 3.4.2
Southern Quebec: physical and economic features

1 (a) Explain why the great majority of Quebec's population is concentrated in the southern part of the province (Figures 3.4.1 and 3.4.2).
   (b) Suggest which peoples inhabit the huge expanses of northern Quebec.

2 (a) On an outline map of Canada showing provincial boundaries, construct histograms (bar graphs) to represent the two data sets in Figure 3.4.3.
   (b) Identifying some of the problems likely to be encountered with these particular statistics.
   (c) How many French Canadians live outside Quebec and what percentage of all French Canadians do they represent?
   (d) Why do French Canadians continue to move to anglophone Canada and the US?

The competition between these two regions – the one francophone, the other anglophone – has always been intense. Indeed, the real significance of Quebec is that its population includes 5 663 000 French-speakers – over 85 per cent of all Canadians whose mother tongue is French.

The Quebec Act (1774) guaranteed religious freedom for the Roman Catholic French Canadians, permitted the continuance of the French code of civil law, and recognised French as an official language in the new British colony. However, following the American War of Independence, these benefits were offset by the arrival in Quebec of many 'United Empire Loyalists'. These were American colonists who, wishing to retain their allegiance to Britain, emigrated from their homes in the newly-formed USA and settled in the Ottawa Valley, Montreal and the Eastern Townships (now known as Estrie) – the area east of Montreal and south of the St Lawrence River (Figure 3.4.2).

Quebec thus gained large numbers of English-speaking inhabitants for the first time. The port of Montreal increasingly became the main attraction for British merchants and by the mid-nineteenth century the city's anglophones outnumbered francophones. The balance has swung back since then; Montreal is now, after Paris, the world's largest francophone city – but its English-speaking minority have continued to control much of Quebec's economic life.

During the nineteenth century, population pressure, mainly from natural increase, caused many French Canadians to move to remoter parts of Quebec where they pioneered farming, lumbering and mining activities. This movement ensured that the province's rural areas became almost completely francophone.

Other French-speakers migrated to northern Ontario and Manitoba and to New Brunswick and other parts of Atlantic Canada, as well as to New York and New England in the adjoining USA. However, once out of Quebec, French Canadians became minorities in anglophone communities where French culture had little chance of survival. Today, only in New Brunswick can the francophone minority be regarded as substantial (Figure 3.4.3).

Figure 3.4.3
Location of French-speakers, by province and territory, 1991

	Number of French-speakers* (thousands)	Percentage of total population
Quebec	5663	82.1
Ontario	505	5.0
New Brunswick	244	33.6
Alberta	58	2.3
British Columbia	52	1.6
Manitoba	51	4.7
Nova Scotia	37	4.1
Saskatchewan	22	2.2
Prince Edward Island	6	4.5
Newfoundland	3	0.5
NW Territories	2	2.5
Yukon	1	3.2
CANADA	6643	24.3

* Having French as mother tongue.

Political developments

In 1867 the British North America Act included Quebec as one of the four founding provinces of modern Canada (the others were Ontario, Nova Scotia and New Brunswick). The new nation was established as a parliamentary democracy based on the British model, and with the British monarch as head of state. Not surprisingly, French Canadians were not wholly comfortable with this

Figure 3.4.4
Quebec's provincial parliament, Quebec City

situation. However, Quebec – then containing a third of the entire Canadian population – was able to insist on a federal system of government and that the province retain its distinctive institutions, especially the legal protection given to the French language.

Despite these concessions French-Canadian nationalism became a permanent feature of both federal politics (in the Ottawa parliament) and Quebec provincial politics (in the provincial parliament, the self-styled 'National Assembly', in Quebec City) (Figure 3.4.4).

Since World War II, calls for the political secession of Quebec from Canada have become increasingly strident. Notable developments towards separation have included the following:

- 1948: The introduction of a new provincial flag emphasising Quebec's historic links with France (Figure 3.4.5).
- 1953: The emergence of a terrorist group, the *Front de Libération du Québec* (FLQ), whose campaign of sporadic violence culminated in the murder of Quebec's Minister of Labour in 1970.
- 1963: The 'nationalisation' of all electricity production and distribution in the province. Hydro-Québec became Canada's biggest producer of electricity, much of which was to be sold to neighbouring New York state, the proceeds being earmarked to finance Quebec's future political independence.
- 1967: The official visit to Canada of President de Gaulle of France, who uttered the cry *'Vive le Québec libre'*, delighting Quebec nationalists but offending most other Canadians.
- 1970: The formation of the *Parti Québécois* (PQ) to campaign democratically at provincial level for the separation from Canada on social and economic grounds, not just for historical and cultural reasons.
- 1976: the PQ won control of Quebec's National Assembly and a year later banned the use of any language other than French in public places throughout the province, including shopfronts, advertisements and public transport. In addition, all immigrant children had to attend francophone schools, and French became the compulsory language of business in all offices, factories and workplaces throughout Quebec.
- 1980: the PQ held a referendum on the separation issue but, rather surprisingly, Quebecers voted 60–40 against any immediate attempt to secede from Canada. It was clear, however, that the issue had not been finally resolved: *'On revient'* ('We'll be back') was the PQ's defiant message.

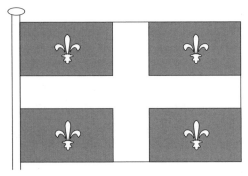

Figure 3.4.5
Quebec's provincial flag

The banning of English became a major issue in Montreal (which contains most of the province's 18.5 per cent non-francophones). It resulted in many footloose companies relocating in Toronto and the departure of around 250 000 English-speaking Quebecers in the 15 years after 1976. Quebec's language laws were modified in 1993 after rulings by the UN's Human Rights Committee and Canada's Supreme Court. English (and other languages) may again be used on shopfronts, etc. as long as an accompanying French sign is 'markedly predominant' – i.e. twice as big!

Meanwhile, Canada's federal government had moved officially to give French equal status with English throughout the country. Even before the Official Languages Act of 1969 both languages had been used in the Ottawa parliament and in government documents. Moreover, many French Canadians have risen to the nation's highest positions; in 26 of the 28 years preceding 1997, French Canadians served as the country's prime minister, and in 1997 they held nearly a third of the 25 cabinet posts in Jean Chrétien's Liberal government in Ottawa. Even so, it remains true that to enjoy such success, francophones have to become fluent in English while their anglophone compatriots have never been under the same pressure to become proficient in French.

Other federal attempts to appease Quebec's separatist tendency foundered at constitutional conferences at Meech Lake in 1987 and Charlottetown in 1992. Proposals to recognise Quebec as a 'distinct society' with special rights within the Canadian confederation were vetoed first by the other provinces and then in a nation-wide referendum.

Following these rebuffs, a new political party, the *Bloc Québécois* (BQ), was formed at the federal level to press for separation from Canada. The BQ won nearly three-quarters of Quebec's seats in the Canadian general election – enough to make its leader, Lucien Bouchard, the official 'Leader of Her Majesty's Loyal Opposition' in the Ottawa parliament! Equally ironically, Canada's prime minister (1997), Jean Chrétien – another French Canadian from Quebec – was totally dedicated to preserving Canada's unity and keeping the province within the confederation.

Quebec's 1995 Referendum

In September 1994, Quebec held a provincial election. As expected, the PQ, now committed to holding another referendum on the separation issue, was returned to power with 77 of the National Assembly's 125 seats but with only 44.7 per cent of the total votes cast. (Like Britain, Canada operates the 'first past the post' system of voting in both federal and provincial elections).

More than a year of bitter political debate preceded the promised referendum. The issues included:

- the right of Quebec's government, elected by a minority of the province's voters, to commence moves to secede on the basis of a simple majority of votes (50 per cent plus one);
- Quebec's relatively poor economic condition (despite a disproportionately large share of federal contracts and other government funding) would probably deteriorate further after independence. Surveys showed that most Quebecers favoured some form of economic association with Canada, including unhindered free trade and the continued use of the Canadian dollar. But most other Canadians objected to such proposals;
- would a breakaway Quebec automatically inherit Canada's membership of NATO, NAFTA, the WTO (see Chapter 13), the OECD, the Commonwealth, and other international bodies? The PQ assured its supporters that Quebec would readily be accepted by such organisations, but no guarantees were ever offered;
- the effects of Quebec's independence on minorities: (a) francophones living in other parts of Canada would probably experience greater discrimination, and (b) Quebec's own minorities, including the native people of northern Quebec, all rejected secession since they feared losing their already limited rights without the protection of federal Canadian law;
- the United Nations Human Development Reports of 1995 and 1996 both rated Canada as the best place to live in the world. The reports considered a range of socio-economic criteria, such as GDP per capita, employment, taxes, education, nutrition, health and life expectancy. Did Quebecers really want to leave the world's best country? Surveys showed that almost 80 per cent of them wanted to remain Canadian citizens even if the province became independent, while nearly a quarter of all Quebecers – including 255 000 francophones – would seriously consider moving to another province;
- the wording of the referendum question itself became an issue. Opponents of secession wanted a clear, unambiguous question, such as: 'Do you want Quebec to separate from Canada and become an independent country? Yes or No?' The question actually put to Quebecers was: 'Do you agree that

QUESTIONS

1 At confederation in 1867, why did Quebec insist on a federal system of government for Canada?

2 Argue the case for and against Quebec's language laws from the standpoint of (i) a French-Canadian nationalist; (ii) an anglophone descendant of Empire Loyalist settlers; (iii) a newly-arrived immigrant from Italy.

3 What evidence is there to support the federalists' assertion that since the British conquest of New France little more could have been done to recognise the 'French fact' of Quebec and to accommodate the province within the Canadian family?

Quebec should become sovereign after having made a formal offer to Canada for a new economic and political partnership, within the scope of the [National Assembly] bill respecting the future of Quebec and of the agreement signed [by the main separatist leaders] on June 12, 1995?'

Referendum result and aftermath

The referendum on Quebec sovereignty was held on 30 October 1995. With a very high 93 per cent turn-out at the polls, Quebecers voted as follows:

'NO' to sovereignty 50.6 per cent (2 350 657 votes)
'YES' to sovereignty 49.4 per cent (2 301 917 votes)

This extremely narrow victory for the federalists brought Canada back from the brink of imminent break-up, but it was scarcely an affirmation of national unity. PQ leaders blamed the defeat on 'money' (i.e. Montreal-based big business) and 'the ethnic vote' (i.e. Quebec's non-francophone minorities) – declaring that the separatist struggle would continue and another referendum would be put to Quebecers 'at the right time – a winning one'.

The federal government immediately drew up a plan to meet the Quebec separatists' three main demands:

- formal recognition (yet again) of Quebec as a distinct, French-speaking society;
- giving Quebec a veto on constitutional change (balanced by similar vetoes for Ontario and, as regions, the four Atlantic provinces and the four western ones);
- transferring responsibility for manpower training from the federal to the provincial level.

Only time will tell whether these measures will be enough to satisfy Quebec's separatists. Before the 1995 referendum, people from all over the country demonstrated to express their affection for Quebec and their desire to see it remain part of the Canadian family. However, others in English-speaking Canada are becoming increasingly disillusioned with Quebec and its antics as 'the spoilt child of confederation'. If Quebec does eventually decide to become an independent country, not all Canadians would regret its departure.

QUESTIONS

1 (a) Discuss the general advantages and disadvantages of the 'first past the post' voting system.
  (b) Proportional representation (PR) is often suggested as a fairer system. What are the strengths and weaknesses of PR?
  (c) What difference might PR have made in Quebec's provincial election in 1994, and therefore in the subsequent referendum?

2 Why was the referendum question that was put to the Quebec electorate in 1995 severely criticised by opponents of secession?

3 Partition has been proposed as a radical solution to the Quebec problem. All areas which recorded more than a 60 per cent 'Yes' vote in the 1995 referendum would be united into a new independent Republic of Quebec; the remainder would comprise a smaller Province of Quebec that would remain part of Canada. Taking account of similar situations elsewhere in the world, discuss the advantages and disadvantages of the partition proposal for Quebec.

4 (a) Present a rational, dispassionate case for Quebec remaining part of Canada.
  (b) Why do so many francophone Quebecers continue to demand their own separate country?

4

Urban Patterns and Problems

4.1 Continental Overview

Historical perspectives

When the first US Census was taken in 1790, the American people were overwhelmingly rural in terms of their residential location. The country's largest urban places, the ports that developed along the Atlantic coast in colonial times, were little more than small towns or large villages. With a population of 42 520, Philadelphia was still the largest of these settlements, though it would soon be overtaken by New York (33 131), which was destined to become the continent's primate city. Even smaller were Boston (18 038), Charleston (16 359) and Baltimore (13 503). Below these five largest towns, the urban hierarchy fell away dramatically.

By 1870, despite the opening up of the west and the development of industry in the north east, the United States was still a rural nation. Only 21 per cent of the population lived in towns of more than 8000, although big cities already contained 12 per cent of the total population. By far the largest was New York (942 000), which together with Brooklyn and the towns on the New Jersey shore, formed a conurbation of well over 1½ million, clearly the USA's major metropolis. Left trailing in second place was Philadelphia (674 000), while Chicago (300 000) and St Louis (311 000), the regional centres of the Midwest, were now larger than Boston (250 000) and Baltimore (267 000).

The majority of Americans remained rural dwellers until well into the twentieth century (Figure 4.1.1). Since then, the balance has shifted increasingly towards urban living, with urban growth rates exceeding that of the population in general and, until recent decades, the biggest cities growing fastest of all. Today, 75 per cent of the United States population is classified as urban (Figures 4.1.1 and 4.1.2), one of the highest proportions in the world.

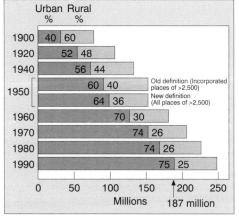

Figure 4.1.1
US population: urban–rural ratios, 1900–1990

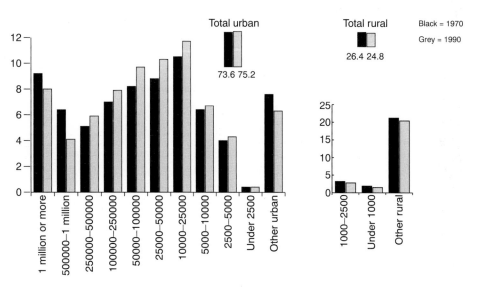

Figure 4.1.2
USA: distribution of urban and rural populations of place size 1970 and 1990

Figure 4.1.3
USA: Urban population (%), by division

Division	1910	1950	1990
New England	73.4	74.8	74.4
Middle Atlantic	70.2	75.6	80.5
East North Central	52.7	66.3	74.0
West North Central	33.1	49.9	66.3
South Atlantic	23.1	41.6	69.4
East South Central	18.6	35.5	56.2
West South Central	22.2	53.1	74.5
Mountain	35.8	49.1	79.7
Pacific	56.8	63.6	88.6

Canada

In Canada, the urbanisation process was slower and on a smaller scale. In 1851 only 13 per cent of the total population was recorded as urban, the balance tilting towards towns and cities only in the late 1930s. The 1941 Census indicated that there were still only 21 towns of more than 25 000 inhabitants and only eight cities of more than 100 000. The intervening half-century has seen the pace of urbanisation quicken, and in 1991 three out of four Canadians were classified as urban, putting the country on a par with its southern neighbour.

Regional patterns of urbanisation are quite distinctive. The provinces of Ontario, British Columbia, Alberta and Quebec are all more urbanised than Canada as a whole, while the Atlantic region is markedly less urbanised than any other except the Northwest Territories. Paradoxically, Prince Edward Island, Canada's most densely populated province, is also its least urbanised.

Although the historical trend has been towards increased urbanisation, there are now signs that the urban–rural ratio is stabilising: between 1986 and 1991 the proportion of Canada's population living in urban areas showed an increase of only one-tenth of a percentage point.

Figure 4.1.4
Major metropolitan areas of North America

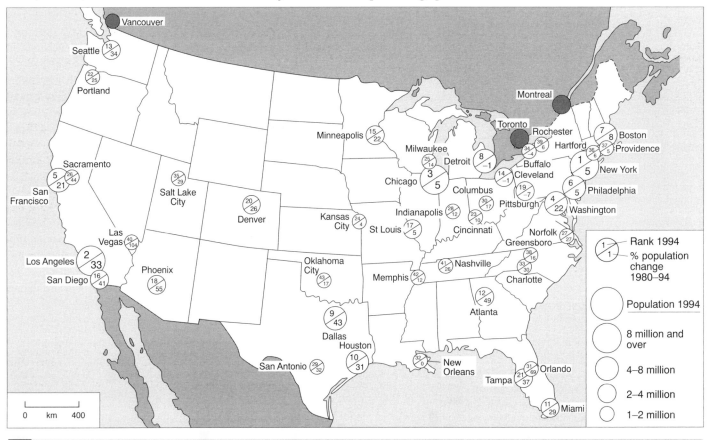

1 (a) In which census year was the United States confirmed as having become predominantly urban? (Figure 4.1.1). How did World War I hasten this change?

(b) 'Over the last few decades the *relative* increase in the United States urban population has been small, but the *absolute* increase has been large.' Explain and elaborate this statement (Figure 4.1.1).

2 Why do so many North Americans live in towns and cities?

3 (a) What is meant by the terms 'primate city' and 'urban hierarchy'?

(b) Explain why the number of cities in the different size categories varies markedly (Figure 4.1.4).

4 Referring to Figure 4.1.2, identify, and suggest reasons for, the main trends in the distribution of population by size of place.

5 Describe and account for the contrasting regional patterns of twentieth century urbanisation (Figure 4.1.3) and recent city growth (Figure 4.1.4) in the US.

Problems of definition: city-metropolitan area

Nowadays most North Americans who live in big cities reside not in the city proper (i.e. the legally defined area within the boundaries demarcating the limits of the rule of the city government; also known as the central or core city) but in the surrounding suburbs and dormitory communities. These satellite communities are linked to the city by economic ties (e.g. employment and shopping) and social activities (e.g. leisure and recreation).

Acknowledging these functional bonds, in both Canada and the USA official statistical units have been created to cover both cities and their adjoining suburbs. Such units are termed 'metropolitan areas' and normally include whole counties, parts of which may not be urban at all, but farmland, forest – even desert.

In Canada such units having a population of over 100 000 are known as 'Census Metropolitan Areas' (CMAs). The 1991 Census indicated 25 CMAs, together containing 62 per cent of Canada's population; the three largest – Toronto, Montreal and Vancouver – accounted for nearly one-third of all Canadians (Figure 4.1.5). All the CMAs with population growth rates above the national average between 1986 and 1991 were located in Ontario and the two western provinces of Alberta and British Columbia, Vancouver recording the nation's second highest rate.

Within all Canadian CMAs the populations of the suburban areas are growing faster than the cities they serve. For example, the city of Toronto's population grew by 3.8 per cent between 1986 and 1991, but the population of the Toronto CMA grew by more than 13 per cent; some outlying suburban areas, such as Richmond Hill and Vaughan, had growth rates exceeding 70 per cent.

In the United States the urban scene is both larger and more complex. Apart from the city proper no fewer than three types of metropolitan area are recognised: the Metropolitan Statistical Area (MSA) and the Primary and Consolidated Metropolitan Statistical Areas (PMSAs and CMSAs). A typical MSA contains a core city of at least 50 000 population and at least one adjoining suburban county meeting the appropriate functional criteria. The 71 PMSAs designated in the 1990 Census comprise constituent parts of the largest urban units, the CMSAs, of which 20 currently exist. Thus the 1990 Census recorded the population of New York in the following ways:

City: 7 323 000
PMSA: 16 938 000
CMSA: 19 342 000

In 1990 nearly 133 million Americans (53 per cent of the entire population) lived in the 40 'million-plus' areas shown on Figure 4.1.3. However, one virtually universal trend in the distribution of population within US metropolitan areas is notable: core cities have lost population over recent decades while suburbs have gained, often dramatically.

Figure 4.1.5
Canada's ten largest CMAs, 1995

Population (1000s)

Toronto	4338	Calgary	829
Montreal	3328	Winnipeg	695
Vancouver	1827	Quebec	677
Ottawa-Hull	1027	Hamilton	642
Edmonton	883	London	413

QUESTIONS

1 (a) With reference to Figure 4.1.3 and an atlas, how many of North America's 'millionaire' cities (metropolitan areas) are located (i) on the coast; (ii) beside major rivers or lakes; (iii) beyond 55°N?
  (b) How have these factors influenced city locations? What other factors are, or have been, important?
  (c) What broad geographical pattern characterises the distribution of large cities in North America? Suggest reasons for this pattern.

2 Draw an annotated sketch map to illustrate the structure of a typical North American metropolitan area, emphasising the relationship between the core city and, in this case, three adjacent suburban counties.

3 Compare and contrast the United States and Canada, regarding (i) trends in urbanisation; (ii) the characteristics of metropolitan areas.

Cities: characteristics and trends

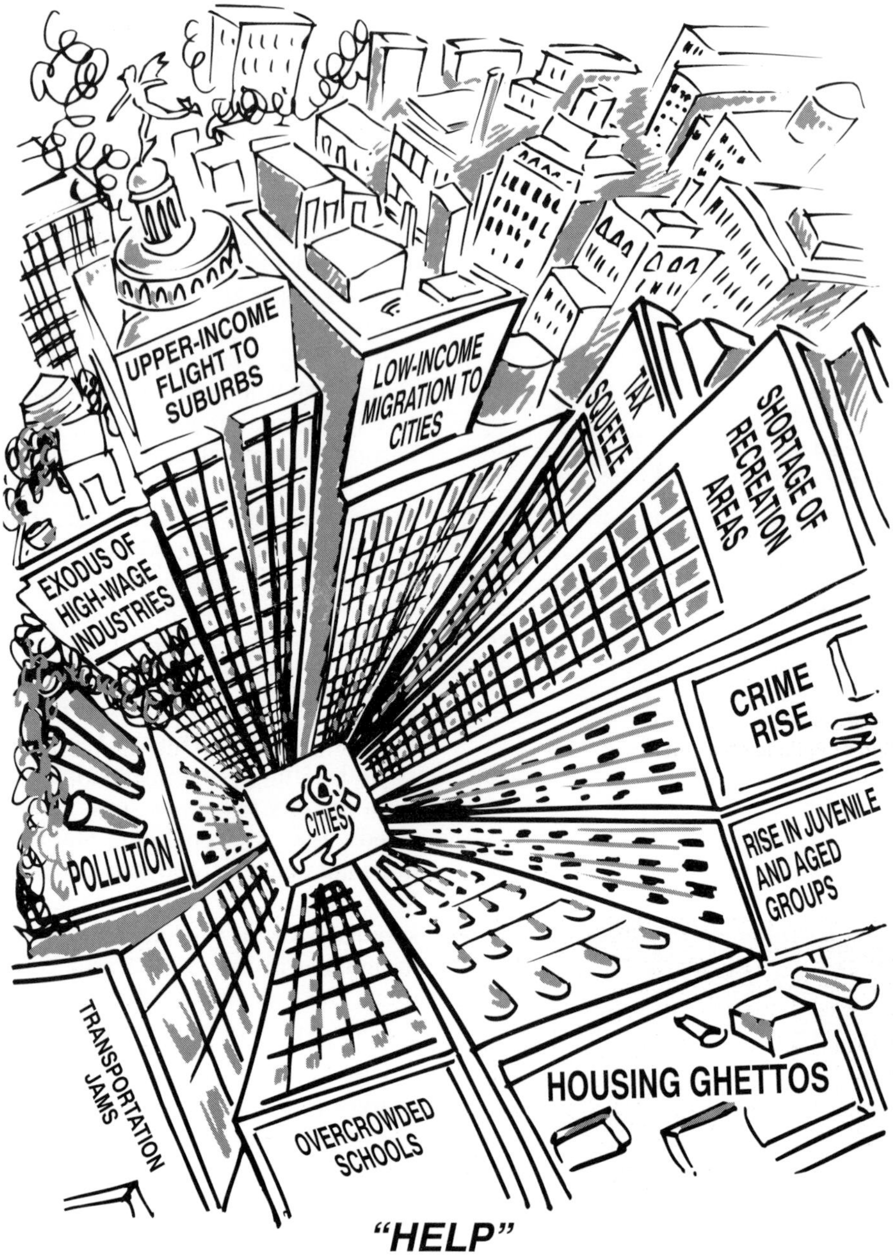

Figure 4.1.6
American cities in distress

Infrastructure pressures

Throughout the twentieth century not only has the continent's urban population increased more rapidly than the rural but, until recently, the bigger the city the faster it grew. Moreover, unlike cities in present-day developing countries, those in North America were historically able to match rapid population growth with the provision of proper housing, employment opportunities and the whole range of services commonly expected in developed countries. Now, however, there are serious concerns that the further expansion of cities may generate insuperable problems of water supply (especially, but not exclusively, in the arid and semi-arid West) and of waste disposal (see Section 4.4). In addition, continued population growth will place extra strain on the social infrastructure such as education and medical services; this is already a problem in Los Angeles and other Western cities.

Preservation of older buildings

The rapid construction of North American cities has been equalled by rapid transformation. Old buildings and whole urban areas are torn down and replaced by modern ones to meet new needs and requirements. The impermanence of the built environment may contribute to the instability, even hostility, apparent in many residents, who lack pride in, and cannot identify with, their home city. Two recent trends are combining to prevent the destruction of some older urban quarters:

- A conservation movement has been successful in preserving and restoring important elements of urban heritage; e.g. decaying waterfront areas, such as Baltimore's Inner Harbour, have been transformed into 'festival marketplaces', while old mills and warehouses have been recycled in Boston and San Francisco to accommodate high-technology industry.
- The 'gentrification' of many run-down inner city areas by 'yuppies' has given a new lease of life to neighbourhoods that would otherwise have fallen into further decay.

Street plans

Initially, European colonial towns in North America evolved haphazardly without any formal urban plans. In 1681, however, William Penn laid out Philadelphia on a strict rectilinear grid principle, with straight streets aligned north–south and east–west and intersecting at right angles (Figure 4.1.7). This gridiron pattern became the normal layout for new cities and extensions of older ones, especially after 1785 when the United States government chose a rectangular survey for all public lands west of Pennsylvania. In due course much of Canada followed suit. This chequerboard street plan is perhaps the most distinctive feature of the North American city.

Suburban sprawl

The great personal mobility afforded by high levels of car ownership and car use has resulted in the characteristic suburban sprawl of most cities in North America (Figure 4.1.8). Los Angeles is, of course, the quintessential example (see Section 4.5), but since World War II others have followed where southern Californians led. Consequently, the sheer areal extent of the continent's cities is another highly distinctive feature. Apart from the distances that now separate suburb from city and suburb from suburb, other disadvantages of horizontal sprawl have emerged; for example, the problem of supplying services to such dispersed communities, not least the provision of financially viable public transport links.

Central business district (CBD)

If horizontal growth characterises the cities of North America, so too does vertical development, at least within the relatively small area of the central business district (CBD). In this area of high business levels and (originally, at least) greatest accessibility, the need to maximise use of the high-value land at the centre of the city resulted in a dense cluster of skyscrapers, which in the public imagination epitomise the American city. Originating in Chicago in the 1880s, and utilising a framework of structural steel on which the rest of the building was supported, the skyscraper can now reach dizzying heights; Chicago's Sears Tower, with 110 storeys rising 442 metres above street level (Figure 4.1.9), is the current North American record holder. The skyscrapers of the CBD house concentrations of corporate offices, banking, finance and insurance operations, representing what has been termed 'the transactional city'. The 'downtown' area also contains high-class shops, hotels and places of entertainment such as theatres, restaurants and night-clubs, along with other enterprises which can

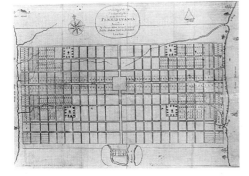

Figure 4.1.7
Plan of Philadelphia, 1682

Figure 4.1.8
New York City: suburban sprawl

afford the high costs of this location. Normally, the CBD's residential population is numerically very small. The recent drift to the suburbs not only of upper- and middle-class residents but of the businesses that employ or depend upon them, has raised a question mark over the future of the CBD, which is in danger of becoming a 'dead eye' to the city.

Ghettos

Levels of ethnic diversity (and tension) are higher in North American cities than in any others in the world. This is due mainly to the constant flow of immigrants from all over the world who have headed for the city as the most likely place to find employment in their adopted country. Today, Los Angeles and Miami have replaced Toronto, Vancouver, Boston and New York as the principal magnets for immigrants. The ghetto is a particularly notable feature of the continent's urban geography. An inner-city area of old, low-quality, over-crowded housing, the ghetto originally served a useful purpose in 'processing' European immigrants into North Americans who could then become assimilated into the mainstream of national life, graduating after a generation or so to better jobs and higher-grade housing in the suburbs. For some groups, however, such free exit from the ghetto has been denied. The Chinatowns of New York, San Francisco and Vancouver are early products of this denial, while more recently the same process has 'trapped' Puerto Ricans in New York's Spanish Harlem and Cubans in Miami. But above all, African Americans have suffered from this form of urban apartheid, and the black ghetto seems likely to remain a permanent feature of most American cities.

Finance problems

As racial and ethnic tensions have increased and the city's socio-economic problems have multiplied, those residents and businesses able to desert the city have done so. This 'white flight' to the suburbs has been accompanied by the removal of capital and of tax revenues. This erosion of the tax base has impoverished many cities, which despite falling incomes still have to maintain their services, such as the police and fire brigade, and support an increasing concentration of lower-income groups and welfare recipients. This problem is exacerbated by the per capita basis of federal funding for cities: falling populations mean cuts in this source of finance too. New structures of local government and of federal assistance seem to be the only solution to this basic and increasingly acute difficulty; otherwise many cities will be bankrupt in the not-so-distant future.

Figure 4.1.9
Chicago: Sears Tower and the CBD

Figure 4.1.10
Diagrammatic section through a typical US city

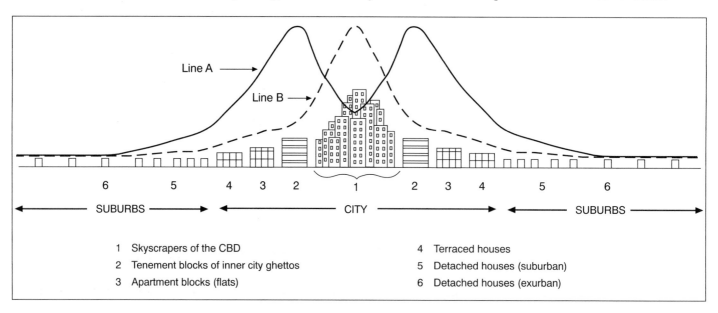

1	Skyscrapers of the CBD	4 Terraced houses
2	Tenement blocks of inner city ghettos	5 Detached houses (suburban)
3	Apartment blocks (flats)	6 Detached houses (exurban)

Manufacturing industry

Manufacturing industry, aided by rail transport, largely created the nineteenth century North American city. More recently, the switch to road and air transport has liberated industry from urban locations and encouraged companies to seek new suburban sites, often in specially designated industrial parks. Such locations offer not only modern premises but many other advantages vis-à-vis the old, often cramped, congested and inaccessible sites near the city centre. This relocation of manufacturing industry, where it survives at all, emphasises the modern importance of the suburbs, which have attracted away from the city proper not just people but a wide range of businesses, retail outlets (usually clustered together in attractive 'shopping malls') and other service activities. Such 'edge city' developments have resulted from what might be termed the suburbanisation of North America, which has replaced urbanisation as the continent's main direction of growth.

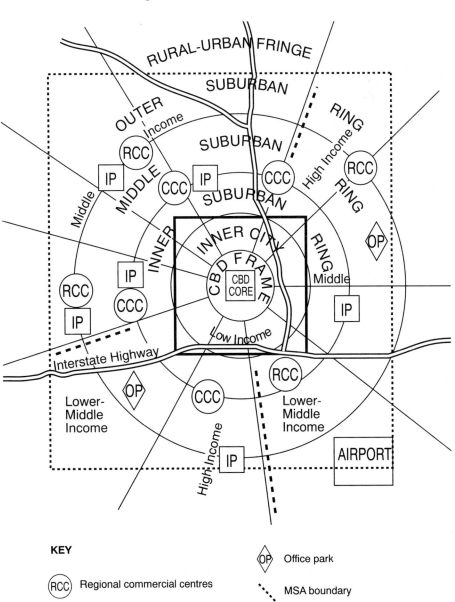

Figure 4.1.11
The spatial structure of the contemporary North American city

KEY

(RCC) Regional commercial centres

(CCC) Community commercial centres

IP   Industrial park

(OP) Office park

∴ MSA boundary

━ Central city boundary

∴ Commercial 'strip' development

1 (a) Referring to Figure 4.1.7, explain the advantages and disadvantages of gridiron urban layouts, especially for modern traffic movements.
  (b) Why, when cities became more extensive, were diagonal streets often superimposed on the gridiron pattern?

2 (a) Suggest why suburban layouts (Figure 4.1.8) often differ from that of the city proper.
  (b) What housing characteristics contribute to suburban sprawl?

3 (a) The exit from modern ghettos is said to be blocked by a combination of social and economic forces. Elaborate this statement, and assess the likelihood of dispersal from black ghettos in the future.
  (b) What action needs to be taken to facilitate this objective?

4 Identify the variables indicated by Lines A and B on Figure 4.1.10 (N.B. The height of buildings reflects the Line B variable).

5 With reference to Figure 4.1.11, answer the following questions:
  (a) To what extent can earlier models of urban structure be identified in the diagram?
  (b) Explain why (i) the CBD occupies a central location; (ii) the CBD may become a 'dead eye'.
  (c) Account for the difference in (i) land use; (ii) vertical profile (height of buildings) in the CBD core and the CBD frame.
  (d) List six differences that you would expect to find between residential environments in low-income inner city areas and high-income neighbourhoods in the middle suburban ring.
  (e) List six types of commercial activity that you would expect to find in the regional and community commercial centres.
  (f) Account for the location of industrial and office parks.

4.2 Megalopolis: an Overview

Many cities, many people

The first conurbations appeared in nineteenth century Europe, and many examples occur in Britain, the Low Countries and Germany as well as in Japan. However, the aptly named 'Megalopolis' – the super-city that stretches along the USA's north eastern coast – is unrivalled in size and complexity anywhere on

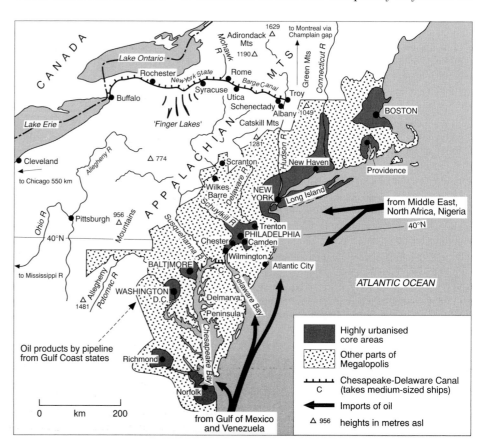

Figure 4.2.1
Megalopolis: position and extent

earth. An immense chain of cities, suburbs, towns and satellite areas forms the world's largest, wealthiest and most productive urbanised region, where about 50 million people (approximately one-fifth of the total American population) occupy less than two per cent of the nation's land area. Originally identified as extending between Boston and Washington, over the last few decades Megalopolis has expanded steadily southwards and is now widely regarded as embracing Richmond, Norfolk and the other towns around the mouth of Chesapeake Bay (Figure 4.2.1).

By North American standards the urbanisation within Megalopolis is intense. The region contains many of the US's largest urban clusters, including four CMSAs: Boston (1994 population: 5 497 000), Philadelphia (5 959 000), Washington–Baltimore (7 051 000) and New York (19 796 000). Yet the built-up areas are often separated by extensive rural and semi-rural landscapes, and the overall population density is lower than Britain's. What binds Megalopolis together as a single entity is the formidable network of transport and communications systems, the interdependence of industrial and commercial linkages, and the overlap of spheres of influence or urban hinterlands. It is thus difficult to determine where one city's dominance ends and another's begins. Residents of dormitory towns in central New Jersey, for example, may commute northward to New York or southward to Philadelphia, daily passing those heading in the opposite direction.

In terms of overall population increase, Megalopolis has long been the national leader. However, over the last quarter-century or so the core cities have all suffered net losses of population, though their metropolitan areas have all recorded increases. Clearly, as elsewhere in urban America, those (mostly white, middle-class) residents of Megalopolis who desire and can afford homes in the suburbs have 'voted with their feet' on conditions in the deteriorating cities. In this continuing movement of people Megalopolis may still be seen as a pointer to trends occurring throughout the country.

Development factors

The unprecedented development of Megalopolis can hardly be ascribed to the wealth of its natural resource base, for its soils are generally of poor quality and its mineral deposits are usually of little value. The region's rise to pre-eminence is more easily explained in terms of its history and location. It was here, confined between the Atlantic and the Appalachians, that the original British colonies grew in economic and political strength; indeed, an embryonic pattern of population clustering along the north eastern seaboard was already evident by the mid-eighteenth century. It was here, too, that the United States emerged as an independent country and first formulated its national institutions and identity. Such circumstances ensured initial advantages that proved decisive in the subsequent development of the region and its cities.

A favourable geographical position also underpinned the rise of Megalopolis. Facing Europe and situated in middle latitudes, the region is ideally located for handling the international trade which still contributes to its commercial importance. Wide valleys, flooded at the end of the Ice Age, became the great inlets and estuaries which allow ocean-going vessels to penetrate deep inland at all seasons. Apart from Washington, a latecomer to the region's urban scene (see Section 4.3), all the tidewater cities developed into major ports which have competed for trade and industry since colonial times. They also served as points of entry for millions of immigrants and as gateways to the rich hinterlands beyond the Appalachians.

Apart from its industrial and commercial importance, Megalopolis contains both the financial capital of the United States (New York's Wall Street) and the nation's political capital (Washington, DC). At the same time it remains –

1 From Figure 4.2.1 calculate the approximate extent of Megalopolis, north–south and east–west.

2 (a) Explain why Megalopolis has been termed 'Boswash' and 'Bosnywash' in the past.
   (b) Suggest a new description, taking account of its present extent.

3 On Figure 4.2.1, what is the significance of the oil imports information in terms of the region's energy needs, port industries and coastal air quality?

4 Parts of Long Island, New Jersey ('The Garden State') and the Delmarva Peninsula specialise in truck farming.
   (a) Why is it so-called?
   (b) Why is this activity important in Megalopolis?

despite competition from the West Coast – a major cultural centre which, in such realms as art, fashion, publishing, advertising, the media and education (all the famous Ivy League universities and many others are located here), continues to lead and influence many aspects of life in America – and in the rest of the world. In short, Megalopolis can be aptly described as 'Main Street USA' and 'The General Headquarters of Enterprise America' – the control centre of the national economy.

Megalopolis can also claim, through the United Nations headquarters in New York, to contain the diplomatic capital of the world, further underlining the region's global importance.

4.3 Washington, DC: Planning the US Capital

Planning in the past

Figure 4.3.1
Early Washington

The District of Columbia

Washington, DC (District of Columbia) has been the capital of the United States since 1800. Unlike most other cities it did not evolve gradually: it was planned from the outset. The need for a federal capital for the 13 newly-independent states was obvious once they had decided to combine as one country. However, the proposed capital's location became a major political issue as one suggestion after another was rejected due to rivalry between the northern and southern states. Eventually a compromise was agreed: the capital would be located on the Potomac River, the traditional boundary between North and South. George Washington, the first US president, was invited to select the exact site of the city that would bear his name.

In 1790 the president chose a 10-mile-square (260 km²) 'diamond' of Maryland and Virginia, the two states ceding this area for the new federal Territory (later District) of Columbia (Figure 4.3.2). Although predominantly a swampy wilderness, the area contained three established settlements: Hamburg, a riverside village in that stretch of the Potomac valley traditionally known as Foggy Bottom;

Alexandria, in the section of the District returned to Virginia in 1847 (Figure 4.3.3); and Georgetown, a thriving tobacco market at the head of navigation on the Potomac, whence a canal (the Chesapeake and Ohio, completed in 1850) would reach out to exploit the 'western' frontier lands then being opened up. Indeed, it was originally envisioned that the new city would become an important commercial centre – 'the Emporium of the West' in George Washington's own words – but it was never able to compete successfully with the long-established port cities to the north.

City planning

Much more significantly, Washington, DC was conceived primarily as a political, diplomatic and governmental centre, and as a kind of national shrine. In 1791 the task of planning the new capital was entrusted to Pierre-Charles L'Enfant, a French architect and engineer who had served in the American Revolutionary War. L'Enfant's basic scheme – which drew upon several European examples of town planning, including Versailles and Sir Christopher Wren's plan to reconstruct London – combined geometry with geography. On the Potomac basin's uneven topography of ridge, ravine, swamp, scrub and forest he created a series of wide avenues radiating from the Capitol and the White House, which were set on natural rises in the terrain to emphasise their importance. Another key element was The Mall; originally intended to be a broad, tree-lined boulevard like the Champs Élysées in Paris, it has become an elongated lawn interspersed with occasional buildings and cross-roads (Figure 4.3.4).

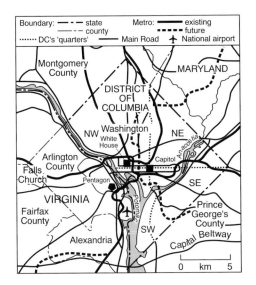

Figure 4.3.2
The political Geography of the Washington region

Figure 4.3.3
Washington, DC, capital of the USA

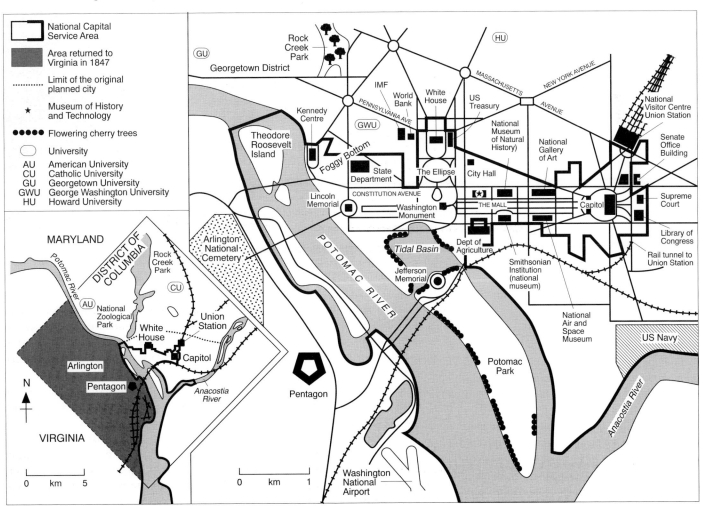

Figure 4.3.4
Aerial view of Washington

L'Enfant's pattern of diagonal avenues (named after states) was complemented by a gridded infill of streets which, with the Capitol as the axis, were lettered (A, B, C, etc.) from north to south, and numbered from east to west. At the intersections of streets and avenues, numerous squares, circles and triangles provided opportunities to create open spaces and small parks where monuments and fountains could be placed. The intervisibility of these open spaces greatly contributed to Washington's character as 'a city of magnificent distances'. Although L'Enfant was dismissed in 1792, his plans were closely followed in completing the layout of the city, and in October 1800 the federal government moved from its temporary quarters in Philadelphia to take possession of the new capital.

Acceptance as the capital

However, Washington's early years were not easy. In the days before modern communications, the city's location was regarded as remote and inaccessible; it was described as 'Wilderness City', 'Capital of miserable huts' and 'a mud-hole', and by 1808 its population was still only about 5000. Not surprisingly, there was constant agitation in Congress and the press to relocate the capital. In 1814, Washington was temporarily abandoned when a British force entered the city during the War of 1812–14 and burned all the public buildings, including the Capitol and White House. Paradoxically, this event changed American attitudes towards Washington, ending the pressure to move the seat of government and firmly establishing the city as the nation's capital in fact as well as in name. Later, Confederate threats to the federal capital during the Civil War further consolidated this sentiment; after 1865 Washington emerged as the undisputed capital city of a newly confirmed union of states.

In 1895, Georgetown, the last self-governing area within the District, was legally annexed by the city; thus the City of Washington and the District of Columbia became areally identical for the first time. Within DC the federal government has direct control over the National Capital Service Area (Figure 4.3.3), the historic city centre with its important government buildings and neo-classical skyline of domes, temples and monuments which embody much of the nation's history and spirit. This is the Washington that attracts over 20 million visitors every year – more than anywhere else in the entire United States.

QUESTIONS

1 Identify and comment on the principal features of the land-scape and economy of early Washington, as depicted in Figure 4.3.1.

2 Why did Washington fail to become an important centre of trade and industry as originally intended?

3 In terms of the city's role as the capital of the United States, how would you assess Washington's location (i) in 1790; (ii) today?

4 Referring to Figure 4.3.3, identify the buildings A–G; the streets H and I; and the state J on Figure 4.3.4.

5 With reference to Figure 4.3.3, outline a sightseeing route that you would recommend for a fit, energetic visitor to Washington, justifying your choice of places of interest. (N.B. Most museums, etc. make no charge for entry).

6 With reference to Figure 4.3.3, list the evidence that Washington is
(a) a 'national shrine'.
(b) an important centre of (i) culture; (ii) higher education; (iii) international finance.

7 Name two other planned capital cities built on 'greenfield sites' like Washington. What was the rationale for their construction and location?

Modern Washington: the company town

Metropolitan Washington is still a company town. People are drawn to the city and its suburbs because the federal government is there, because it is a prime centre of political power.

Businessmen, like everybody else, are coming to the Washington area to display their goods and services to the government. Specifically, they display their research and development. Battalions of lawyers, lobbyists, journalists and other hangers-on still prowl the city and its environs as they always did. But these days their ranks are reinforced by phalanxes of physicists, chemists, biologists, electrical engineers, data-processors, software supermen and salesmen. Close behind comes the ever-ready commissariat, proffering fast-food and other essentials to the front-line troops.

These newcomers settle in the District of Columbia, in suburban Maryland, in northern Virginia. The nearer companies can be to the government department that they have, in one way or another, to deal with, the easier life is for them. In Maryland, the life sciences cluster round the National Institutes of Health; in Virginia the death sciences cluster round the Department of Defence.

Other magnets draw businessmen, research workers and scientists to the region. They come for "the quality of life": green space, clean air, fine schools. Culture, universities, good food and other urban blessings are accessible just round the corner. An educated labour force waits to be employed. All good reasons to pick on Washington—but less important than the site of government.

The federal government no longer dominates the metropolitan area as a direct employer (though in the District itself about one-third of all workers are federal employees). The government still employs roughly the same number of people as it did 15 years ago, but the new jobs are going to private industry. In the splurge of growth since 1983, the government has accounted for only 3% of the new jobs; in the big-government days some 20 years ago it accounted for around 30% of the growth.

Yet Washingtonians continue to live on money from the federal government. The difference is that less of it comes directly in the form of salaries and pay cheques. What matters nowadays is not government jobs but government contracts, sub-contracts and procurement. Companies come to Washington to service the government and to sell things to it. Increasingly the government contracts out jobs that it once did in-house. About half the region's total sales of technology are now government-related.

Nothing much is made in and around Washington. It was never a manufacturing zone. But most of the *Fortune* 500 (largest American) companies are represented in the area and many of them find Washington a sensible place to set up their corporate or regional headquarters. Mobil Oil, for instance, is in the process of moving its headquarters from super-expensive New York to less expensive Merrifield, Virginia.

The classiest industry these days is research—into aerospace, weapons systems, energy, biotechnology, agriculture, telecommunications—much of it done on contract for the government.

The Economist,
16 April 1988

Figure 4.3.5
Modern Washington: the company town

1 What is meant by the terms 'company town', 'death sciences' and 'in-house'?

2 (a) What, in general, characterises the 'frontline' industries which have recently located in the Washington area?

  (b) What special attractions does Washington offer to such industries?

3 Since the *Economist* article was first published, dramatic political developments have occurred on the world scene. Assess the likely economic impact of these developments on the Washington area generally and on northern Virginia especially.

Planning for the modern metropolitan area

In planning terms, the metropolitan area, or greater Washington, generally means the District and an inner ring of suburbs consisting of Montgomery and Prince George's counties in Maryland, and Fairfax and Arlington counties in Virginia, plus the independent boroughs (locally 'cities') of Alexandria, Herndon, Fairfax and Falls Church (Figure 4.3.6). For census purposes, the Washington Metropolitan Statistical Area (MSA) has been extended to an outer ring of suburbs: Charles, Calvert and Frederick counties in Maryland, and Prince William and Stafford counties in Virginia.

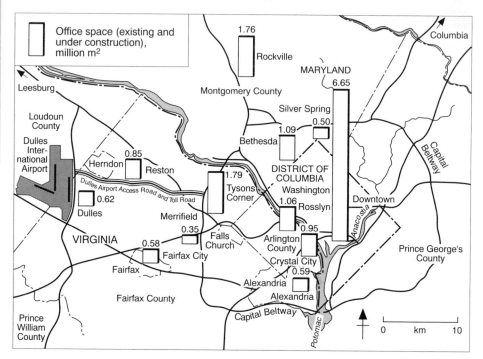

Figure 4.3.6
Emerging cities in the Washington metropolitan area

Washington is the richest metropolitan region in the United States, but the wealth is very unevenly spread between core city and suburbs. Nowadays, for every new job in the District, 18 are added in the suburbs, so that two-thirds of all jobs in the MSA are located in suburban areas. Moreover, by 2010, while an estimated 200 000 new jobs will be created in the District, about one million will appear in the suburbs.

'Edge city' developments in Washington's suburbs

Most of the Washington MSA's frenetic growth – actual and predicted – focuses on 14 centres, all but one of which are located in the suburbs (Figure 4.3.6). Such foci are known locally as 'emerging cities', each of which should meet the following criteria: at least 0.45 million m² of office space, at least 54 000 m² of retail space, more jobs than bedrooms, more people commuting in than commuting out, and a public image of mixed land use. More generally in the USA, such developments have acquired the description 'edge city'.

Some 'emerging cities' are long-established towns (e.g. Alexandria and Bethesda), others are suburbs which have developed into towns (e.g. Rockville and Silver Spring), while Columbia and Reston are genuine new towns. Still others, most notoriously Tyson's Corner, seem to be little more than a jumble of hotels, office blocks, shopping malls and snarled-up traffic (Figure 4.3.7). Visually, some are so alien to their surroundings that they could have been dropped from outer space.

Figure 4.3.7
Tyson's Corner, Fairfax County, Virginia

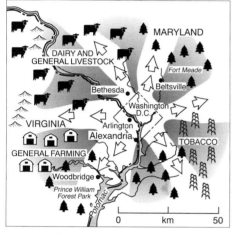

Figure 4.3.8
Open wedges and developed corridors of the Year 2000 Policies Plan

The 'wedges and corridors' plan

Now virtually out of control, metropolitan Washington's 'emerging cities' were once intended by the region's planners to be components of a series of small towns and community centres extending along urbanised corridors of growth. These corridors were to be separated by wedges of open countryside – thus preserving valuable amenity areas for the people of the metropolitan area (Figure 4.3.8).

This 'wedges and corridors' plan (more formally, 'A Policies Plan for the Year 2000') was drawn up by the National Capital Planning Commission in 1961, and can still be faintly detected as the blueprint of metropolitan growth. Most of the economic development has taken place along the radials designated in the plan. Maryland, by and large, has been able to preserve parks and green belts. In Virginia growth has spread into those precious open-space wedges. And, as land values rose, office parks and shopping malls crowded out the planned houses and flats, schools and libraries, post offices and community centres. Builders hate to fuss with a house if a large profitable office block is the alternative. And, with no houses, there is no call for schools or libraries.

Metrorail (Washington's underground/suburban railway) was built, as planned, along several of the spokes, with intense patches of development at its stopping-places. The Capital Beltway ring road that encircles Washington, about 20 km from the centre, was completed in the mid-1960s. But earlier plans for an 'outer beltway' came to nothing; people discovered it might run through or near their property, disrupting their lives, and lowering the value of their homes.

The new development areas have not worked out quite as the original planners hoped. But geography, and state laws, have helped the plan to work better in Maryland than in Virginia. Wedges and corridors fit the topography of Maryland; the configuration of the land in Virginia turns the straight lines into concentric circles.

And Maryland state law gives the counties more authority over development than does Virginian law. This is a north–south distinction. Maryland is a border state, neither northern nor southern. But Virginia is an integral part of the old South, where the laws of the land favour the landowner.

QUESTIONS

1. (a) What role did lines of transport and communication play in designating the radial corridors shown on Figure 4.3.8?
   (b) To what extent can an earlier model of urban development be identified?

2. Public resistance to the planned 'outer beltway' (ring motorway) could be described as an example of 'the NIMBY factor'. Explain this acronym.

3. (a) Briefly describe the countryside around Washington (Figure 4.3.8).
   (b) To what extent was it worth attempting to preserve?

4. Explain in your own words why the Policies Plan for the Year 2000 has been more effective in Maryland than in Virginia.

5. The National Capital Planning Commission was established in 1926 to coordinate planning throughout the Washington metropolitan area, but its efforts have never been wholly successful. What makes planning for this region particularly difficult? (See Figure 4.3.6).

The forgotten District: black Washington

The end of the Civil War saw some 40 000 freed slaves move into Washington, virtually doubling its population and establishing the racial diversity that has been an enduring feature of the city ever since. By 1910, the black population had risen to 94 000 (28.5 per cent of the total) and by 1960 blacks had become the majority within the District. A record high 538 000 (71.1 per cent) was reached in 1970 but since then the black population has been declining both absolutely and relatively (Figure 4.3.9).

Figure 4.3.9
Washington: total and black population

Total Population (1000s)

Year	D.C.	% change	Suburbs	% change	MSA	% change
1970	757	% change	2283	% change	3040	% change
1980	☐	−15.7	2613	☐	3251	6.9
1990	607	−4.9	3317	26.9	3924	20.7

Black Population (1000s)

Year	D.C.	% change	Suburbs	% change	MSA	% change
1970	538	% change	166	% change	691.7	% change
1980	449	☐	405	☐	☐	☐
1990	399	−11.1	643	☐	1042	22.0

Even so, the District still contains a large swathe of black residential areas from Rock Creek Park in the north-west to the Anacostia River in the south-east (Figure 4.3.3). Within this broad crescent exists a great variety of housing, from detached, upper-income types to zones of deteriorating houses and tenements of the worst kind. In such areas there has been little reconstruction and few new developments. Here up to 10 000 people, including 2000 children, are homeless. Where 'yuppies' have moved into some inner city areas to create improved, 'gentrified' housing, the displaced blacks are even worse off than before, adding to the massive homelessness problem that no agency of government seems able to tackle.

While average family incomes in the MSA as a whole are 50 per cent higher than the national average, well over 100 000 Washingtonians, mostly black, live at or below the official poverty line. The District's infant mortality rate is 21/1000 – twice the US average, which itself is considerably higher than rates in most other advanced countries. The inner city's schools are notoriously substandard, and many children are caught up in the drug-related gang warfare that afflicts much of the District but especially along 14th Street in the South-East quarter. The murder and mayhem is of such frightening proportions that the criminals' automatic and semi-automatic weaponry exceeds anything available to the police.

Small wonder that those who can escape from such conditions usually do so. Whites were the first to head for the comparative safety of the suburbs, and blacks have followed in increasing numbers, especially to Prince George's County. While Washington's black suburbanites typically earn less than their white counterparts their median incomes are higher than those of whites living outside the MSA. More encouragingly, young blacks in Washington's suburbs are better educated than young whites elsewhere.

Nevertheless, when the total metropolitan picture is reviewed, the most striking feature remains the great contrast between the 'two Washingtons': the mainly white suburbs with a highly educated, skilled workforce with above-average incomes; and a core city (the District) where a majority of residents are black, educational and income levels are low, and unemployment rates are way above the national average (e.g. 40 per cent for black males in DC is not unknown).

All this is depressing, but the 1990 census revealed a new phenomenon for Washington: 5.4 per cent of the District's population was Hispanic. This includes a growing community of illegal immigrants from Mexico and Central America whose housing conditions and general quality of life are even worse

1 (a) Complete the blank spaces in Figure 4.3.9.

(b) Describe and attempt to explain the trends shown in the tables.

(c) Suggest two techniques that could be used to represent the data diagrammatically.

2 (a) Construct another table to show the black population as a percentage of the total population in the years and areas indicated.

(b) Comment on the results obtained.

3 (a) What do you understand by the term 'gentrification'?

(b) What are the advantages and disadvantages for cities like Washington?

4 (a) Why might the District's unemployed blacks find difficulty in getting jobs in the suburbs?

(b) Suggest practical solutions to this problem (i) in the short term; (ii) in the longer term.

than those endured by the city's blacks. Needless to say, few visitors to the federal capital are concerned about such matters, simply because few are even aware that such problems exist in the show-place planned by L'Enfant as a celebration of the United States. Here, of all places, contrasts between image and reality are particularly acute.

4.4 New York City: The Big Apple

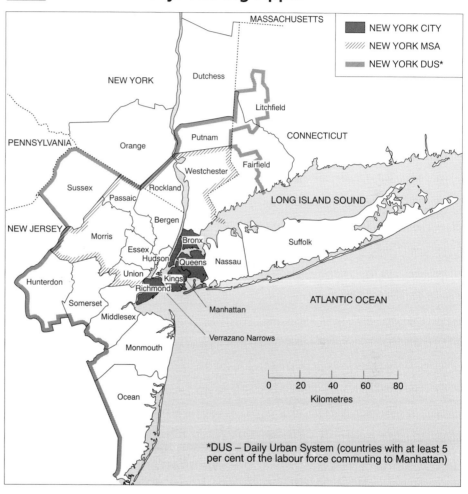

Figure 4.4.1
The New York metropolitan region

The development of New York: some key dates

Pre-European contact: The area's first inhabitants were Algonquin Indians who called it *Man-a-ha-ta* ('heavenly place').

1524   New York harbour was discovered by the Italian navigator Giovanni di Verrazano (commemorated in the Verrazano Narrows between Staten Island and Long Island and the suspension bridge linking them).

1609   Henry Hudson, an English navigator working for the Dutch, explored New York harbour and sailed up the Hudson River as far as Albany, claiming the territory for The Netherlands.

1626   The Dutch purchased Manhattan Island from the local Indians for beads, buttons and trinkets worth 60 guilders ($24). New Amsterdam was built on Manhattan's southern tip; the settlement's northern boundary, a defensive wall, is marked by modern Wall Street.

1653	New Amsterdam became the capital city (population 800) of the province of New Netherland (an area covering what became New York City as well as much of Connecticut, New Jersey and Long Island).
1664	The English seized New Amsterdam and renamed it New York.
1785–90	New York served as the US capital until the federal government moved first to Philadelphia and then, in 1800, to Washington, DC.
1800	New York (population 70 000) became the largest city in North America, a position it has held ever since.
1818	The White Star Line began the first regular transatlantic shipping service, between New York and Liverpool. Massive influx of European immigrants began.
1825	Completion of the Erie Canal, along the only lowland routeway through the Appalachian barrier, gave New York access to a vast hinterland and thus a huge advantage over its Atlantic coast rivals. (The canal was enlarged in 1918 and renamed the New York State Barge Canal (Figure 4.2.1)).
1898	The five boroughs of Manhattan, Bronx, Kings (Brooklyn), Queens and Richmond (Staten Island) merged to form the modern New York City (Figure 4.4.1).
1914–18	Influx of black Americans from the Southern states as World War I prevented further European immigration.
1918–present	New York – the 'Big Apple' of opportunity – has continued to attract both Americans and foreigners in large numbers; about 40 per cent of the city's residents are either foreign-born or have foreign-born parents.

Today, New York City is both the hub of Megalopolis and a route centre with access to some 160 million people living within a 1200-km radius – the biggest and richest market in the world. The city's own economy is huge; with only three per cent of the US population, its gross product is about the same as the GDP of Australia. However, during the past two decades or so, New York has encountered difficulties which have removed some of the Big Apple's shine. The following sections present contrasting aspects of this fascinating city.

Manhattan: the heart of New York City

Figure 4.4.2
Downtown Manhattan

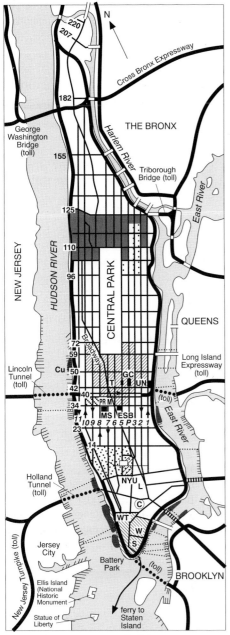

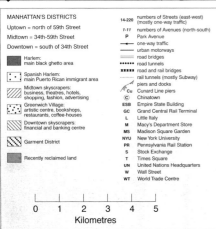

Figure 4.4.3
Manhattan

Although the world-famous skyscrapers are symbolic of New York (Figure 4.4.2) they occupy only two relatively small areas of Manhattan, just one of the city's five boroughs; moreover the city is only one part of a much larger metropolitan region extending into three states (Figure 4.4.1).

Despite its comparatively small area (74 km²), most of the activities which contribute to New York's national and international importance are crowded on to Manhattan Island. Amidst the ceaseless hustle and bustle, Manhattan offers both New Yorkers and visitors a wide range of world-famous museums and art galleries, together with dozens of theatres and cinemas along the 'Great White Way' of Broadway (originally an Indian trail, now the only major thoroughfare to deviate from the island's rigid grid-iron layout (Figure 4.4.3). For the shopper, fashionable Fifth Avenue and the huge department stores of the Midtown area are irresistible attractions. More generally, the contrasting ethnic neighbourhoods and famous landmarks on or accessible from Manhattan add to the island's tourist allure.

Manufacturing industry

Competition for Manhattan's limited space has made its land the most expensive in the world. Once an important centre of manufacturing industry, Manhattan's rise in the twentieth century as the core area of the core city of the core region, first of the nation and then of the world, has removed most of the workshops and factories which used to cluster on the island. As the control and decision-making functions of Manhattan's offices expanded, so too did the space – horizontal and vertical – needed to house them and the myriad other services that Manhattan now contains. Consequently most manufacturing activities have been ousted to mainland locations, especially the New Jersey shore, while the chain of industrial cities which arose along the Hudson–Mohawk valley from Albany to Buffalo (Figure 4.2.1) have long symbolised this trend for manufacturing to decentralise away from the core. The clothing industry located in the Midtown Garment District (Figure 4.4.3) exemplifies the survivors: small-scale operations making high-value products and offering employment to low-waged, low-skilled and (often) new-immigrant workers.

Employment

Nowadays Manhattan's commercial, financial and other service interests are much more important than its manufacturing industry, not least in terms of jobs (Figure 4.4.4) 'Manhattan's business is business' aptly summarises the island's *raison d'être*. Its two skyscraper clusters contain the greatest concentration of office space in the world – even if these days vacancy rates may be 15–20 per cent. The Downtown Financial District of Lower Manhattan dominates the world of American banking, insurance and finance, with Wall Street's Stock Exchange having global influence. Six km to the north, many big American and foreign corporations have main or branch offices in the Midtown area.

Midtown

In several specialised services Midtown is virtually unchallenged; for example, its publishers account for about 75 per cent of all American books while a similar proportion of US advertising agencies have Madison Avenue addresses. This second business node developed after the only two main-line railways to gain access to the island built passenger stations (Grand Central and Pennsylvania) at 42nd Street and 34th Street respectively. This later node is not only larger than the original Downtown area but, with its hotels, shops and theatres, much more diverse. New York is thus unusual in having its central business district (CBD) divided so distinctively, both functionally and geographically.

Figure 4.4.4
Employment trends in New York City

Figure 4.4.5
Location of Fresh Kills landfill

Manhattan's skyline and general cityscape are forever changing, new high-rise buildings appearing where older ones have been demolished and where new land has been reclaimed along the island's shores (Figure 4.4.3). The twin towers of the World Trade Centre are currently the city's highest buildings (110 storeys, 411 metres). Approximately 50 000 people work in the Centre and it receives another 80 000 visitors daily, its main business being international shipping.

New York in crisis: garbage disposal

Americans are prodigious generators of waste materials. In an average lifetime they produce around 4000 times their own bodyweight in waste – four times as much as Europeans and 27 times as much as people in the developing world. Of the 180 million tonnes of garbage annually disposed of in the United States, 72 per cent goes to some 6000 active landfills at a yearly cost of $20 billion.

With landfill sites reaching capacity and few new ones being approved, the United States is in a widely proclaimed 'garbage crisis'. The problems of collecting, transporting and disposing of such huge quantities of municipal waste are amply exemplified by New York City, which produces five million tonnes of rubbish every year – twice the national per capita average.

In New York the problems begin on the streets. Apart from the difficulties created by the City's own unionised refuse workers who collect domestic waste (about half the total), the rest – from about 250 000 commercial establishments such as offices, hotels and restaurants – must, by law, be handled by over 300 private companies who work as a cartel at fixed prices. This arrangement, plus the fact that two of the city's five Mafia families have controlled the industry for decades, creaming off millions of dollars in the process, results in the excessively high cost of rubbish collection in New York – three or four times that in Los Angeles or Chicago.

The Fresh Kills landfill

Half of New York City's rubbish ends up at the Fresh Kills landfill on Staten Island, 23 km from Manhattan (Figure 4.4.5). Opened on tidal marshland in 1948, it receives 17 000 tonnes of refuse by barge every day, six days a week (Figure 4.4.6). This unceasing stream of valueless cargo has made Fresh Kills the world's biggest rubbish tip and one of the largest-ever human-made monuments. It already covers 1250 hectares (almost four times the area of Manhattan's Central Park) and occupies 72 000 m³ of space – 25 times the volume of the Great Pyramid at Giza.

Several myths exist about the contents of landfills like Fresh Kills, reflecting the 'convenience' aspects of modern American life. For example, the US public

Figure 4.4.6
Garbage barge heading for Fresh Kills

believes that fast-food packaging accounts for 20–30 per cent of total landfill volume, polystyrene foam (including disposable cups) for 30–40 per cent, and disposable nappies for 25–45 per cent; since plastics are thought to account for another 30 per cent, such beliefs are clearly statistically ill-founded. Recent scientific excavations of Fresh Kills by archaeologists have revealed its actual contents (Figure 4.4.7).

Figure 4.4.7
Fresh Kills landfill: archaeological analysis (by volume)

Major Category	% of total	Principal Contents
Paper	50	Newspapers (18%), magazines, telephone directories, mail order catalogues, packaging (0.25%)
Construction and demolition materials	15	Bricks, stone, plaster, timber, mortar (often used to construct roads on landfills)
Organic	13	Garden waste (10%), wood, food scraps
Plastics	10	Milk jugs, pop bottles, food containers, bin liners, polystyrene foam (0.9%)
Metal	6	Aluminium and steel cans for food and drink, iron and steel products (including engine blocks!)
Glass	1	Beverage bottles, food containers, cosmetic jars
Miscellaneous	5	Tyres, textiles, disposable nappies (0.8%)

Another common belief is that landfills behave like big compost heaps, i.e. that material in them biodegrades quite rapidly. But this is usually not so; refuse is seldom shredded, large quantities of fluids are prohibited and air circulation is nil. Thus most micro-organisms cannot flourish and materials do not decay; even 30-year-old newspapers remain legible when brought to the surface.

1 Construct a flow diagram to represent New York's system of refuse collection, transport and disposal.

2 How does the scientific evidence in Figure 4.4.7 contrast in detail with the widely held beliefs about the contents of American landfills?

3 Although plastic is much more commonly used in the United States than 25 years ago, it still represents a constant 10 per cent of landfill volumes. Suggest explanations for this apparent paradox.

4 Study Figure 4.4.7. What actions can ordinary individuals and families take to reduce the volume of materials going to landfills, without necessarily reducing their own levels of consumption or living standards?

However, at Fresh Kills the stream that flows between the landfill's two sections affects the refuse. Within 7–12 metres of the stream level all the refuse is moist; another five metres further down and the refuse is saturated; below this level only grey slime and a few baulks of timber exist. The explanation is that the site's tidal wetlands support the bacteria which produce methane and become degraders under anaerobic conditions.

The real problem with landfills is not what goes in but what comes out. Emissions of methane gas can be hazardous if uncontrolled, but at over 100 American landfills the gas is recovered to produce about 2.5 billion m^3 a year – a small but useful contribution to total US gas production. The best, state-of-the-art landfills are lined with impervious clay and/or heavy plastics to prevent leaching, and are located where they lack the potential to leak toxic residues into aquifers, rivers, lakes or seas.

Fresh Kills is currently the only landfill located within the boundaries of New York City; it is filling up rapidly, and its tipping fees are increasing steeply. Much of the rubbish that is not taken to Fresh Kills is transported by road and rail as far as Ohio and Indiana where tipping fees are only one-seventh those in New York. The city has also attempted to 'export' its rubbish by sea (Figure 4.4.8).

Don't Be a Litterbarge

No one wants the wretched refuse of New York's teeming shore

In the Gulf of Mexico last week the *Mobro 4000* searched in vain for a friendly harbour. Southern ports had good reason for turning away the bereft barge: it was loaded with 3250 tonnes of rancid, fly-infested trash from New York.

The *Mobro*'s ill-starred journey resulted from a deal made by Alabama entrepreneur Lowell Harrelson, founder of National Waste Contractors, Inc., to haul away a massive batch of refuse, more than half of it from New York City.

Harrelson had a destination in mind. After he convinced the commissioners of Jones County, N.C., that they could make a fortune by extracting methane gas from the trash, they agreed to let him dump 76 000 cubic metres of rubbish into their local landfills. But when the barge with its festering cargo pulled into Morehead City, Harrelson found that the deal was off. "We told him

to get the boat the hell out of North Carolina waters," said Stephen Reid of the state's solid and hazardous waste management bureau. "We have enough garbage of our own. We didn't need New York's."

The stinking scow then headed for a dump site just outside New Orleans. When the *Mobro* arrived at the Delta town of Venice, Louisiana's department of environmental quality sent inspectors to examine the gunk. Although the bales consisted primarily of paper, investigators discovered hospital items—syringes, bedpans—that could pose a health hazard. The bales were also beginning to ooze, and inspectors feared the scum would leak into the river. Governor Edwin Edwards half jokingly threatened to deploy National Guardsmen on the levees with orders to shoot if the barge tried to dock. As the vessel meandered about the Gulf of Mexico, Louisiana Attorney General William Guste trashed

federal authorities. "The Environmental Protection Agency was not watching it at all," he said. "If that barge tips or runs into a storm, the garbage could end up on our shores."

Goaded into action, EPA officials commandeered a Coast Guard cutter last week and were en route to a rendezvous with the barge when they got word that the *Mobro* was headed for Mexico. But there too it was turned away: the Mexican navy even went on alert to repel the Yanqui rubbish.

Having been rejected by six states (North Carolina, Mississippi, Alabama, Florida, Texas and Louisiana) and three foreign countries (Mexico, Belize and the Bahamas), the ill-fated *Mobro* eventually returned to New York City and its cargo, termed 'the most famous 3000 tonnes of garbage in the history of the universe', was dumped locally.

Time, 4 May 1987

Figure 4.4.8
Don't be a litterbarge

4.5 Los Angeles: A Transportation Palimpsest

Although founded as a Spanish *pueblo* in 1781, Los Angeles (1994 population: city 3 449 000; metropolitan area 8 863 000) is essentially a twentieth century American city (Figure 4.5.1). With a sunny climate and an oilfield as its only initial advantages, it grew rapidly as the centre of developments such as petroleum refining, film making, the processing of farm produce and, later, defence, aerospace and other high-tech industries. Los Angeles is now California's main industrial area, the economic capital of the Pacific region, and the United States' second largest population cluster (15 302 000 in 1994).

Figure 4.5.1
Percentage of 1960 population level attained during several eras in selected cities

	Pre-1830	1830–1870	1870–1920	1920–1960
Los Angeles	0	0+	15	85
Boston	9	18	48	25
Chicago	0	8	47	45
Philadelphia	9	15	38	38
New York	3	11	44	42

As late as 1880 Los Angeles was still little more than a small country-town of only 11 000 people. Its subsequent growth and extraordinary lateral expansion across the plains, into the valleys, and up the foothills of the Los Angeles Basin (Figure 4.5.2) can be ascribed to two main factors: first, to the desire of Angelenos to live in one-family, detached homes, in low-density residential areas; and secondly, to a succession of transport developments – not just, as commonly believed, to freeways and the motor car. The still-discernible imprint on the region's urban morphology made by each stage in this succession of dominant transport systems has been termed a 'palimpsest'.

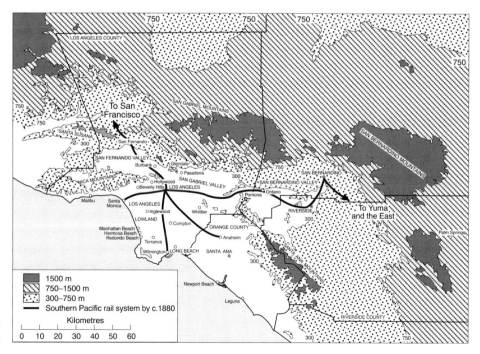

Figure 4.5.2
Topography of the Los Angeles Basin

Rail systems encouraged suburban sprawl

The first of these transport developments was the Southern Pacific Railroad, which from the 1870s not only provided Los Angeles with a vital transcontinental link, but also created its first, star-shaped commuter network (Figure 4.5.3) each line generating new satellite communities along its length.

Thus by 1880 the steam railways had established a skeletal outline of the shape and extent of the future city, and had initiated the pattern of commuter movement that was to become its hallmark.

In 1887 the horse-drawn streetcars which had been established during the previous decade in Los Angeles and neighbouring communities like Pasadena, Santa Monica and San Bernardino, were superseded by the Los Angeles Railway (LARY), the region's first electric railway, a commuter service operating out of down town Los Angeles. Similar suburban and inter-urban railways soon spread over a wide area of the Los Angeles basin, eventually being consolidated under the control of one company, the Pacific Electric (PE).

By the 1920s, PE's trains of 'Big Red Cars' were running over one of the most extensive rail networks in America, interconnecting Los Angeles – itself a wide-spread city – and a conglomeration of independent cities and far-flung communities which comprised the metropolitan region (Figure 4.5.3). Indeed, PE's route map practically defined Greater Los Angeles as it exists today. Since PE was also heavily involved in the real estate business, the company contributed greatly to the development of the new communities served by the rail system.

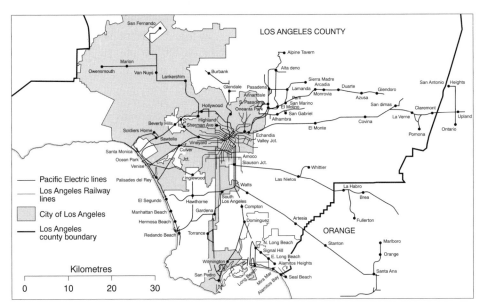

Figure 4.5.3
Electric railway systems and surburban development, 1925

The advent of the automobile

Further dispersion of Los Angeles' suburban communities resulted from the introduction of the motor car. By the mid-1920s massive numbers of cars began to appear throughout the metropolitan area. In Los Angeles County alone, car registrations had increased nearly fourfold between 1918 and 1923 from 110 000 to 430 000, while car ownership in the city itself had always run well ahead of the national average. Car ownership gratified Angelenos' desire for individual freedom and offered a wider choice of residential location than that allowed by the fixed-route rail system.

With steep rises in both the numbers and the use of cars, congestion on downtown Los Angeles' narrow streets – originally laid out in the horse and buggy era – became increasingly hazardous as cars, lorries and streetcars jostled for space. One reaction was to decentralise some of the CBD's functions to sub-urban locations but, by the late 1930s, shoppers and commuters were driving almost randomly across the metropolitan area. The resultant confused pattern of movement was described by the Automobile Club of Southern California as 'a million automobiles moving in a million different directions, with their paths of travel conflicting at a million intersections a million times a day'.

Freeways advocated

In 1937 the Automobile Club suggested a dramatic solution: a network of traffic routes for the exclusive use of motor vehicles with no cross-roads and no interference from adjacent land use activities. Such a system of highways, from which rail traffic was excluded, uninterrupted by traffic lights or intersections, would allow high-speed automobile movement throughout the metropolitan region (Figure 4.5.4).

These roads, first called 'motorways' by the Automobile Club, would later be known as freeways and were hailed as the answer to the decades-old problem of congestion. Not only would the freeways allow residents to live in one part of the region and work in another, but they would also provide easy access throughout what had become a multi-centred metropolis whose principal CBD would now be first among near-equals rather than absolutely dominant.

An official study, involving all the communities in the Los Angeles metropolitan area, accepted the Automobile Club's conclusions and in 1939 published a map of the proposed route network, which after the Second World War became the basis of today's freeway system (Figure 4.5.5). In fact, a short stretch of motorway, the Arroyo Seco, America's first freeway, was completed in 1939, later being incorporated into the Pasadena Freeway.

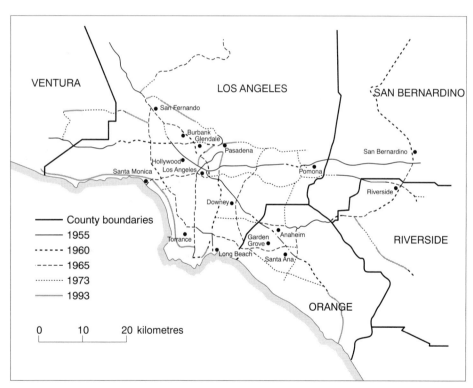

Figure 4.5.4
Los Angeles: development of the freeway system, 1955–1993

The Second World War temporarily reprieved the region's railways, but as the freeway-building programme began in earnest in the late 1940s the railways plummeted into irreversible decline, the last train running in 1961. Buses have replaced the Big Red Cars as the main public transport system in the Los Angeles region but are widely regarded as unsatisfactory. A recent survey of bus passengers indicated that the great majority would prefer to use cars if they could. Indeed, it has been said that the only people regularly using the city's buses are the old, the poor, the infirm and the otherwise disadvantaged.

A metropolis on wheels

Today, life in Los Angeles is based on the automobile to a greater extent than anywhere else in the world. Over 70 per cent of the downtown area is devoted to roads, parking lots and other demands of the car. There are almost as many cars as people of driving age, over 90 per cent of all journeys are made by car, and 160 million km a day are driven on a 2000-km network of freeways that have become virtually indispensable to movement around the metropolis (Figure 4.5.5).

However, despite all the early expectations, in recent years the freeways have proved increasingly unable to cope with the huge volumes of modern traffic. Every new freeway generates ever more traffic and rarely eases congestion. Consequently, Los Angeles is notorious for its traffic jams and parking problems. 'Rush hour' lasts from 5 a.m. to 7 p.m. and, on average, it takes 1–1½ hours to travel 30 km. Not surprisingly, an end to freeway construction in the region has been announced; the last was completed in 1993.

Figure 4.5.5
Part of the Los Angeles freeway system

QUESTIONS

1 (a) Convert the data in Figure 4.5.1 into a multiple bar chart.
   (b) Taking 1920 as the start of the 'automobile age' in the United States, compare and contrast the urban growth of Los Angeles (measured by population) with that of the other cities listed.
   (c) Which city matured most before the onset of mass motoring, and which after it?

2 With reference to Figure 4.5.2 and other relevant maps, explain the relationship between Los Angeles' physical setting, transport systems and urban/suburban sprawl.

3 Explain in detail why Los Angeles can be described as a 'palimpsest' in terms of its transport systems, past and present.

Los Angeles and the smog problem: yearning to breathe free

Figure 4.5.6
Downtown Los Angeles in the smog season

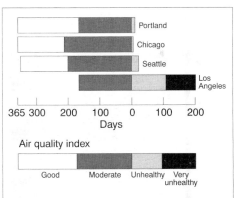

Figure 4.5.7
Annual summary of air quality for selected cities

Year-round sunshine, warm temperatures, and light winds, especially in summer and autumn, are among southern California's most attractive climatic features. However, these same climatic elements combine to produce atmospheric conditions poorly suited to receive the emissions of a motor vehicle-dependent city like Los Angeles, which has long experienced some of the worst air pollution in America (Figure 4.5.7).

Under normal conditions, as the sun warms the earth's surface, the air at ground level expands, becomes less dense and rises, carrying pollutants with it to disperse at high levels. Fresh, cool air then flows downward to replace the polluted air. But when air descends from the North Pacific sub-tropical anticyclone it causes compressional heating of the lower atmosphere and – especially when a layer of dense, cool air flows in from the sea – a temperature inversion is formed. Such low-level inversions, at 500 metres or less, restrict vertical diffusion of pollutants while the blocking effect of the surrounding mountains prevents lateral diffusion (Figure 4.5.8).

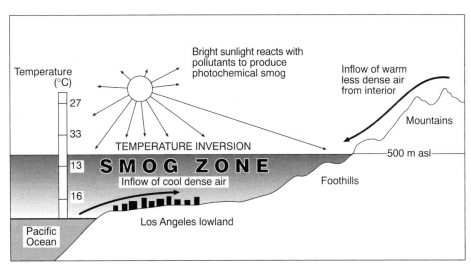

Figure 4.5.8
Smog conditions in the Los Angeles Basin

This serious problem is compounded by the region's bright sunshine which energises the reaction that transforms gaseous wastes into photochemical smog – a brownish haze that reduces visibility, damages vegetation, irritates eyes and generally threatens human health.

Pollution control

Smog became a problem in Los Angeles in the early 1940s and official attempts to reduce atmospheric pollution date from 1947 when the Los Angeles County Air Pollution Control District was formed. Its regulations include the banning of open refuse burning, the installation of control devices in industrial chimneys, and the burning of natural gas or low-sulphur oil in power stations – coal is banned completely. However, in 1971 four million vehicles burning 30 million litres of petrol every day were responsible for nearly 90 per cent of the 14 645 tonnes of pollutants released *daily* into the air of Los Angeles County alone. Despite limiting the olefin content of petrol sold in the area, and restrictions on exhaust emissions through the introduction of catalytic converters from 1966, the problem continued.

Smog distribution

The geographical distribution of smog in the Los Angeles region is extremely complex; each chemical constituent of the toxic brew – hydrocarbons, nitric oxide, carbon monoxide, sulphur dioxide – has a different distribution. Their concentrations are originally related to varying sources of pollution, but their distribution is soon affected by the prevailing winds – generally from the coast to inland areas. Pollutant volumes also vary both hourly and seasonally, July and August being the worst months for smog.

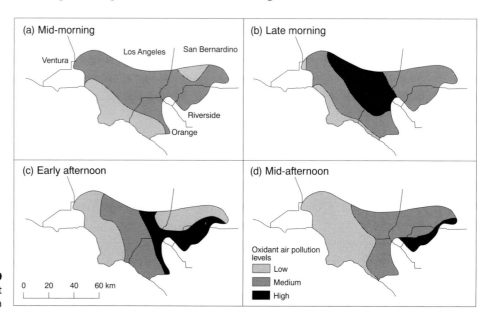

Figure 4.5.9
Changing daily distribution of oxidant pollution in the Los Angeles Basin

The stationary sources of pollution – power stations, oil refineries, factories – are located mainly in the south and south-west coastal area and in the east San Fernando Valley. These plants produce considerable quantities of sulphur dioxide and nitric oxide, the latter being formed during the combustion of all fuels, including domestic heating oil. Pollutants produced by car exhausts reach their peak during the morning rush hour. While this pollution is widespread, it is also heavily concentrated in the industrial areas and in the CBD. The original contaminants react photochemically with atmospheric elements to form secondary contaminants such as ozone and nitrogen dioxide which have their own distribution patterns. Ozone is colourless but nitrogen dioxide's light-scattering properties cause the smog's typically brownish or yellowish-brown tinge. Oxidants – the end-products of the reactions described above – have a pattern that is the most complicated of all (Figure 4.5.9).

Technofix attempts

Motor vehicles, especially cars, remain the major cause of Los Angeles' air pollution problem, yet any proposal to limit their use, even in the main smog season (for example, by rationing petrol) has always resulted in a public outcry. Instead, Angelenos seem to prefer the 'instant technofix'. Since the 1950s, suggested technofixes have included: removing the temperature inversion by using ground-based fans, helicopters or thermal means; eliminating the sunlight which triggers the photochemical reactions by using aircraft to spread a huge parasol of white smoke over the city; extracting the smog through tunnels in the surrounding mountains, using huge fans; and chemically spraying the atmosphere to neutralise the smog.

In reality, lasting technofix solutions seem very remote, but steady progress is being made by more conventional means. The interior of the Los Angeles basin is still America's smoggiest place, but in recent years there have been big reductions in the most dangerous components of smog, including lead, carbon monoxide and, above all, ozone. In the 1970s, during the May–October official smog season there used to be average of 120 Stage-1 smog alerts each year (triggered when the ozone level exceeds 0.20 parts per million (ppm)); nowadays 40 a year is the norm, while Stage-2 alerts (when the ozone tops 0.35 ppm), warning children and old people to stay indoors, are virtually unknown (Figure 4.5.10).

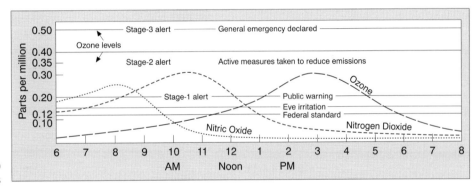

Figure 4.5.10
A typical smog day in Los Angeles

The emission controls on vehicle exhausts, first introduced in 1966, have been steadily tightened ever since. So successful have these measures been that the two million cars in the region in the early 1950s produced more smog than today's $8\frac{1}{2}$ million. These vehicles are, however, still responsible for 60 per cent of the smog in the Los Angeles Basin, and the aim of the pollution authorities is to comply with the federal standard of 0.12 ppm or less of ozone in the region's air by 2010. To achieve this, new cars sold by the year 2000 will be required to emit less than one-fifth of the pollutants pumped into the atmosphere by today's cars. The introduction of electric vehicles will also be beneficial. By then, according to the regulators, smog alerts in the Los Angeles region will be a thing of the past.

A new rail network for Los Angeles

In an attempt to provide an alternative to cars and reduce both congestion and pollution, Los Angeles is building a new rail network, albeit much less extensive than the old one (Figure 4.5.11). In July 1990, the city opened the Blue Line, its first urban railway since 1961, a 30-km track from Long Beach to downtown

1 (a) What exactly is photochemical smog?
  (b) Explain (i) why May–October is the main smog season in Los Angeles; (ii) why smog remains a serious problem throughout the winter.

2 With reference to Figures 4.5.9 and 4.5.10, describe and explain the changing daily distribution of oxidants in the Los Angeles Basin.

3 In 1991, motor vehicles caused 60 per cent of all air pollution in the Los Angeles region, compared with nearly 90 per cent 20 years earlier.
  (a) How has the improvement been achieved?
  (b) What further measures could be taken to make still greater reductions?

Los Angeles. Funded mainly by a local sales tax agreed in a referendum, the whole rail system is scheduled for completion by 2010, but the Red Line, tunnelling towards Hollywood, encountered serious delays caused by earthquake fault lines, jeopardising the federal subsidy. In any case, a relatively sparse rail web in such an extensive metropolitan region can provide only marginal relief. Most commuters will still have to drive to the station to take the train, or endure long journeys by bus.

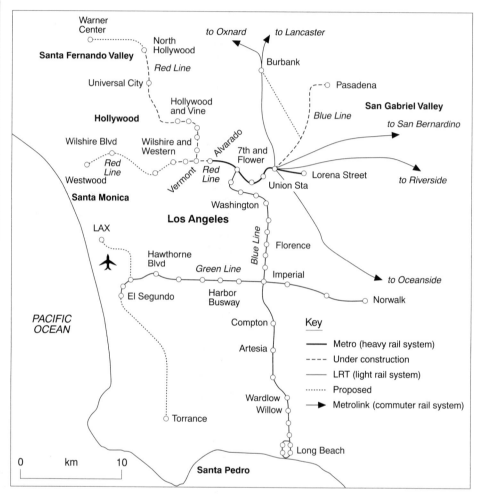

Figure 4.5.11
Los Angeles rail routes

4.6 Urban Planning: Toronto

For over forty years the citizens of Toronto have been the subject of planning as have no others on the North American continent. Having apparently avoided many of the serious problems encountered by its US contemporaries in particular, Toronto is regarded worldwide as a model for major metropolitan development.

Location and character

Toronto, the Huron Indian word for 'meeting place', lies on the north west shore of Lake Ontario (Figure 4.6.1) and is one of the busiest Canadian ports on the Great Lakes. As Canada's largest metropolitan area (population 4.3 million in 1995) it is the chief manufacturing, financial and communications centre in the country. About a third of Canada's manufacturing industries are within 160 km of the city.

The city of Toronto covers 97 km² while the Municipality of Metropolitan Toronto extends for 630 km² (Figure 4.6.2). About a quarter of Ontario's people and nearly one-tenth of the Canadian population live in the Municipality. The Greater Toronto Area (GTA) includes Metropolitan Toronto and the regional municipalities of Halton, Peel, York, and Durham. The GTA covers an area of approximately 7000 km², over 5500 of which are rural. It is thus an area of considerable contrast.

Metropolitan Toronto also lies within the GT Bioregion. The boundaries are defined by the Niagara Escarpment to the west, the Oak Ridges Moraine to the north and Lake Ontario to the south. This area comprises six watersheds, all of which drain south to Lake Ontario. The municipalities within the GT Bioregion share its natural resources and are mutually responsible for the health and integrity of the natural ecosystem.

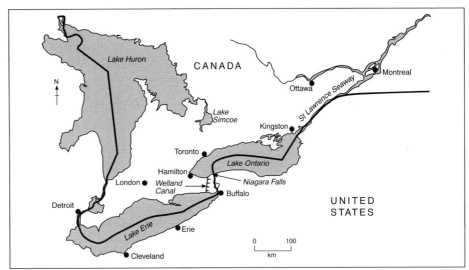

Figure 4.6.1
The location of Toronto

A high density legacy

An important part of Toronto's twentieth century success is a legacy from the previous century, when the working quarters of the city, lying east and west of the core were largely constructed. They were made up mainly of three storey semi-detached houses on lots no wider than fifteen feet, separated by a narrow firegap. Although house style was to change, the legacy of residential compactness did not and large areas of the city's east and west ends were filled, in the 1920s, with the ubiquitous Tuscan porch houses, characterised by narrow side-yards and setbacks. The pattern was punctuated, along the streetcar routes, by contiguous architecture and dense walkup apartment buildings. Thus by 1945, with a population of almost 600 000, Toronto had a density of over 7000 people per km², extremely high by North American standards.

Figure 4.6.2
The Greater Toronto Area (GTA)

Infrastructural spending

Another factor responsible for the relative compactness of Toronto and the well-being of the inner city compared to its US counterparts has been the direction of government spending. Throughout most of this century the balance of investment from upper-tier governments for social infrastructure and hard infrastructure has differed between the USA and Canada. Canada has provided more for health, education and social services and less for highways, waste treatment facilities, and other elements of hard infrastructure. There can be little doubt that such targeting has had significant beneficial effects on Canadian cities in terms of compactness and the liveability of inner areas.

The establishment of Metropolitan Toronto

Toronto managed to sidestep the largely unregulated low density urban sprawl that engulfed most US and some Canadian cities through the 1950s, 1960s and much of the 1970s. This was to a large extent due to the foundation of the Municipality of Metropolitan Toronto in 1954, becoming North America's first metropolitan government federation. The Municipality, commonly called Metro Toronto, consisted of Toronto and twelve of its suburbs. The Ontario legislature created the federation to provide a way for Toronto and the suburbs to solve various problems they had in common. In 1967, the 13 members of the federation were merged to form Toronto and five boroughs; East York, Etobicoke, North York, Scarborough, and York. Since 1979, all but East York have become cities but remaining in the federation. Each of the federation's six units has its own government to handle local needs.

The new metropolitan government was given jurisdiction over the provision of regional infrastructure, including highways, transit systems, water supply and sewage treatment facilities. Metro Toronto thus had considerable power to control the pace, density and location of urban development at the supra-municipal scale. The metropolitan authority's powers were reinforced by the Ontario Planning Act which linked approval of all subdivisions to the supply of public facilities. Thus the process of 'leapfrogging' which characterised suburban growth in so many North American cities was largely avoided in Toronto.

The 1959 Metro Plan

In its foundation year Metro Toronto started the first urban planning exercise in North America which was truly metropolitan in scale, a task completed in 1959. Although this plan was never formally adopted it was consistently used as an unofficial guide to decision-making. The remarkable content of the 1959 plan owed much to concern about the centrifugal trend that was then beginning to fracture US cities. Metro Toronto had unique powers, if not to reverse this trend, then at least to reduce its impact. As a result the plan focussed on three related objectives:

- **A balanced transportation system**
  'Individual' and 'collective' transport were viewed as playing complementary roles. While a new expressway grid was needed to complete the partial network then in place, the importance of a comprehensive public transport system was stressed, not only because it was socially justifiable, but because of its role in affecting urban shape and density. The Toronto subway opened in 1954 to become Canada's first underground rapid transit railway. The first line ran under Yonge Street. Several new rapid transit lines and extensions opened during the 1960s and 1970s while a line serving Scarborough opened in 1985. Metro Toronto was the first city in the world to put a computer-controlled road traffic system into operation.

- **A strong central city**
  The economic vitality of the City of Toronto was seen as essential to the health of the wider metropolitan region. The Metro Plan could only aid this objective indirectly, by allocating gross residential densities and favouring central accessibility through regional transportation decisions. Additionally, in the 1950s, the composition of Metro Toronto's membership reflected a population distribution that favoured the interests of the City of Toronto, so policies favouring a strong centre were generally unopposed. A major benefit was that provincial money paid for much of the transportation infrastructure that focussed on the centre in accordance with the 1959 Metro Plan.

- **A compact metropolitan form**

The Metro Plan advocated a 1980 population of 2.3 million inside the metropolitan boundaries to give a density of about 4000 per km², roughly double the continental average. In 1990 Metro Toronto had a population density of just over 3500 persons per km², significantly higher than many of its counterparts elsewhere in the continent (Figure 4.6.3). Such compactness has had important implications for public transport.

By the late 1970s virtually all of the land within the metropolitan boundary had been developed. However, the suburban municipalities had reduced the purpose of subdivision control to limiting the incidence of negative externalities. The result was a binary landscape with large expanses of low density housing separated by high-rise commercial and residential structures occupying arterial routes and their intersections. About 80 per cent of all development fell into one of two categories: either lower than two storey or higher than fifteen.

The 1980 Metroplan

The new Metroplan which was adopted by Metro Toronto and approved by the Province took its lead from a number of European cities, which had adopted the idea of polycentrism as a way of alleviating central area congestion, while bringing life and employment to the suburbs. The objective was to base commercial, retail, cultural and housing development into a number of new high-density suburban nodes that would be accessed primarily by rapid transit.

The Metroplan seemed the ideal instrument to combine the often divisive interests of the centre and periphery. However, much depended on the ability and will of the suburbs to channel office growth into the designated centres. Between 1970 and 1985 Metro Toronto's commercial and service employment doubled, while that of manufacturing fell by almost half. However, while the bulk of new office jobs were suburban the majority of these were locating outside the designated centres, either in office parks close to the freeway system or near the arterial intersections, causing continued congestion on the suburban arteries. During the same period, capital works investment in the mass transit system was negligible, so there was little chance to consolidate the polycentric plan objectives.

The Liveable Metropolis

The latest official plan for the Municipality of Metropolitan Toronto was published in December 1994 and approved in early 1995. It replaces the 1980 plan and the amendments to it which followed. The Liveable Metropolis provides a long-term planning strategy to improve the quality of urban life through an integrated approach to land use planning and management of environmental, economic and social change. The policies in the plan are intended to shape the future within Metropolitan Toronto's boundaries and influence the future in the whole Greater Toronto Area.

The Metropolitan Structure

Urban sprawl is recognised as not being a sustainable solution. The Metropolitan Corporation cannot look to the outlying parts of the GTA to absorb all the new population growth. However, unlike other areas in the GTA, most of the land in Metro Toronto is already developed; thus reurbanisation is required. Reurbanisation is a process of redevelopment and reinvestment that involves taking advantage of the opportunities and facilities that already exist across the urban area. Fundamental to the reurbanisation strategy are reductions in: urban sprawl, reliance on the car, natural resource use and public servicing costs associated with development.

Figure 4.6.3
Population densities (persons per km²) of major cities, 1990

Metro Toronto	3531
City of Toronto	6300
North York	3150
Outer GTA [urban]	2100
Montreal [city]	5700
Edmonton	860
Winnipeg	1040
Vancouver	3800
Sydney	1800
Chicago	2000
Los Angeles	2000
Hong Kong	30 000
London	5600
European Capitals	4700–6000

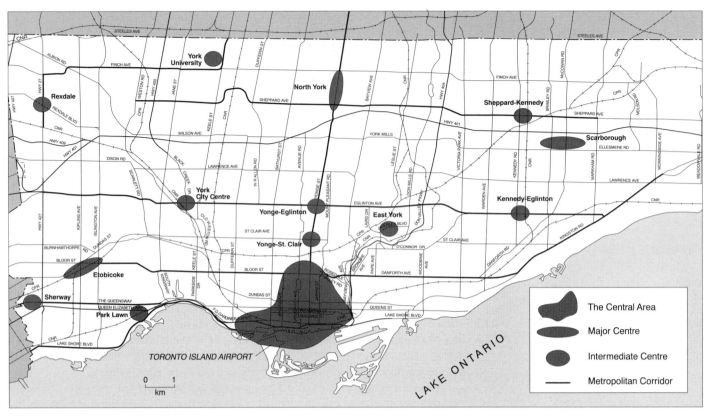

Figure 4.6.4
Toronto: Metropolitan Centres and Corridors

The objectives for Metropolitan Toronto in terms of structure are:

- to create and maintain a structure of Centres and Corridors (Figure 4.6.4) that:
  uses land, infrastructure and other services efficiently;
  concentrates employment and population in areas well served by transit;
  promotes living close to work, walking and cycling and the use of transit and other high-occupancy vehicles;
  facilitates social interaction, public safety and cultural and economic activity;
  strengthens the vitality and identity of Metropolitan Toronto.
- to maintain a sufficient supply of industrial lands and a diversity of employment necessary to enhance the economic competitiveness of the city.
- to plan and manage the Metropolitan Green Space System (Figure 4.6.5) and abutting lands in terms of ecological functions and recreational needs.

The Plan anticipates a minimum population of 2.5 million people and 1.7 million jobs by the year 2011. It is policy that investment in transit and other services is required to achieve the desired structure of Centres and Corridors. However the Plan also recognises that Metro Toronto contains many local centres, not designated as Metropolitan Centres by the Plan, but which are important to the neighbourhoods and districts in which they are located. The development of such local centres should be encouraged providing that this does not significantly detract from development in Centres and Corridors. The Plan also stresses the importance of supporting and strengthening the central area to enhance the overall development of both Metro Toronto and the GTA.

It is policy to improve and expand the rapid transit network by implementing the Metropolitan rapid transit network shown in Figure 4.6.6. The implementation of such transit facilities shall be considered only after detailed analysis of their overall costs and benefits. Rapid transit reviews will be conducted at least every five years to reassess network needs and establish priorities.

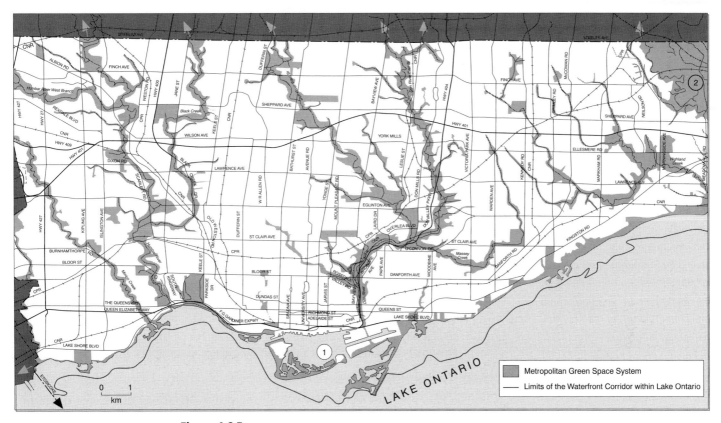

Figure 4.6.5
Toronto: Metropolitan Green Space System

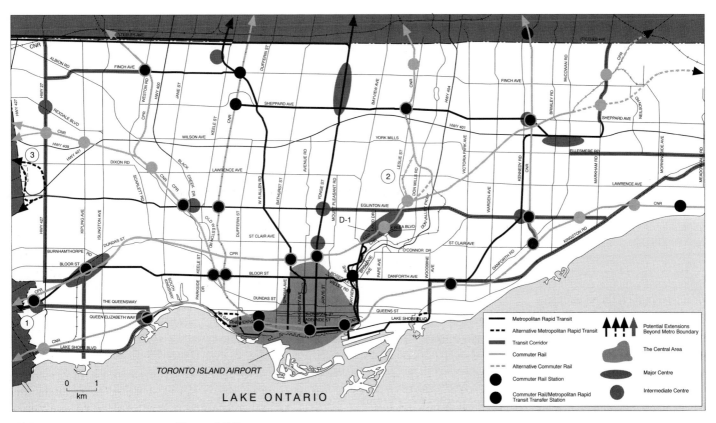

Figure 4.6.6
Toronto: rapid transit system

Sustainable community development

Social, economic and environmental issues are intrinsically linked with land use planning and urban structure. For example, the pattern of development across the municipality affects access to open space, to jobs, and to a suitable labour force, and influences the degree to which opportunities exist for social interaction and integration. The Plan sets out the following objectives for sustainable community development:

- to maintain and enhance Metropolitan Toronto's role as the economic and employment focus of the GTA, to achieve a diverse, competitive, adaptable economic environment, and to support a broad range of employment opportunities;
- to ensure the availability of an adequate supply and mix of housing to meet the full range of housing needs, and to attract and accommodate population growth;
- to promote cultural expression and to provide opportunities for social and leisure activities that reflect the lifestyle needs and interests of residents and visitors;
- to achieve strong communities where residents have equitable access to services and opportunities;
- to conserve, protect and enhance the integrity of the natural systems so that they may benefit the health and well-being of current and future generations.

Implementation

The Plan provides a guide for public and private decisions that shape the future for Metro Toronto. Effective implementation requires collaboration by the Metropolitan Corporation with the community, Area Municipalities within Metropolitan Toronto, GTA municipalities, the Province, the private sector, the non-profit sector, the Metropolitan Toronto and Region Conservation Authority, and other public agencies. The Plan sets the strategic framework for managing growth and change. Considerable reliance is placed on the Area Municipalities to implement many of the policies and objectives. It is intended that the official plans and zoning by-laws of the Area Municipalities shall conform with, be complementary to, and elaborate on, the policies of the Plan.

QUESTIONS

1 Describe the location of Toronto and suggest why it has become the largest city in Canada.

2 Explain the distinction between (i) the City of Toronto; (ii) Metropolitan Toronto; (iii) the Greater Toronto Area.

3 (a) Why did Toronto develop at a higher population density than its US counterparts?
  (b) To what extent has this proved beneficial for contemporary planning?

4 When and why was Metropolitan Toronto established?

5 Discuss the main elements of the 1959 Metro Plan. How influential was it in the development of Toronto?

6 How did the 1980 Metroplan build on its predecessor?

7 Attempt to justify the system of Centres and Corridors which are fundamental to the latest plan, the Liveable Metropolis.

8 (a) Describe the distribution of Green Space in the metropolitan area.
  (b) Why is the management of the Green Space System a significant element of the plan?

5

Energy and Mineral Resources

Continental Overview

Energy

US 'Vulnerability' to rising imports

North Americans use huge amounts of energy each year (Figure 5.1.1) with consumption per person over twice that of western Europe. Both the USA and Canada have been fortunate in having large deposits of fossil fuels and many good sites for hydro-electric power (HEP) plants. The two countries have also achieved the affluence and technology to exploit these resources and to invest in newer forms of energy.

For most of its history the continent was able to supply its net energy needs. However, by the 1950s demand for oil had risen to such a high level in the USA that significant net imports were required, rising to over 47 per cent of domestic consumption in the late 1970s, and exceeding the 50 per cent mark in the early 1990s. Canada became a net importer of oil in the late 1970s but new discoveries in the 1980s restored the country's position of net self-sufficiency and also allowed renewed export to the USA. In total, North America holds only 4 per cent of the world's proven oil reserves.

The demand for energy increased because of rising industrial production, an increasing population and lavish per capita usage due to a combination of high average incomes and cheap energy prices. However, in the early 1970s North Americans were forced, for the first time, to think really seriously about energy use. In the USA many oil and gas fields were drying up, with oil production reaching a peak in 1970, followed by natural gas in 1973. International politics made the situation worse. After the 1973 Arab–Israeli war the Arab members of OPEC, who were and remain large-scale producers, reduced exports of oil to the USA and other countries.

The oil embargo was short-lived but OPEC took the opportunity to sharply increase the price of oil on the world market. Such a severe increase served as a warning about the need for the sensible management of resources. Interest in renewable energy increased considerably, leading to substantial investment in research and development. However, increasing global production of oil in the 1980s caused a significant reduction in price and declining interest in renewable energy. By late 1992 Americans were paying less in real terms for their petrol than they had done at any time since the 1973 oil embargo. This was due to a combination of a very low petrol tax and a subdued world market price. Such a low price encouraged consumption and thus importation (Figure 5.1.2) rekindling US fears of vulnerability to supply discontinuities. If nothing is done, experts warn, the USA will need to import two-thirds of its oil by the year 2000. In 1994 the oil import bill amounted to almost $50 billion, a major item in the nation's huge trade deficit.

Energy supply: a changing situation

In 1993 fossil fuels contributed 88 per cent to US energy supply, a reduction of less than 8 per cent since 1950 (Figure 5.1.3). Nuclear power, a non-contributor in 1950, has steadily increased production but at a much slower pace than the

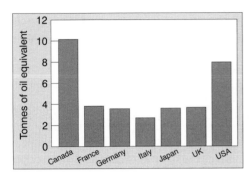

Figure 5.1.1
Energy use per capita in G7 countries, 1994

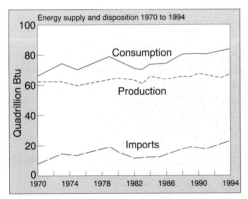

Figure 5.1.2
US energy supply and disposition, 1970–1994

Figure 5.1.3
US energy consumption by source (%)

Type of fuel	1950	1970	1993
Oil	39.7	43.9	40.2
Natural gas	18.1	32.8	24.8
Coal	38.0	18.9	23.3
Nuclear	0.0	0.3	7.7
HEP	4.2	3.9	3.7
Other	0.0	0.1	0.2

optimistic forecasts of the 1960s and early 1970s predicted. Over the same period the developed installed capacity of HEP increased almost fourfold, but its relative contribution actually declined slightly. Biomass has dominated other forms of renewable energy in the USA.

In Canada, 85 per cent of consumption was met by fossil fuels in 1991, the remainder coming from HEP and nuclear power. About 75 per cent of Canada's primary energy is used in final consumption such as petrol for cars or fuels for heating. Industry is the largest single energy user, followed by transportation. Most of the remaining 25 per cent of primary energy is used to convert energy from one form to another. In terms of electricity generation, 62 per cent of Canada's total supply comes from HEP, 22 per cent from thermal plants and 16 per cent from nuclear stations.

Frontier and offshore exploration

With declining production from traditional oil resource regions, frontier and offshore locations have attracted increasing attention in recent decades. The North Slope of Alaska is a classic 'resource frontier' with rapidly rising oil production from the late 1970s significantly offsetting declining production in other states. However, in recent years Alaskan oil production has been in decline, adding to US concerns about further import dependence.

Not long ago, the Gulf of Mexico, the first offshore exploration zone in North America was known as the 'dead sea' within the industry. But by 1993 the Gulf was back in favour. It currently produces about 25 per cent of US gas and 12 per cent of its oil. Would-be drillers have always been nervous about the Gulf's very salty bed, which is difficult for seismic detectors to penetrate, making exploration very unpredictable. However, with new techniques, such 'subsalt prospecting' is becoming more fruitful. Another boost has come from better technology for deep-sea production. Most of the Gulf's existing output comes from shallow offshore wells, which typically work at depths of less than 100 m. In the 1980s a few wells were set up at depths of up to 450 m. Now production in water twice as deep appears to be profitable.

The US Geological Survey has estimated that 32 per cent of the still undiscovered oil resources and 27 per cent of undiscovered gas resources are located offshore. Apart from Alaska, with its severe environmental difficulties, offshore leasing has attracted great interest in the Santa Monica Basin, off California, and the Georges Bank, off Newfoundland.

The North America energy market

Both electricity and primary energy are transferred daily between the USA and Canada. Much of this transfer is from Canada to the USA in the form of gas, oil and electricity. The USA also imports significant amounts of oil from Mexico, reaching a record 317 million barrels in 1993. In the same year Canada exported 330 million barrels to the USA, the highest since 1973. In contrast, since 1985 Mexico has imported increasing volumes of natural gas from the USA. Before that year the flow of gas had always been in the opposite direction.

Environmental impact and conservation

Economic activity can have a severe impact on the environment and as the continent has grown rich, environmental destruction has increased. Energy production and use has caused much of this damage. As concern grew in the 1960s and 1970s, both the USA and Canada introduced new legislation to limit the impact of energy activities in terms of both production and end use. For example:

- where land is strip mined for coal it now has to be restored to a usable condition after mining has ceased;

- the Clean Air Acts of both countries place strict limits on the amounts of pollution released into the atmosphere by industry and power stations;
- the production of renewable energy is encouraged through financial incentives.

All of the above strategies concern energy supply. However, conservation or demand strategies are also of considerable importance. In 1992 US electric utilities spent 1.3 per cent of their total revenues on demand-side management (DSM) programmes (Figure 5.1.4). In return these programmes cut annual electricity sales by 1.2 per cent and peak demand by 6.0 per cent. Relative to economic growth, the US has reduced energy consumption significantly. In 1970 the country used 23 400 BTU/$ of GDP, reduced to 16 200 BTU/$ in 1994. From the initial 'oil shock' in 1973 through to the 1980s, seven times more 'new' energy was obtained through efficiency than from all other sources of supply.

DEMAND-SIDE MANAGEMENT BY U.S. ELECTRIC UTILITIES

Electric utilities in the United States have been world leaders in taking the initiative to improve the efficiency of electricity use by consumers because so often an investment in efficiency improvement shows a better payoff than an investment in additional supply capabilities. From among a great many, illustrative examples include the following:

- The Smart Light program of the Burlington, Vermont, Electric Department provides energy-efficient compact fluorescent light bulbs to customers who pay a small leasing fee as part of their monthly electric bills. This program is part of the more general Neighbor $ave program, which has an estimated lifetime savings of more than 45 gigawatt-hours of electricity supply.

- The Peak Corps Air Conditioner Load Management program of the Sacramento, California, Municipal Utility District (SMUD) allows the utility to reduce peak power supply requirements by cycling customers' air conditioners. SMUD also offers rebates and financing to encourage the purchase of more efficient end-use equipment, and it offers incentives to builders to make new construction more energy efficient.

- The Good Cents New and Improved Homes program of Waverly, Iowa, Light and Power offers reductions of 10 percent for 10 years in the energy portion of the electricity bill for homes that qualify for "Good Cents" designation for being energy-efficient. The program includes financing programs to make efficiency improvements more accessible to customers. The utility also provides appliance rebates, free energy audits, and innovative rate structures.

- The Multi-Family Conservation program of Seattle City Light (SCL) provides grants for the installation of energy conservation measures in low-income multifamily residential units. SCL also offers rebates, weatherization programs, and incentives for efficient building construction.

- The Energy Star Homes (ESD) program of the Austin, Texas, Environmental and Conservation Services Department rates the energy efficiency of new homes. ESD also offers audits, rebates, loans, and weatherization services. Cumulative commercial savings from all of ESD's programs were estimated in 1992 as the equivalent of 23 megawatt-hours of electrical supply.

Environment, 36(9), November 1994

Figure 5.1.4
Demand-side management by US electric utilities

QUESTIONS

1 What are the main reasons for the variations in energy use per capita shown in Figure 5.1.1?

2 (a) Why is the concept of 'vulner-ability' so high on the US energy agenda?
   (b) How can energy vulnerability be reduced?

3 Discuss the costs and benefits of frontier and offshore energy pro-duction.

4 With reference to Figure 5.1.4 outline the benefits of DSM to both producers and consumers of energy.

In 1973 the Canadian government began an incentive programme to get older industrial plants to save 31 per cent of their fuel use by 1993, with usage in 1973 as a base line. As of 1990, a 26 per cent saving had been achieved.

Canada has become a world leader in constructing energy-efficient buildings. In 1973 a typical office building used about 600 kwh/m². In 1992, an experimental office building using underground heat storage and solar panels used only 20 kwh.

Non-fuel minerals

The same basic dichotomy between the USA and Canada exists for the non-fuel sector as it does for fuels, in that the intense pressure of demand has necessitated rapidly increasing imports into the USA while Canada remains a very significant net exporter.

Excluding fuels, the remaining range of minerals can be divided into metals and non-metals. Although concentrations of very high activity are apparent, mining is undoubtedly a truly national industry in both countries, affecting to some degree every state, province and territory.

The United States

The total value of the non-fuel mineral sector was about $35 000 million in 1994. Despite its apparent rich endowment of mineral deposits (Figure 5.1.5), the USA imports large quantities of essential metallic minerals, and the country's share of several metals, including iron, copper and zinc has declined in recent years. The temporal pattern of metals production is very irregular, reflecting the volatile nature of market prices for metals, particularly precious metals. In contrast the value of the non-metallic sector, which includes clays, sand and gravel, stone and salt has increased in almost every year over the last three decades. The USA is self-sufficient in all but a few non-metallic minerals.

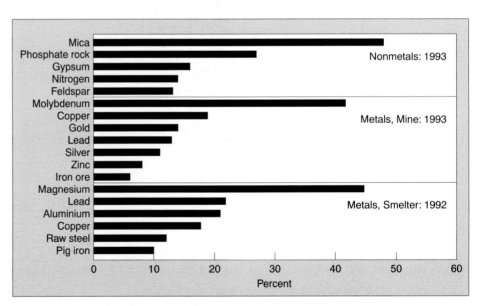

Figure 5.1.5
US mineral production as a percentage of world production

Government regulations, particularly those relating to the environment, have affected the mining and mineral processing industries significantly. For example, demand for platinum has grown because it is needed for catalytic converters, pollution control devices on motor vehicles. Almost 50 per cent of domestic consumption of platinum is now used in these devices. The Environmental Protection Agency is actively involved in the regulation of mining waste, and monitoring emissions from base-metal smelters and acid levels in lakes.

Lands owned and administered by the federal government contain significant amounts of minerals. For example, 16 per cent of the value of phosphate production comes from federal and Indian lands. However, many such areas, including national parks, are subject to various levels of protection and currently off-limits to the mining companies. The latter argue that the development of these resources will reduce US dependence on foreign sources for many strategic minerals and bring employment and much-needed revenue to local communities.

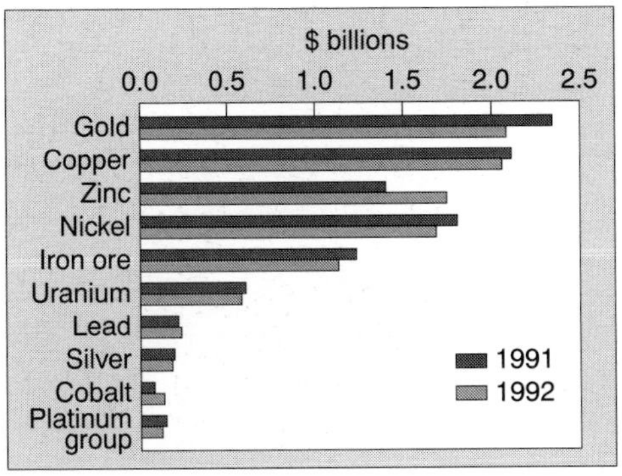

Canada

Canada is the third most important producer of non-fuel minerals after the US and the countries of the former USSR. It is the world's No. 1 in total volume of mineral exports. 80 per cent of production is destined for export markets, with about two-thirds of this going to the USA. In 1992 exports of all non-fuel minerals totalled $23 billion, accounting for over 15 per cent of Canada's total exports. Canada paid $13.2 billion for imported non-fuel minerals, resulting in a trade surplus of more than $9.8 billion. The location of production by value in 1992 was as follows: Ontario 32.3 per cent, Quebec 18 per cent, British Columbia 12.7 per cent, Saskatchewan 8.6 per cent, Manitoba 7.2 per cent, with the remaining 21.3 per cent spread among the other five provinces and two territories.

Canada is the world's largest producer of uranium and zinc and the second largest producer of asbestos, nickel, potash, sulphur and gypsum. However, gold and copper are the leading minerals by value (Figure 5.1.6). The Canadian Shield is particularly rich in metallic and non-metallic ores and as the northlands become increasingly mapped and surveyed in detail, further resources will undoubtedly be found. The technology employed in exploration matches the most sophisticated in the world (Figure 5.1.7). However exploitation in some of these remote areas is opening up a whole series of economic, environmental, political and technical issues.

Figure 5.1.6
Canadian production of leading metals

FROM ROCKS TO RICHES

In the past, much Canadian mineral exploration was carried out by prospectors, many of whom were rugged individualists equipped only with a pick, a pan, a canoe or a horse, and a lot of optimism. Most of today's explorers are a new breed of professionals who hold one or more university degrees in geology, geophysics or geochemistry and work for mining and exploration companies. Modern exploration involves the use of fixed-wing aircraft, helicopters, 4-wheel drive vehicles and air and ground operated geophysical instruments such as magnetometers, gravimeters and various electromagnetic instruments that are essential for detecting buried orebodies. Scintillometers and gamma-ray spectrometers are used both from aircraft and on the ground to discover deposits of uranium.

Mining companies search for geophysical and geochemical anomalies and for geological clues that may indicate orebodies. They use geophysical and geological surveys and maps and various other advanced methods of mineral exploration. For example, geochemical surveys are used to locate anomalous concentrations of elements of economic interest in soils, lake and stream waters and plants. Such concentrations may indicate the presence of an orebody. In special cases, geobotanists may search for indicators such as species of vegetation specific to elements of possible economic interest.

Canada Year Book, 1994

Figure 5.1.7
From rocks to riches

QUESTIONS

1 For which of the minerals listed in Figure 5.1.5 does the USA account for more than 10 per cent of world production?

2 Use Figure 5.1.7 to make a list of high technology techniques that are used in mineral exploration.

3 Why has it become more difficult to cost mineral development projects?

Aboriginal land claims and new parks have shrunk the amount of land available for exploration. A joint industry-government task force recently estimated that at least 12 per cent of Canada is now 'out-of-bounds'. The costing of mineral development is now a more complex issue than it once was, with environmental delays and variations in provincial taxes high on the list of factors that mineral companies have to take into account.

5.2 The Bingham Canyon Copper Mine, Utah

Figure 5.2.1
Bingham Canyon Mine

Early developments

Bingham Canyon is located in Utah's Oquirrh Mountains, 35 km south west of Salt Lake City (Figure 5.2.1). The presence of low-grade copper ore was first reported in 1860, but when gold, silver and lead were discovered a few years later, mining – of the conventional underground type – remained largely devoted to these more valuable metals until near the turn of the century. Eventually, however, the commercial potential of Bingham Canyon's porphyry copper was recognised, and in 1906 open-pit operations began. Ore that contained only two per cent copper was hauled to processing plants near the Great Salt Lake by predecessors of the Kennecott Corporation, one of the largest mining companies in the United States and the present owner of the Bingham Canyon Mine.

In 1927 Copperton, a planned residential community, was built by the company at the mouth of Bingham Canyon. Similar company towns and squatter communities sprang up on the north slope of the Oquirrhs, near the mills and smelter at Magna/Garfield which processed the Bingham Canyon ores. With time, all these settlements – which were as ethnically diverse as those in the Canyon – gave way to expanding copper operations. In 1956 the last residents were relocated by Kennecott.

In the early days the very idea of attempting to exploit Bingham Canyon's copper was ridiculed; few believed that mining such poor-quality ores would ever be profitable. However, subsequent operations have transformed what was once an imposing mountain (Figure 5.2.2) into the world's first open-pit mine, the largest artificial excavation, and the most productive copper mine, which has been dubbed 'the richest hole on earth'.

Large-scale operations

From the outset large-scale use of power machinery has been the norm: steam shovels for digging the ore and steam trains for removing the material gave way in the 1920s to electric-powered machinery, while in 1963 the in-pit trains were replaced by huge diesel trucks capable of carrying 190 tonnes of material up steeper gradients without resorting to the traditional 'switch-backing'. Approximately five billion tonnes of material have been removed from the mine since 1906. Currently 100 000 tonnes of overburden and 107 000 tonnes of ore are hauled from the pit every day.

These gargantuan operations have resulted in a huge inverted cone consisting of numerous terraces or 'benches'. This conical shape not only reflects the configuration of the ore body but is necessary to bank the mine walls at an angle at which the slope will be stable, so that material does not fall from one terrace to the next. Stability depends both on the rock type and the effect of rain or floodwater. Such matters have to be carefully calculated as excavations go ever deeper. The mine is now over 800 metres deep; the USA's tallest building, the Sears Tower in Chicago, would reach only half-way up the side of the mine (Figure 5.2.3). Its other dimensions are equally impressive: the mine covers almost 800 hectares and is over four km from rim to rim.

Figure 5.2.2
Bingham Canyon, 1903

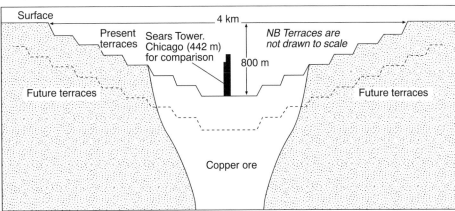

Figure 5.2.3
The Bingham Canyon Mine: diagrammatic cross-section

To date, Bingham Canyon has yielded approximately 14 million tonnes of copper metal whose cumulative value exceeds eight-fold the yields of the Comstock Lode, Klondike and California gold rushes combined. Nowadays Bingham Canyon ore contains an average of only 0.59 per cent copper, and it takes a tonne of ore to produce five kg of copper. In 1993 output was approximately 345 000 tonnes of copper (about 15 per cent of the USA's newly-mined total), together with valuable by-products such as 14 510 kg of gold, 123 750 kg of silver and 8 100 tonnes of molybdenum. Proven and probable reserves are estimated at 1200 million tonnes of ore – enough to ensure the continuation of the mine's productive life well into the twenty-first century.

Investing for the future

Maintaining the economic viability of the mine has not always been straightforward. After many years of production, the profitability of the operations began to deteriorate in the 1970s. By the early 1980s, the mine had high production costs, low productivity and outdated facilities. These handicaps, combined with an economic recession, necessitated the complete closure of the mine between 1985 and 1987.

Since the mid-1980s, Kennecott has undertaken a massive programme of investment at Bingham Canyon. The company has spent $625 million on modernising and expanding the mine and concentrator, including $80 million on specific measures to protect the environment. Another $880 million was invested in a new smelter and refinery modernisation project, completed in 1995 (Figure 5.2.4).

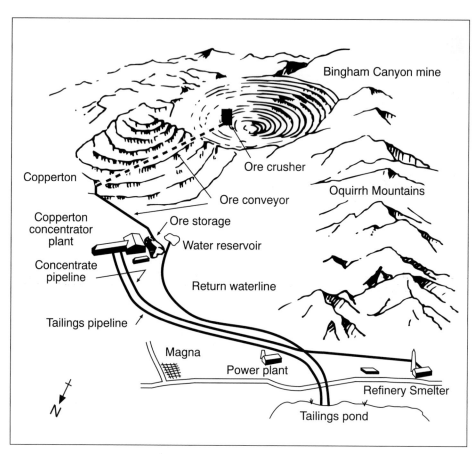

Figure 5.2.4
Modern production facilities at Bingham Canyon

The new smelter incorporates a Finnish state-of-the-art technology called 'flash converting', which eliminates the open-air transfer of molten metal, a process which contributed significantly to air pollution. It is also able to refine all Bingham Canyon's copper concentrate, whereas 40 per cent had previously been exported, mostly to Japan. All this work created 1100 direct jobs over three years, with another 2200 in support and service operations.

Finally, Kennecott plans to spend a further $500 million on expanding the tailings impoundment facilities to provide for the disposal of smelter waste at Magna/Garfield up to the forecast end of mining operations. This ten-year programme of investment will complete the modernisation of the company's integrated facilities at Bingham Canyon. All told, the Bingham Canyon Mine employs directly about 2300 people, with another 260 at Kennecott's office headquarters in Salt Lake City. It is notable that since 1987 the mine's ultimate corporate owners have been British. In that year British Petroleum (BP) acquired Kennecott as part of BP Minerals America, and in 1989 the company was sold to the Rio Tinto Zinc (RTZ) Corporation. The fortunes of the Bingham Canyon Mine are thus reflected directly on the London Stock Exchange – an example of large-scale foreign investment in the modern US economy.

Figure 5.2.5
Cumulative production at Bingham Canyon since 1906 (tonnes)

Copper	14 000 000
Molybdenum	405 000
Silver	4200
Gold	480

1 (a) With reference to Figure 5.2.3, why is it necessary to remove so much overburden (waste material) to extract the ore?

(b) Why will the cost of exploiting the ore increase with depth?

2 (a) Calculate the date when the mine is due to come to the end of its productive life.

(b) What factors may affect the forecast?

3 Before 1988 an extensive rail network transported the ore from the mine to the concentrator, smelter and refinery (Figure 5.2.4). What new systems have replaced the railway?

4 (a) What waste products result from the mining and other processes at Bingham Canyon?

(b) How are such waste products dealt with or disposed of?

5 The minerals in Figure 5.2.5 are ranked by cumulative weight of output. Rearrange the ranking to reflect their cumulative values.

6 (a) List the most common uses of (i) copper; (ii) molybdenum.

(b) Where are Kennecott's principal industrial customers likely to be located?

(c) Why has most of Bingham Canyon's copper always been processed near the mine rather than near the markets for the finished products?

5.3 Coal Mining

Mining almost one quarter of the global total, the USA is one of the world's great coal producers. However, in recent years it has fallen to second place, after China, in the production league table. In 1994 US production was just over 940 million tonnes while Canadian production topped 40 million tonnes. After prolonged US domination of the world market in coal exports, Australia nudged ahead in 1984 and the two countries have vied for the top position since then (Figure 5.3.1). Canada is also a significant exporter of coal.

The United States

More coal is now mined in the USA than at virtually any time in the past, but as other energy sources have been developing at a faster rate, coal now accounts for only 23.3 per cent of total energy supply compared with 36 per cent in 1950. Coal's share of the energy market has fallen because:

- it is a heavy and bulky solid, costly to store and transport;
- the mining and burning of coal can cause considerable environmental problems which have resulted in increasingly strict government legislation;
- other popular fuels are available at very competitive prices;
- new technology often excludes the use of coal. For example in 1945 the railways bought a quarter of all coal mined in the USA. This market has totally disappeared as all trains are now either diesel or electric powered.

Coal is mined in half of the states in the USA but the 'big five', Wyoming, West Virginia, Kentucky, Pennsylvania and Illinois, account for 65 per cent of production (Figure 5.3.2). The Appalachian coalfield is the oldest and still the most important of America's coalfields. Until recent decades the eastern coalfields produced nearly all of US coal. However, since 1970 mining in the western states, particularly Wyoming, has expanded rapidly due to two main factors.

The 1970 Clean Air Act placed strict limits on the amount of sulphur released into the atmosphere. Most eastern coal has a high sulphur content and the new legislation meant that expensive scrubbers had to be installed in smokestacks if such coal was to be burned. Most western coal has very little sulphur and could be burned directly without breaking the law.

Most western coal is found at or near to the surface, allowing strip mining which is much cheaper than underground mining (Figure 5.3.3). Less labour is required and huge coal cutting machines can be used. Many eastern mines are underground and where strip mining occurs the thickest and most accessible seams have already been cut.

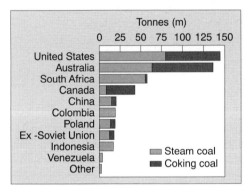

Figure 5.3.1
Major coal exporting countries, 1992

Figure 5.3.2
Coal production by leading states [in millions of tonnes]

STATE	1970	1994
Alabama	19.1	21.8
Illinois	59.2	48.2
Indiana	20.0	28.2
Kentucky	113.8	149.4
Montana	2.7	38.2
Ohio	50.1	27.3
Pennsylvania	82.0	56.4
Virginia	31.9	33.7
West Virginia	131.1	147.4
Wyoming	6.4	215.7
Other States	41.5	176.5
Total	557.8	940.8

Figure 5.3.3
Strip mine in Hanna, Wyoming

But the western mines have faced some problems. The coal has a lower calorific value than coal mined in the east. It is also distant from the concentration of coal-fired power stations in the central and eastern states which purchase the bulk of US coal. Transport costs can add 60 per cent to the price of coal at the mine. However, the advantages of western coal generally outweigh the disadvantages, enabling it to compete effectively with other sources in its market area. Undoubtedly, more and more of US coal will be mined in the west in the future.

The coal industry has for some years been faced with a high degree of uncertainty. After the energy crisis of the early 1970s it was suggested that production should be increased to between 1500 and 2000 million tonnes a year, a measure which would greatly reduce oil imports. However, such an increase in output would have created significant problems, some of which have been discussed above.

The industry has been faced with further challenges since the mid-1980s. Prices on the Rotterdam spot market, which peaked at almost $80 a tonne in 1981, were stuck at about $35 in 1994, prompting coal companies to try to push production costs down even further. Only in a few areas of the world, such as the western states of America, where six of the world's ten most profitable coal mining companies own large mines, have firms been able to marry first-class geology to efficient labour and a good infrastructure.

Steel industry

In a flat market for steel, two trends are causing concern in the coal industry: an increasing proportion of scrap is being used in electric arc furnaces; and pulverised coal injection, allowing steel makers to replace half their coking coal with cheaper steam coal. Bethlehem Steel has predicted that metallurgical-coal demand in the USA could be as much as halved between 1988 and 2000. However, it must be remembered that the steel industry currently takes only 4 per cent of US coal output. In contrast 75 per cent goes to electric utilities and over 10 per cent is exported.

Environmental concerns

Coal with a sufficiently low sulphur content to meet federal air pollution control standards (emitting less than 0.5 kg of sulphur per million BTU's of energy content) is known as 'compliance coal' and commands a premium price. The location of such coal, in huge quantities, has been the major influence on the spatial foci of investment in recent years. Today the twenty largest mines in the US are all west of the Mississippi River, and twelve of these are in Wyoming, particularly in the Powder River Basin. The use of new technology in such large mines has enabled the industry to steadily raise production and productivity. Average productivity in the USA increased from 1.9 tonnes per miner per shift in 1981 to 3.7 in 1991. However, falling coal prices have resulted in the closing of many marginal mines and the exit of small producers.

Great environmental damage was caused in the early years of coal mining. The evidence is widespread in the Appalachian Mountains where mountain tops and ridges were scalped of vegetation and topsoil to reach the coal below. Slopes were gouged out to build haul roads and trees and soil were bulldozed over hillsides. Rain falling on exposed coal seams carried coal dust into nearby rivers, where the sulphur in the coal made the water acidic. Once the coal was mined no reclamation was attempted and the mining companies moved to other areas to repeat their havoc yet again. Fortunately the Surface Mining and Reclamation Control Act of 1977 did much to change the situation, requiring mine operators to use effective techniques to restore stripped land to its original condition, capable of supporting whatever functions prevailed before mining (Figure 5.3.4).

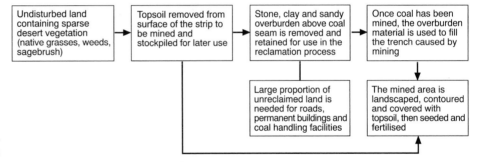

Figure 5.3.4
The main stages of strip-mine reclamation

Recent Clean Air Act amendments now require all new coal-burning power plants in the USA to use 'scrubbers' (chimney filters), irrespective of the sulphur content of the coal used. While environmental groups applaud, the coal industry worries that their major market will gradually be eroded further as some utilities increasingly diversify their sources of electricity generation. A linked issue and a specific pressure point that is attracting more and more attention on the international scene is the emission of CO_2. Coal emits nearly twice as much CO_2 per unit of energy consumed as natural gas and 20 per cent more than fuel oil.

Canada

Almost 95 per cent of Canadian coal comes from surface mines in Western Canada. The remainder comes from mines in Atlantic Canada, most of which are underground. Proven reserves total 6 billion tonnes, equivalent to over 80 years' supply at current production rates. In addition some 30 billion tonnes of coal, contained in known deposits, have not as yet been fully delineated.

Canada is a net exporter of coal, with a trade surplus of C\$1.6 billion in 1991. Because of the great length of the US–Canadian border, there are frequent movements of coal across it, primarily to reduce the cost of transporting the coal required. Imported coal from the USA goes mainly to Ontario for use in generating electricity and steel-making. While this coal is less expensive than domestic coal, imports have declined recently as Ontario Hydro gradually switches to low sulphur coal from Western Canada.

The impact of restructuring on an Appalachian mining community.

Rhodell is typical of many small town mining communities in Appalachia which are reeling from the effects of mine closure and other forms of restructuring taking place in the industry (Figure 5.3.5). Coal has been the lifeblood of Rhodell since the town was founded in 1907 and early maps show mines at the end of Main Street. As local production increased the population grew and the town prospered. However, such prosperity depended almost entirely on the circulation of money earned by the town's miners.

THE CLOSING OF THE AMERICAN MINE

Oblivion comes to a West Virginia coal town

The town is Rhodell, and it sits at the center of the southern West Virginia coal seam, one of the richest in the world. Thirteen years had passed since I'd been down here, and as I drove into the coalfield, I entered a landscape I could scarcely recognize. Restaurants I'd eaten in, places I'd stayed were all boarded up. Twenty miles west of Rhodell, the huge tipple at Itmann, where 1,200 men once worked, was rusting in silence; the cluster of mines that fed it, closed. I even found it difficult to pick out the coal camps, once the salient and inescapable feature of the region. The oppressive uniformity of company towns—with their identical houses in precise rows—had been lost within a jumble of mobile homes, junk cars, heavy machinery, and plywood shacks interspersed at random among the vestiges of the older settlements.

I had expected dying towns, but I couldn't reconcile the continued production of coal with the moribund communities I visited. Automation in the mines has been increasing—and employment declining—since the mid-1950s, but only in the last decade has the economic order, once so monolithic it seemed divinely sanctioned, met its demise. And only within the last year or so has it become possible to glimpse the lineaments of the new Appalachian dispensation.

In a Logan County camp, forty miles west of Rhodell, I met a retired miner named Willie Andersen, who explained the changing realities of life in coal country. "The coal companies have done away with the community concept," he said. "They scattered out the miners. In these camps, everybody used to know everybody's son's name, everbody's daughter, everybody's cousin, and if I had a dog, they knew what my dog's name was. Used to be the miners in this camp would fill up the ballpark on a Sunday." Andersen said that the mines I'd been looking for, which I had remembered as being in or near the camps, were now located deep in the mountains, often behind barbed-wire fences patrolled by guards with shotguns. "There are kids in the coalfield today," he added, "that unless they see a lump of coal go down the road on one of these trucks, they don't see a lump of coal."

The primary effect of the new order in coal country seemed to be diffuseness. As I toured the coalfields, I drove through towns with no stores, no restaurants, no bars, no gas stations, once-thriving towns like War, Cucumber, Red Jacket, and Skygusty that had almost ceased to exist. And one Sunday morning, in the steady rain, I stumbled on Rhodell, a town well advanced in the process of unbecoming.

I recalled my first visit to the area in 1981, when a photography assignment took me to the coalfields of McDowell County. At the time, Main Streets in mining country were full

Figure 5.3.5
The closing of the American mine

of new cars, the taverns were packed with paying customers, and you couldn't book a motel room because of all the surveyors who'd come to town to map new coal seams. But the real heyday of West Virginia coal mining had been four decades earlier.

In the last decade, Rhodell has completed the shift from a solidly working-class community to a place where virtually everybody gets some sort of monthly check without hitting a lick. The town's mayor estimates that 98 percent of his constituents are on welfare or a pension. This steady but limited cash flow keeps Rhodell's few businesses barely solvent, but as older pensioners die

off and the generation now in its forties ages, even this will dry up.

Although cheap foreign coal has given West Virginia's product a run for its money, unemployment in the coalfield today is not linked to low demand, thanks to falling wages and steady orders from public-utility companies. Nor are the current woes the result of depletion, for though the best seams have been mined, extensive reserves of high-quality coal remain. Men like Bimbo and Dave are idle because of complex structural changes in the industry. When automation made it possible to operate with smaller crews, huge producers like A.T. Massey and Island Creek turned the actual

production of coal over to local contractors while retaining title to their coal lands and tipples. The trend has accelerated in the last decade, making smaller mines the rule. The contractors get the "easy coal" and fold up, usually within seven to nine years. The larger mines of yore often produced for fifty years. And mines that once employed 500 men now required fewer than 100. Statistics reflect these changes neatly. In the 1950s, coal accounted for 25 percent of West Virginia's jobs. Today, the figure is 4 percent—yet more tonnage is mined now than forty years ago.

Harper's Magazine
December 1994

Some coal mining towns in West Virginia have had their fate sealed by mine closure. In contrast Rhodell's decline has been due to what miners see as more insidious trends in the industry. As mining companies sought to drive down costs their methods brought them into increasing conflict with the United Mine Workers (UMW). After a series of unsuccessful disputes the general feeling was that the union had lost much of its clout. In a bid to reassert its authority the UMW launched a series of decentralised 'selective' strikes, a tactic that many miners blame for fragmenting the workforce and producing a succession of weak contracts. In the new 'weak-union' climate the companies have increased production with fewer miners working in generally poorer conditions. As mining jobs were lost in Rhodell the impact on services quickly became apparent as shop after shop closed leading to the inevitable flow of outmigration.

As the old order of the industry has fragmented, the impact has been felt throughout the coalfield. The handful of large producers has given way to a constantly shifting proliferation of small contractors (the larger producers retaining title to the coal lands), who bear responsibility for paying the miners UMW benefits. Today in Rhodell, coal moving in large quantities is no longer an index of local prosperity. With only four residents now employed in mining the town struggles to contain two disparate poles of the American experience – raw energy and irreversible decline.

QUESTIONS

1 Why has coal's share of the North American energy market fallen?

2 Outline the essential differences between coalmining in the eastern and western states of the US.

3 Compare the impact of stripmining in earlier years in Appalachia with stripmining in the western coalfields today.

4 (a) Describe the outward signs of decline exhibited in Rhodell.
   (b) Examine the reasons for Rhodell's demise.

5.4 Alaskan Oil: Difficulty and Disaster

Physical and economic background

Bought from Russia in 1867 for the equivalent of £2 million, Alaska was administered as a US territory until 1959 when it became the 49th state. In area, Alaska is by far the biggest state in the Union, its remote and detached position tending to conceal the fact that Alaska is equal to about one-fifth of the 48 contiguous states' entire area, or nearly seven times larger than Britain.

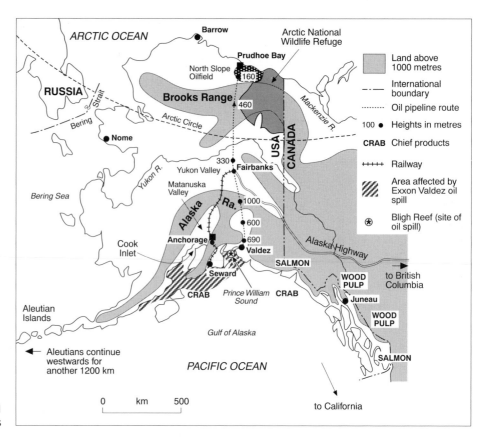

Figure 5.4.1
Alaska: principal features

Physiographically, Alaska consists of two mountain ranges (extensions of the Pacific Coast Range and the northernmost Rockies) between which plateau-like areas, most notably the Yukon Valley, dip towards the Bering Sea (Figure 5.4.1). Climatically, the Pacific-facing slopes and valleys in the south make up a distinctly warmer area that contrasts with the inhospitable central and northern parts of the state (Figure 5.4.2).

Figure 5.4.2
Alaska: selected climatic statistics

	Temperature (°C)		Precipitation (cm)		Annual total Precipitation (cm)
	January	July	January	July	
Barrow (North Slope)	−26	4	0.5	2.3	10.3
Fairbanks (Yukon Valley)	−11	15	2.0	3.9	30.0
Juneau (Pacific Coast)	−2	14	29.0	8.9	231.0

Long dependent on primary activities such as mining (e.g. the Klondike gold rush of the 1890s), salmon and crustacean fisheries, and lumbering (especially in the southern 'panhandle' area), Alaska's economic development remained at best patchy until well into the latter half of the twentieth century. By 1960 the state's population, including its native peoples (Indians, Inuit and Aleuts), still numbered only 226 000, mostly concentrated in the more favourable south.

Then, in 1968, oil was discovered at Prudhoe Bay on Alaska's North Slope (Figure 5.4.1). From that moment the state's economic prospects were transformed. In 1969 a group of oil companies, including British Petroleum (BP), paid the state government nearly $1 billion in oil-land revenues, but before production could commence several major problems had to be overcome:

Figure 5.4.3
Oil worker, North Slope, Alaska

- supporting the oil-drilling crews in what one BP geologist described as 'a mean, nasty, unforgiving place to work' (Figure 5.4.3);
- moving the oil to market in the 'Lower 48' states – initially California;
- the opposition of a powerful conservationist lobby who argued that Alaska, America's last great wilderness, would be despoiled by the oil development.

Apart from the low air temperatures, windchill and winter darkness, the main environmental problem focused on northern Alaska's tundra zone and its permafrost. The tundra's surface is stony or marshy with an intermittent vegetational cover of mosses, lichens, dwarf shrubs and occasional stunted birches and willows. During the brief summer season, the surface snow cover melts and the land surface thaws out to produce waterlogged soils and marshy hollows which become breeding grounds for an abundance of insect life, including the mosquito – so large and voracious that it has been dubbed Alaska's unofficial state bird! Alaska's 600 000 caribou (reindeer) are the most notable of the tundra's large mammals.

Figure 5.4.4
Permafrost in Alaska

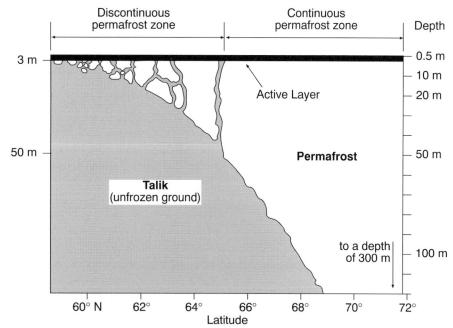

Associated with the tundra zone is permafrost (Figure 5.4.4). Below the surface in the north, the ground is permanently frozen to great depths, and has been since the last Ice Age. At the surface, however, a narrow 'active layer' thaws in summer to a depth of about 30 cm, but this layer deepens towards the south. Continuous in the north, the permafrost area becomes discontinuous in the south; here a series of permafrost 'islands' occurs, dependent upon such local factors as vegetation cover, aspect and relief. Human activities, including any construction work, must therefore reckon with a surface layer that is rock-hard in winter but usually soft and waterlogged in summer (Figure 5.4.5).

The conservation versus development debate

The conservationists argued that developing the North Slope oilfield would irreparably damage the delicately balanced tundra ecology, just as had other works like the Alaska Highway (built for military purposes during the Second World War) which still has problems of waterlogging and instability.

And how would the oil be transported? In 1969 a trial run by the 150 000-tonne *Manhattan*, a specially strengthened American supertanker, attempted to

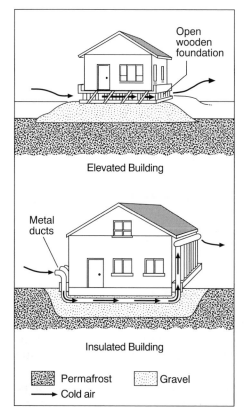

Figure 5.4.5
Two solutions to construction problems in the tundra zone

force a route through the North West Passage. Unable to make headway through the international waters of McClure Strait, the Manhattan turned south towards the mainland and into Canadian territorial waters before reaching Prudhoe Bay. The *Manhattan*'s voyage raised the spectre of a potentially disastrous oil spillage in the Arctic Ocean with scarcely imaginable environmental consequences. Since a 40-strong fleet of 200 000-tonne tankers would be needed to carry the oil, and Canadian objections were so strong, the North West Passage route was abandoned.

The only alternative was a pipeline running southward across Alaska to the ice-free port of Valdez. Again the conservationists expressed concern. Constructing the pipeline would not only inflict environmental damage similar to that on the North Slope, but survey teams had already accidentally diverted caribou off their traditional migration tracks and had blocked salmon streams when excavating gravel for foundations.

Moreover the 1280-km pipeline itself would be an environmental threat. The oil would have to be heated to 86 °C to keep it flowing in sub-zero temperatures and any escaping heat would thaw the underlying permafrost. This could cause the pipe to fracture, leading to a disastrous outpouring of oil. Most alarming of all, the pipeline would have to cross one of the world's most active earthquake zones. In 1964 Anchorage, the largest city in the whole of the continent's northland fringe, was severely damaged by violent tremors.

Why, asked the conservationists, risk damaging one of the world's last wilderness areas when Alaska's supply of oil to the United States will last only an estimated 20 years?

The oil companies replied that every care would be taken to protect the vulnerable Alaskan environment. The 1.22-metre diameter pipeline would be buried for over half its length and so would not interfere with the migration routes of caribou and other animals. In the most critical sections the pipeline would be insulated by brine-filled jackets to prevent any heat loss that could damage the permafrost. Further south the pipeline would be carried above ground level (Figure 5.4.6) and the route would be carefully chosen to avoid adverse effects on forests and animal populations.

Figure 5.4.6
The Trans-Alaska pipeline

Additionally, the economic case in favour of developing the oilfield and constructing the pipeline was very powerful. The United States was running short of oil. Its other reserves were showing signs of depletion while demand was ever-increasing.

Such problems could be at least partly offset by developing the Alaskan oilfield, which contained perhaps one-third of all US proven reserves, the biggest that the nation possessed. Most Alaskans were certainly in favour of exploiting the oilfield and building the pipeline, which offered them the best-ever chance of long-term growth and prosperity based on oil revenues.

By 1973 the mounting oil crisis created by steep OPEC price rises and an Arab embargo on oil exports to the United States ended the debate over Alaskan oil. Quite simply, it had become vital to the American economy. In November 1973 the US Congress passed the law authorising the pipeline's construction, which began the following year. On 20 June 1977 the pipeline came on stream, moving North Slope oil to Valdez for transfer to supertankers. These huge vessels then carry it southward for refining, mostly in the port cities of southern California, where much of it is marketed locally; the remainder is moved eastward via long-established pipelines. For more than a decade the Alaska pipeline operated without major mishap – then disaster struck in perhaps the least expected place and fashion.

QUESTIONS

1 Study Figure 5.4.2. Identifying the three climate stations on Figure 5.4.1, account for any important differences in temperature and precipitation.

2 (a) Apart from the Arctic fringe of North America, name three contrasting world locations where oil is exploited in physically challenging conditions.
   (b) Identify the main hazard in each case.

3 (a) With reference to Figure 5.4.4, explain the gradation of depth of the active layer, and the transition from continuous to discontinuous permafrost zones.
   (b) Why is the tundra surface waterlogged in summer?

4 Explain how the techniques illustrated in Figure 5.4.5 overcome the permafrost problem, and why all services – such as heating ducts, water pipes, electrical wiring and sewage pipes – have to be installed above ground in insulated boxes called 'utilidors'.

5 (a) From the information on Figure 5.4.1, draw a labelled cross-section of the trans-Alaska pipeline route.
   (b) The BP oil company, which built 380 km of the pipeline, claims that it 'remains one of the greatest feats of engineering ever undertaken'. What evidence supports this view?

6 (a) Complete Figure 5.4.7.
   (b) With particular reference to Alaska, comment on the trends revealed.

Figure 5.4.7
US petroleum production

US petroleum production (million barrels): four major states

	Average 1966–70	1980	1985	1991	Percent change 1970–1991
Alaska	53	592	666	656	☐
Texas	1142	977	889	683	−40.2
Louisiana	804	469	508	147	−81.7
California	365	357	424	320	−12.3
US Total	3929	3146	3274	2707	−31.1
Alaska as percent of:					
Texas total:	4.6	☐	☐	☐	☐
US total:	1.3	☐	☐	☐	☐

For more than a decade the Alaska pipeline operated without major mishap – then disaster struck in perhaps the least expected place and fashion.

The Exxon Valdez disaster

On 24 March 1989 one of the world's largest oil spills occurred. The tanker *Exxon Valdez*, heading south and laden with Alaskan crude, ran aground on Bligh Reef and 38 000 tonnes of oil (about 40 million litres) poured into Prince William Sound. The worst fears of the conservationists had been realised: a massive tanker disaster, but in the waters of the Gulf of Alaska rather than the Arctic Ocean.

The spillage affected an area of some 2400 km² and 2400 km of coastline. Along the shore the spilt oil was over 10 cm thick, and five years after the accident less than 10 per cent had been removed. The effect on local wildlife was devastating. The death toll includes 3500–5500 otters, 200 harbour seals and nearly 400 000 birds. In 1990, the first run following the spillage, 45 million salmon returned to their ancestral rivers; by 1992, this number was down to five million.

Oil kills fish not only because it is toxic but because it coats their gills, while oil on the surface inhibits oxygenation. When oil sinks to the sea bed it immobilises fish sperm and reduces fertilisation rates. Studies by the Alaska Department of Fish and Game indicate that even minute quantities of residual oil are capable of killing salmon eggs and new-born fish, and of causing long-term genetic problems, especially for benthonic fauna (sea-floor dwelling animals) (Figure 5.4.8).

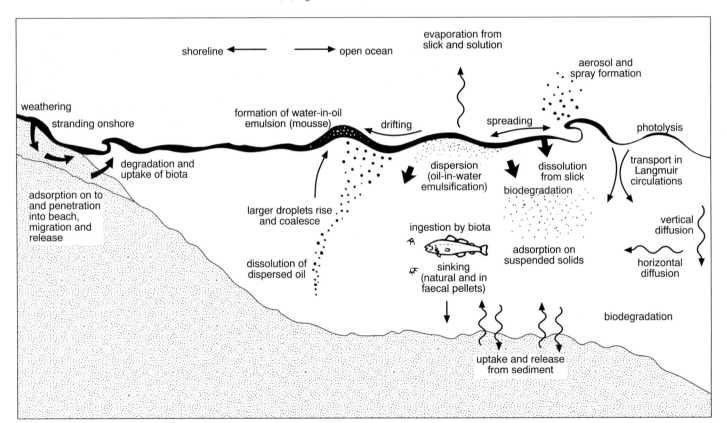

Figure 5.4.8
The behaviour of oil released into the sea

Oil affects birds by coating their feathers, reducing buoyancy and insulation, and causing pneumonia. Oil also affects the permeability of birds' eggshells and reduces levels of food supply as fish life dies. In the same way, mammals and other higher life forms are adversely affected when food chains are disrupted.

At sea, oil-spill pollution is both pervasive and persistent, although in the long term it does disperse and dissolve. In addition, oil slicks can be tackled by artificial means, including:

- burning the oil (warplane attacks have proved effective);
- drawing off the oil with high-capacity pumps;
- scraping up the oil with machines called adhesion collectors;
- using carbosand or other sinking agents to settle the oil on the sea bed;
- spreading emulsifiers and dispersants to create droplets and thus aid bacterial degradation;
- applying biological agents (e.g. desulfovibrio and desulfomaculum), organic compounds which feed on the oil.

These techniques, however, cannot readily be applied to shoreline pollution where the clear-up task is much more difficult (Figure 5.4.9).

Figure 5.4.9
Cleaning up after the *Exxon Valdez* disaster

Aftermath of the Exxon Valdez disaster

After five years of legal proceedings involving the Exxon oil company, the State of Alaska and the federal government, in June 1994 a jury finally ruled that the skipper of the *Exxon Valdez* had behaved recklessly in allowing the tanker to run aground (he had a history of consuming alcoholic liquor whilst on duty) and that Exxon, his employers, had been equally reckless in allowing him to captain the tanker since they knew of his drink problem.

Exxon have already paid $2 billion to clean up the polluted beaches, $1.1 billion in fines and $2.4 billion in state and federal criminal penalties. In addition, 10 000 people who fish and hunt and live around Prince William Sound must still be compensated. All told, the company faces a bill of $16.5 billion – $1.5 billion in compensation, the rest as punitive damages. However, Exxon has appealed against the latter, opening up the prospect of years of litigation in the US Supreme Court.

An important upshot of the *Exxon Valdez* tragedy was the federal Oil Pollution Act of 1990, which aims to make the punishment fit the future crime. Before the accident the definition of those entitled to compensation from spillages was too narrow, but the new Act has been criticised for making the definition too wide. Those eligible for future compensation will include not only those who enjoy the wilderness through direct personal contact but also those who never intend to visit the wilderness but who nevertheless place some abstract value on its mere existence; this legal notion is termed 'contingent valuation'.

The future

For Alaska's 603 000 residents (1994) the oil industry remains extremely important. It finances 85 per cent of the state budget, and Alaskans pay neither sales tax nor state income tax, which was abolished in 1980. This makes Alaskans the least taxed of all Americans; indeed, the state government pays every Alaskan an annual stipend of nearly $1000 and spends over four times as much on each resident as any other state. Until the mid-1980s Alaskans enjoyed the highest per capita personal incomes in the whole of the United States.

However, North Slope oil production peaked in 1988 and revenues have been falling since then. By the early 1990s each Alaskan was being supported by the proceeds of only three barrels of oil a day compared with four barrels in 1988, and the state had dropped to 13th in the personal income rankings. Worse still, the state government was considering restoring income tax and dipping into its $15 billion 'permanent fund' of oil revenues in order to balance the budget.

In 1993, Atlantic Richfield, the largest oil company operating in Alaska, announced big new discoveries in Cook Inlet (in the south) and at Kuvlum (in the north). These will partly compensate for the declining North Slope. Even there, new diagonal drilling techniques developed by BP can reach offshore oil cheaply from land-based rigs, thus prolonging the life of Alaskan reserves allocated to the company.

It is clear that plenty of oil still exists in Alaska, but only one per cent of the state's huge area is privately owned. The federal government owns approximately 80 per cent, including 20.5 million ha in national parks, 31.6 million ha of wildlife refuges, huge tracts of native-controlled land, offshore oil leases and the corridor along the trans-Alaska pipeline. State and local governments account for the remainder. While so much land remains under official control, the pro-development inclinations of most Alaskans will be kept in check. Thus the vast 48 million ha Arctic National Wildlife Refuge (Figure 5.4.1), with its caribou, polar bears and migrating geese, is closed to drilling even though it is probably underlain by an oilfield at least as big as the Prudhoe Bay Field to the west.

This is the great dilemma for Alaskans. They have only their state's natural resources to sustain them, so hunting, fishing, forestry, mining and oil-drilling are virtually second nature to them. But at the same time, Alaskans are the guardians of some of the world's last great wilderness areas. Only time will tell if they succeed in striking the right balance between exploitation and conservation.

1 Discuss the various ways in which people are affected by major oil spills.

2 Exxon is one of America's most successful companies. Its annual turnover in 1993 amounted to $111 billion (three times the GNP of Ireland) and its profits were $5.3 billion. In the light of these figures and the information presented in the account, argue the case for and against the scale of the financial penalties imposed on the company for causing the *Exxon Valdez* disaster.

3 Outline your views on (i) the 'contingent valuation' method of assessing damages under the Oil Pollution Act of 1990;

(ii) the proposal that the price of US motorists' cheap petrol should be raised to include a 'pollution levy' to pay for future *Exxon Valdez*-type disasters.

4 World oil prices (in 1993 dollars) declined from $43 per barrel in 1983 to $10 in 1994.
(a) Account for this downward trend.
(b) Explain the likely effect on exploration for new oilfields in Alaska.
(c) Describe and explain any significant trends in the price of oil since 1994.

5 (a) Present the arguments for and against drilling for oil in the Arctic National Wildlife Refuge.
(b) What is your own recommendation on this issue?

5.5 Hydro-Electric Power

The continental scene

While hydro-electric power (HEP) is the third most important energy source in Canada, contributing almost 16 per cent to total supply in 1991, its relative position in the USA is much less significant, accounting for only 3.7 per cent of energy supply in the same year. However, in absolute terms total HEP capacity in the USA is 74 000 MW in comparison to a Canadian capacity of 98 000 MW. For HEP there is a finite limit to development set by physical and climatic considerations which cannot be exceeded unless extremely complex and costly construction and technology is used. Thus, once a region has utilised all its available sites, it must of necessity look to other sources of power to satisfy rising demand. This has been the general pattern of development in the USA in the past and, as a result, the relative contribution of HEP has steadily declined. More recently this situation has been repeated in Canada where in 1960 HEP accounted for 93 per cent of all electricity production, falling to 69.5 per cent in 1977 and 62 per cent in 1991.

In some regions where relief is very subdued HEP has never, even in the early stages of power development, played an important role. The East North Central division with only 1200 MW is a case in point (Figure 5.5.1). In contrast the Pacific division has an installed capacity of 38 300 MW. In total the USA has an estimated undeveloped capacity of 71 000 MW. However, even if it is decided that a substantial amount of this capacity should be developed on economic grounds, the increasingly strict array of environmental regulations placed on the statute books in recent years will prove a significant obstacle. An interesting trend in the past decade or so has been the refurbishment and reactivation of a number of retired low-head, low-capacity hydro facilities located mainly in New England and the Pacific division.

Figure 5.5.1
Hydro-electric power in the US

Division	Developed installed capacity, (millions of kilowatts)			Estimated undeveloped capacity, (millions of kilowatts)
	1950	1970	1995	1995
New England	1.2	1.5	1.9	4.4
Middle Atlantic	1.7	4.3	4.9	4.9
East North Central	0.9	0.9	1.2	1.7
West North Central	0.6	2.7	3.1	3.1
South Atlantic	2.8	5.3	6.7	7.2
East South Central	2.7	5.2	5.9	2.3
West South Central	0.5	1.9	2.7	4.6
Mountain	2.3	6.2	9.5	18.8
Pacific	6.0	23.9	38.3	24.0
US	18.7	52.0	74.2	71.0

In Canada, Quebec dominates HEP production although all provinces have developed this energy source except Prince Edward Island where there are no major rivers. Recent advances in extra-high voltage transmission techniques have given impetus to the development of HEP sites previously considered too remote.

The James Bay project

The world's second largest HEP scheme, at present ranked only behind Itaipu in Brazil, is located in the remote James Bay region of Quebec (Figure 5.5.2). The project, which began in 1971, has been one of the most keenly debated environmental issues in North America. As the project expands to cover more of the region, as seems likely, the arguments will become even more intense.

Figure 5.5.2
Position of the James Bay project

The objective of the scheme was to use more of Quebec's abundant water resources to provide low-cost power both for the province itself and for export. Quebec's political leaders saw HEP as a financial windfall, giving the francophone province a major economic bargaining chip. By late 1993 Hydro-Quebec had invested more than US$16 billion in the scheme, including the construction of five wilderness airports and over 8000 km of transmission lines through forest and wetland.

The project was planned in three phases. The initial phase, the La Grande complex, begun in 1971, has flooded more than 15 500 km², diverting and damming six waterways into larger bodies of water. The first power plant, the largest underground generating station in the world, was completed in 1982; four others have since been commissioned, and work is in progress on three more. The latest on-line addition is LG1, completed in 1995, with a capacity of 1368 Mw (Figure 5.5.3).

The rivers flowing into James Bay are powerful, not because of their slopes (which are gentle) but because of the large areas they drain. The greatest of these rivers is La Grande Riviere, its drainage basin covering 98 000 km², more than twice the size of Switzerland. Unfortunately, the La Grande drops only 376 m in its 800 km course. With no natural waterfalls, dams and reservoirs have had to be built to generate electricity. To increase the flow of water in the La Grande and so build large power plants, water has been diverted from two other rivers. Now 87 per cent of the water from the basin of the Eastmain River and 27 per cent of the water from the Caniapiscau basin is channelled into the La Grande. The Eastmain has been parched to a trickle but the La Grande has had its mean annual flow doubled and its winter flow increased by a factor of eight.

When the project began, most Quebecers heated their homes with oil or gas. Within a decade HEP, delivered at favourable rates, had doubled the percentage of all-electric homes in the province to 63 per cent. Government and utility officials also worked hard to attract aluminium and magnesium smelters to the St Lawrence Valley, as both industries require huge inputs of electricity.

Initial environmental impact

When the James Bay project began, Quebec had no environmental legislation of any significance, but as the project progressed a range of protection measures were introduced. However, environmental groups saw such measures as a minimum response, arguing that nothing could be done to compensate for the massive disruption already caused.

The first major tragedy occurred in 1984 when 10 000 migratory caribou died in the artificially swollen waters of the La Grande. The mass drowning happened in the first year the reservoirs were full enough to require controlled releases down the riverway. Hydro-Quebec claimed that, because of torrential rains that year, it was very likely that the caribou would have died anyway. The company also stressed the overall increase in the caribou population in the region since the project commenced. But the indigenous peoples and some biologists insist that the mass drownings would never have occurred if the river had been left alone.

Releases of toxic methylmercury, which have led to limits on the consumption of fish in the region, a staple in the diets of the Cree and Inuit (Figure 5.5.4), have caused considerable concern. A 1984 study of the Cree village of Chisasibi, downstream from the La Grande complex, found that 64 per cent of the residents had methylmercury levels classified as unsafe by the World Health Organisation. Methylmercury is thought to cause numerous birth defects. While Hydro-Quebec acknowledge the problem they have pointed out that through medication and strict limits on the consumption of predatory fish like trout and pike, the incidence of unsafe mercury levels in Chisasibi had reduced to 30 per cent by 1992.

Hydro-Quebec is fighting an uphill battle to improve its image and convince people of its glossy brochure commitment to building 'in harmony with nature'. Challenging this view, Jan Beyea, chief scientist at the National Audubon Society, has concluded that 'in terms of wildlife and habitat, the devastation of James Bay is the northern equivalent of the destruction of the tropical rain forest'. Whatever the differing views on the detail of environmental impact it would

Figure 5.5.3
The LG1 powerplant

Figure 5.5.4
A Cree family in Quebec

be difficult not to acknowledge that the overall landscape change has been staggering. Maps of the James Bay area in 1970 show nothing but string bogs, eskers, rivers, lakes and scattered Indian villages. Current maps show airports, roads, power stations, dams, reservoirs and a network of 735 kV transmission lines, the longest on the continent.

Cree and Inuit Land Claims

The Cree call electricity 'nimischiiuskataau' meaning 'fire that shakes the land'. In 1973 the 10 000 Cree and Inuit living in the region won an injunction to stop construction of the project. However this ruling was overturned by the Quebec Court of Appeal which found that, since construction had already begun, a 'balance of convenience' favoured Hydro-Quebec; that the Cree lacked clear rights in the territory, inasmuch as Charles II of England had granted exclusive rights to the Hudson's Bay Company in 1670; and that hydro development would not harm the environment but probably improve it. Stunned by these findings the native peoples sat down to negotiate and in 1975 the James Bay and Northern Quebec Agreement was signed by the Canadian government, the Province of Quebec, the Cree and the Inuit (whose communities lie mainly to the north of the project area). By signing the agreement the native peoples gave up all claims based on aboriginal rights. In return they got among other things:

- compensation of C$225 million;
- native land reserves and exclusive hunting, fishing and trapping rights;
- a voice in approving or rejecting, on the basis of environmental impact, developments in the region.

Phases 2 and 3

Hydro-Quebec announced its intention to proceed with Phase 2, the Great Whale complex, in 1992. Damming of the Great Whale River is scheduled for 2003, pending an environmental review. Plans also call for the 'eventual' construction and completion of Phase 3 which would involve the diversion of the Nottaway and Rupert rivers into the Broadback River.

The Great Whale complex would take this tumbling 360 km long river and with dams and dykes, convert much of its length into a series of artificial slackwater lakes, submerging more than 1600 km². Three large generating stations are planned, along with an all-season road linking the generating stations and the settlement of Whapmagoostui to the La Grande complex to the south. At present Whapmagoostui can only be reached by air or sea.

Phase 3 will involve a sevenfold increase in the mean annual flow of the Broadback River and the construction of eleven power plants and seven reservoirs.

With all phases completed the project would produce 26 400 MW of which 15 per cent would be available for export. In comparison, all the power stations in the UK have a combined capacity of 55 000 MW.

Further environmental impact

For the Cree and Inuit, the social and environmental impact of Phase 1 has proved to be substantial and greater than most originally feared. It is therefore not surprising that they are firmly against the commencement of Phase 2 (the damming of the Great Whale River) and Phase 3. They argue that their signatures on the 1975 agreement only involved acceptance of Phase 1, fearing magnification of all the current problems but particularly the threat to traditional sources of food.

There is growing concern among environmentalists for the well-being of the birdlife in the region. Because river water flows faster in the winter (to match peak demand) and slower at other times than previously, coastal marshes and tidal flats receive fewer nutrients in spring and autumn, the times of heaviest

QUESTIONS

1 Explain the spatial variations in the supply of HEP in North America.

2 How has the flow of the La Grande River been increased so that more power could be generated?

3 Suggest how the project has had a social impact on the Cree and Inuit.

4 Explain why environmental groups are increasingly appealing to the 'demand side' of the market.

bird migration, when most birds must eat voraciously to prepare for their flights. These migrants may have no substitute habitat if more key sites in James Bay are damaged. More diversions of rivers will also of course result in even more lengths of dry riverbed, with familiar repercussions.

The environmental spotlight has also fallen on the demand side – the purchasers of James Bay power. In 1992 New York state backed out of an agreement to import 1000 new megawatts from Hydro-Quebec's grid (Figure 5.5.5) but other contracts are still in effect with northeastern US customers and these draw off 6 per cent of Quebec's installed hydro capacity. Clearly the future decisions of US customers will have a significant impact on further development in James Bay.

At first glance, the agreement seems like the perfect match of Quebec's supply and New York's demand. By early next century, if the plan goes through, 10 to 12 percent of the state's electricity (compared with about 2 percent today) would come soaring down high-voltage lines from the taiga and into electrical outlets from Massena to Manhattan to Montauk.

Yet opponents have raised a number of disturbing questions that already have forced New York officials to postpone the decision once. Is the importing of "clean" hydropower the salvation of a state that has labored unsuccessfully to comply with Federal clean-air standards? Or is it simply tantamount to exporting environmental and cultural destruction to the taiga, the ultimate not-in-my-backyard solution to New York's energy needs? Will energy demand truly rise by 2 to 3 percent a year, as was forecast a few years ago, before the growth in demand slowed in the face of the current recession? And, finally, should New York not in any case try harder to meet its energy needs through conservation?

By virtue of its financial clout, New York will cast what many analysts say is the decisive vote on Great Whale. Hydro-Quebec officials insist that they will go ahead with the $50 billion to $60 billion project no matter what New York decides. The only effect, they say, will be to delay it for a few years. But less-interested observers doubt that Quebec can proceed without the contract and the capital it represents. "The whole essence of James Bay 2 depends on export sales," says Ihor Kots, managing director of Canadian Bond Rating Service Inc., the largest credit-rating service in Canada. "The electricity demand in Quebec won't be sufficient to take up the capacity for years to come. If there is not demand from elsewhere, it would add considerably to the financial risk."

The New York Times Magazine, 12 January 1992

Figure 5.5.5
Is the James Bay development the right solution?

5.6 Nuclear Power

Spatial distribution

The United States and Canada rank first and sixth respectively in the world nuclear power league (Figure 5.6.1). Together they generated almost 35 per cent of global nuclear electricity in 1993. The 110 reactors in the USA had a gross capacity of 105 276 Mw while Canada's 22 reactors totalled 16 709 MW capacity.

In both countries electricity is produced and supplied by many different companies. Some were very keen on nuclear power from the start. Others were more cautious. In the USA the South Atlantic, East North Central and Middle Atlantic regions have the most nuclear power plants (Figure 5.6.2). Most of Canada's reactors are located in Ontario – eight near Toronto, eight near Kincardine and four near Bowmanville. The other two are located at Gentilly near Trois-Rivières, Quebec, and Point Lepreau near Saint John, New Brunswick. In 1991 nuclear generation accounted for about 16 per cent of Canadian electrical energy, 50 per cent of Ontario's, 34 per cent of New Brunswick's and 3 per cent of Quebec's.

Figure 5.6.1
Commercial nuclear power generation:
Top ten countries, 1993

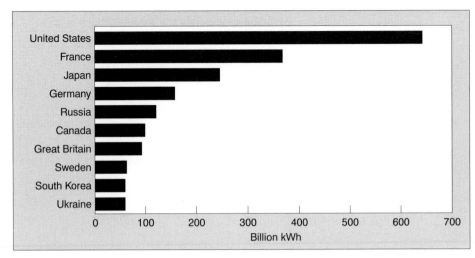

Figure 5.6.2
US nuclear power plants by division, 1992

Division	Number of units	Net generation million kWh
New England	8	38,474
Middle Atlantic	19	105 883
East North Central	23	118 604
West North Central	8	39 895
South Atlantic	27	155 401
East South Central	8	43 225
West South Central	7	46 178
Mountain	3	25 609
Pacific	6	45 509

A sequence of events that would lead to the worst accident possible – the 'China Syndrome': A break on the coolant pipe (1) and failure of emergency cooling (2) would cause the reactor vessel (3) to overheat, melting the fuel (4). This would escape through the concrete slab (5). The containment building (6) would crack, allowing more escapes.

Figure 5.6.3
The 'China Syndrome': the consequences of a nuclear meltdown

There are three main reasons why nuclear power is not more evenly spread over the continent:

- regions with the greatest demand for electricity and regions with limited availability of other power sources were more attracted to nuclear power;
- some companies calculated that nuclear power would be cheaper than the alternatives;
- in some areas opposition to nuclear power was exceptionally strong and companies decided to produce electricity in other ways.

Three Mile Island

At 4.00 am on 28 March 1979 the cooling system at the Three Mile Island nuclear power plant near Harrisburg, Pennsylvania broke down. For a while it was feared that intensely hot radioactive material would burn through its casing. This would contaminate the environment and cause a major disaster. Fortunately engineers brought the reactor under control before this could happen but the clean-up operation inside the plant cost millions of dollars.

If meltdown (often called the China Syndrome in the USA) had occurred (Figure 5.6.3) thousands of people might have died, some immediately, others over many years as the effects of radiation took its toll. The Three Mile Island accident has been replicated many times at a desert testing ground in Idaho to learn more about preventing similar accidents in the future. Since the Three Mile Island incident no new nuclear power plants have been ordered in the USA. In fact, since the early 1970s, over a hundred planned nuclear power plants have been cancelled. However, some plants planned before 1979 are still coming on line.

Environmental pressure

Opposition to nuclear power was already strong before the Three Mile Island accident. However, during the 1950s many environmentalists believed that using nuclear energy to generate electricity would bring cleaner air, reduce strip-mining for coal and offshore drilling for oil and end plans to dam more rivers for hydro-electric power. But in 1974 the Sierra Club (one of the US's major environmental groups) voted to oppose construction of more nuclear power plants. They did so because of growing concern about the unsolved problems of safety, waste disposal, nuclear weapons proliferation and possible theft of nuclear materials. In 1979 after Three Mile Island, the Sierra Club went further and called for the phasing out of all existing nuclear power plants, stating: 'The events at the Three Mile Island nuclear plant reaffirm our concern about the lack of safety of nuclear plants and demonstrate that the possibility of human error dooms the nuclear fuel cycle to unacceptable risks'.

Nuclear waste

The disposal of nuclear waste has been the major issue confronting the industry in recent years. States which at one time felt that the benefits of storing nuclear waste outweighed the costs have changed their minds in recent years.

The Hanford Nuclear Reservation in Washington (Figure 5.6.4) covers 900 km². It is dotted with spent nuclear reactors, dozens of poorly documented sites where chemicals and radioactive waste were dumped into trenches or pits, and 177 big underground storage tanks, many of them leaking.

Figure 5.6.4
Site of the Hanford Nuclear Reservation

2010: America's nuclear waste odyssey

ON 29 JUNE last year, at 3.14 am, a moderate earthquake registering 5.6 on the Richter scale occurred 13 kilometres below Little Skull Mountain. That is about 20 kilometres south-east of the proposed burial site of at least 25 000 tonnes of highly radioactive spent fuel and high-level waste from the US's 120 commercial nuclear power plants, at Yucca Mountain in southern Nevada, 100 miles north of Las Vegas.

Opponents of the project at once claimed that the damage caused to surface buildings in the area showed that the site is unsuitable. But the US Department of Energy—whose research facilities sustained an estimated $1 million of damage in the earthquake—concluded that it actually enhanced the site's suitability, because seismologists were able to verify computer models about the seismological stability of the mountain and its environs they had generated from historical data.

The DOE has chosen Yucca Mountain to be the store for spent fuel and high-level waste with a half-life of 24 000 years. To meet the "10 000 year criterion", set by the DOE, the repository must not release "significant" amounts of its radioactive content into the environment during that period, no matter what changes occur in the local climate, geology or

water table. Earthquakes obviously make it harder to meet that requirement.

Clarence Allen, a member of the Nuclear Waste Technical Review Board, comments that "there are certainly areas in the US with lower seismic hazard than that of south Nevada, although no area is completely devoid of seismicity". Earthquake studies of Yucca Mountain began over 15 years ago, and according to DOE reports have not shown that faulting would break open the mountain to expose radioactive waste.

Meanwhile the NWTRB's 17-strong panel has also studied the earthquake threat to the proposed Yucca repository and concluded that there is no evidence to support the assertion made by opponents—including one former DOE scientist—that the water table has, in the past, periodically risen hundreds of metres, which it would have to do to disrupt the repository from below.

Critics remain unconvinced. Bob Loux, executive director of the state of Nevada's Nuclear Waste Project Office, claims that in the June earthquake "people in Nevada literally felt why Yucca Mountain is an unsafe site for the disposal of radioactive waste". Carl Johnson, a geologist and administrator of the NWPO, says pointedly, "The question that now comes up is—how

big an earthquake and how close does it have to be to disqualify the site?" Recent studies of fault lines along the west coast by seismologists conclude that, for Nevada and California in particular, the odds of a devastating earthquake—say, 100 times more powerful than that of last June—within the next decade have increased significantly.

Nevada senator Richard Bryan, a former state governor, said in the Senate that "the decision to locate a high-level waste dump in an area where 32 active earthquake faults traverse the region defies common sense and logic". Citizens in Clark County seemed to agree. A survey found that after the earthquake 75 per cent of those polled thought the Yucca Mountain site should be "immediately dropped from consideration", and almost 71 per cent believed that the repository would pose "an extreme health and safety threat to the Las Vegas metropolitan area".

Presently the US, like Britain, does not have a single repository for the high-level waste from its civil and military nuclear plants. There is the Waste Isolation Pilot Plant in Carlsbad, New Mexico, which should hold 1.6 million cubic metres of waste left over from nuclear weapons production. But that has not yet opened.

New Scientist

Figure 5.6.5
2010: America's nuclear waste odyssey

QUESTIONS

1 (a) Outline the reasons why many environmental groups first believed that nuclear power was an improvement over existing ways of generating electricity.
  (b) Why are all environment groups now firmly against nuclear power?

2 Account for the regional variations illustrated by Figure 5.6.2.

3 Discuss the assertion that 'the disposal of waste material is now the major issue in the nuclear debate'.

Under a 1989 agreement between Washington state, the Department of Energy (DOE) and the Environmental Protection Agency, the remains of four decades of plutonium were to be eradicated and the site restored to as near pristine a condition as possible. By 1994 over $4 billion had been spent at Hanford but progress has been extremely slow. Clean-up officials describe the site as the most complex piece of polluted land in the United States. However the scope of the clean-up effort is creating a significant local industry with a vested interest in dragging things out for as long as possible. The most buoyant housing market in the state is around Richland, the town adjacent to Hanford, which contains a growing force of clean-up workers (over 19 000 in early 1994).

The long search for a permanent site for US high-level nuclear waste has been narrowed down to one location – Yucca Mountain in Nevada. However, opposition to the plan has been intense, not least because of strong seismic activity in the region (Figure 5.6.5).

The battle for Yucca Mountain has been raging for years and is becoming more urgent for the US's nuclear plants, where the thousands of tonnes of spent fuel are already stored temporarily. By the year 2000 there will be 40 000 tonnes to store. The DOE plans to study Yucca Mountain until the end of the century to see if the site's geological and hydrological setting can isolate the waste sufficiently from the environment. If the studies are favourable, the DOE will in 2001 apply to the US Nuclear Regulatory Commission for a licence. The Commission will then probably take up to four years to review the DOE data. If it backs the DOE report, the repository will be constructed by 2010.

Apart from the plan to receive and store the commercial spent fuel, Yucca Mountain will, if approved, take the high-level waste from the military nuclear sites across the US.

5.7 Energy: Alternatives for the Future

The first major wave of interest in new alternative energy sources resulted from the energy crisis of the early 1970s. However, the oil glut of the 1980s and the withdrawal of tax advantages sent this sector of the energy industry into depression and many renewable energy firms folded. As a result the projected contributions of such energy sources have not lived up to earlier expectations. For example, solar electricity accounts for less than 0.5 per cent of the power generated in the US today, instead of the 2–5 per cent envisaged in the late 1950s.

The tentative rebirth of the renewable energy industry in the 1990s has been due to:

- the 1990 Clean Air Act amendments which provide a financial incentive to electric utilities to purchase renewable energy;
- the 1992 Energy Policy Act which reinstated a modest tax credit for renewable energy of 1.5 cents per kW hour;
- technological advances within the sector.

As *The Economist* stated on 25 September 1993,

'For an industry that once seemed likely to die young, this is a welcome recovery. But the technology of renewable energy is still fraught with problems. It lacks a good way of storing its output; it usually requires a big investment in equipment; and it needs an especially windy or sunny site. Meanwhile, its competitors have not stood still. Small, turbine-assisted generators that use low-polluting natural gas are now cheaper than dirtier fossil-fuel generators. Renewables may yet prove the Fuel of Tomorrow – but they are not the fuel of tomorrow.'

The fate of alternative energy remains tied to government policy. A recent report from the World Energy Council (WEC) predicts slow growth for the industry unless governments take a more active role (Figure 5.7.1). The most important policy envisaged by the WEC would be pricing fossil fuels to include their full environmental costs. Financial assistance for research and development is also important. The renewable energy industry in the USA has long argued that it receives too little in government funding (Figure 5.7.2).

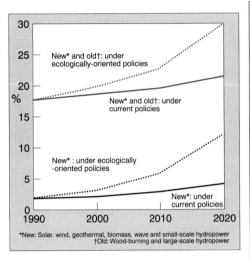

Figure 5.7.1
Renewable energy sources as percentage of world energy supply

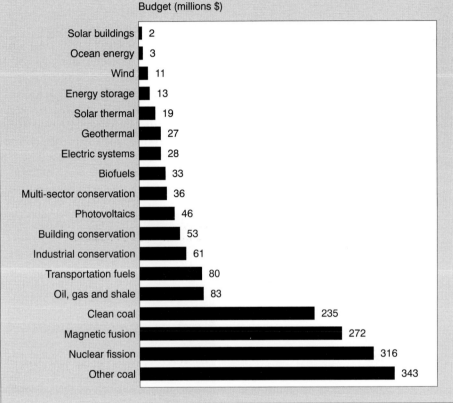

Figure 5.7.2
US Department of Energy: research and development budget, 1991

The USA has been much more active in the alternative energy area than Canada. This is largely because the USA is a heavy importer of energy while Canada has the relative luxury of being a net exporter. Thus energy forms such as solar and wind are still in the developmental stage in Canada where most alternative energy is in the form of biomass.

Solar power

Solar energy is produced by nuclear reactions that take place inside the sun. Every forty minutes the sun delivers as much energy to the Earth's surface as the total population of the planet uses in one year. This resource has been used for heating for some time, but the key to utilisation on a significant scale is solar electricity. Although the USA is the world leader in the field, solar power only contributed 0.2 per cent to total generation in 1993.

Two types of system are currently in operation. The most significant at present are solar furnaces (also known as High-Temperature Collectors). The principle here is to focus the sun's rays on a target. A fluid is pumped inside the target where it is heated.

Nine power plants based on this system are located in the Mojave Desert of southern California. Together they occupy an area of 7.8 km². The Mojave Desert is in the sunniest part of the continent, having more than 300 cloudless days a year. These solar power plants can generate a total of 354 MW, about 2 per cent of the capacity of the local grid, and accounting for about 50 per cent of world production of solar electricity. The rest is largely supplied by hundreds of small photovoltaic systems, most generating less than 1 MW.

The Mojave Desert plants consist of long parallel lines of troughs (Figure 5.7.3), each about 100 m in length and computer controlled to rotate and follow the sun during the course of the day. Each trough is made up of 224 glass mirrors and has a collecting area of 554 m². The parabolic shape ensures that the light from all the mirrors reflects onto a receiver – a horizontal pipe containing oil, that runs above the trough. The solar energy heats the oil to about 390 °C. Pumps then force the oil through a series of heat exchangers that use the recovered heat to convert water into superheated steam for driving a turbine. The largest of the power plants, an 80 MW facility, needs an array of about 900 troughs. The operating company is now looking to build a plant of between 200 and 300 MW capacity in the region.

Photovoltaic cells consist of thin slices of semiconductor material, mainly made of silicon. When the sun shines on a photovoltaic cell, electric current flows from one side of the cell to the other. The strips of semiconductor material are placed on large panels which are computer rotated to benefit from maximum sunlight. The world's first 1 MW size photovoltaic cell unit began operation in southern California in 1982.

Figure 5.7.3
Lines of solar troughs in Texas

The solar furnaces of the Mojave Desert produce power at over twice the cost of coal power but solar proponents argue that the apparent price difference is grossly misleading. They say the cost of oil and coal should be adjusted to reflect:

- price volatility;
- uncertainties over supplies;
- environmental damage (estimates for the pollution costs of coal start at 1 cent to 3 cents per kW hour and range upward).

For the moment power from solar furnaces costs less than that produced by photovoltaic cells. Advocates of photovoltaic cells point out, though, that the gap is narrowing and that photovoltaic cells work in cloudy weather.

The environmental effects of solar generation are mainly confined to the size of the land area required and the visual impact of these facilities.

Wind power

In the early 1900s more than six million small windmills in the US powered water pumps, grist mills and small electricity generators. By the 1940s most had been abandoned as the Rural Electrification Administration wired the country-side. Now the US and Canada are showing renewed interest in wind-powered electricity generation.

The US had more than 1700 Mw of installed wind capacity in early 1995. Most of the 19 000 turbines are located along a few mountain passes in California. However, 16 states have wind energy potential equal to or greater than California's, according to the Department of Energy. While a number of large single turbines operate at various sites over the continent, most recent development has centred on the concept of the 'wind farm' (Figure 5.7.4). Here many turbines operate at a single site and feed power into the local grid. Altamont Pass near San Francisco boasts over 7000 turbines.

Figure 5.7.4
A wind farm, Palm Springs, California

Turbines take toll on wildlife

IN California, between 1989 and 1991, researchers funded by the California Energy Commission searched for bird carcasses at sites on the 7000-turbine wind farm at Altamont Pass. Of the 182 they found, most had collided with turbines, and had suffered amputations. Others had been electrocuted or collided with wires. Almost 120 were birds of prey, including red-tailed hawks, kestrels and protected golden eagles.

Barnaby Briggs of Britain's Royal Society for the Protection of Birds (RSPB) says: "The thing's in the wrong place." Altamont Pass is an important hunting ground for birds of prey, he adds.

New Scientist,
16 July 1994

Figure 5.7.5
Turbines take toll on wildlife

The Kenetech Corporation produces wind power in Altamont Pass for 7 cents per kW hour, compared with 4 cents or less for conventional fossil fuel plants. Kenetech would be out of business were it not for tax incentives and federal and state mandates that have forced utilities to buy its power. Since the early 1980s Kenetech's largest customer, San Francisco's Pacific Gas and Electric, has been required by state law to buy all the wind power generated at Altamont Pass.

New wind generating stations using advanced techniques are planned in a dozen states in the USA. New technology, along with better siting practice, is projected to reduce wind energy cost below 4 cents per kW hour. As a result wind is expected to be one of the least expensive forms of new electrical generation in the twenty-first century. New designs allow remarkably efficient wind capture, eliminating the need for a breaking mechanism to slow the blade in high winds, which causes high stress on earlier generation machines. The new turbines produce power 90 per cent of the time, compared with 50 per cent of the time in the mid-1980s. Industry forecasts suggest that by the end of the decade another 5–10 000 MW will have been installed in the USA alone.

The environmental concerns about wind power centre around the visual impact, the 'hum' of the turbines, and the effect on wildlife (Figure 5.7.5).

Biomass

Biomass is organic matter from which energy can be produced. Living plants generate ten times as much energy each year as people consume. Some developing countries get over 90 per cent of their energy from biomass, mainly by burning firewood. In North America wood is also the major contributor to biomass, but for industrial rather than domestic use. In the USA wood accounts for 84 per cent of biomass energy production. This is mainly in the paper and forest-products industries, which use the technology to meet more than half of their own energy needs. But biomass is also used to generate electricity and to mix with petrol to produce gasohol.

There are already about 1000 biomass-fuelled power plants in the US, although all have capacities of less than 25 MW. Total capacity in early 1994 had reached 6500 MW with a further 2000 MW of refuse-burning plant. The Electric Power Research Institute believes that biomass could be used to supply 50 000 MW of capacity by 2010 and probably twice that amount by 2030.

US says drivers must use more alcohol

Happy days: now the gas guzzlers have gone, gas itself is changing

IN a ruling aimed at cleaning up the air, the US government announced last week that alcohol made by fermenting corn will have to be added to the petrol sold in several American cities.

Car engines do not usually suck in enough oxygen to fully oxidise the carbon in petrol. As a result, their exhaust gases contain high proportions of carbon monoxide. Adding ethanol to petrol helps the fuel to burn more efficiently, reducing emissions of carbon monoxide.

Under a law passed in 1990, petrol sold in cities with severe air pollution must include oxygen-containing compounds known as oxygenates. The agriculture industry wants ethanol, ordinary alcohol, to be used, while the petroleum industry's choice is methyl tertiary-butyl ether. MTBE is made from methanol, which comes from natural gas. The petroleum industry has so far prevailed and MTBE has been the main oxygenate.

Last weeks' decision requires that 30 per cent of the oxygen content of the oxygenated petrol must come from renewable sources. At present, ethanol is the only widely available renewable oxygenate. The remaining 70 per cent of the oxygen is still likely to come from MTBE.

Petrol stations will have to supply the reformulated petrol in the nine cities with the worst air pollution: Baltimore, Chicago, Houston, Los Angeles, Milwaukee, New York, Philadelphia, San Diego and Hartford, Connecticut.

The US Agriculture Department estimates that the regulation will lead to the use of 1.9 billion litres of ethanol a year. In turn, that would need 200 million bushels (about 5 million tonnes) of corn, a small increase compared to the annual US corn harvest of 8.7 billion bushels, but enough to boost the price of corn by about 3 per cent. The Environmental Protection Agency, which made last week's announcement, reckons the reformulated petrol will cost drivers about $25 more a year.

All sides agree that petrol with oxygenates reduces emissions of carbon monoxide, but arguments remain over which additives will cut the most. Daniel Durbin, director of the Motor Vehicle Emissions Laboratory at the University of Cincinnati says that there is little difference in the pollution produced by ethanol and MTBE.

Others claim that the notion of ethanol as a renewable fuel is spurious. David Pimentel, professor of agricultural sciences at Cornell University, New York, says that producing ethanol takes 70 per cent more energy than is contained in the ethanol itself. And usually, the energy used to produce ethanol comes from fossil fuels.

Deborah Gordon, a transport specialist at the Union of Concerned Scientists, questions the whole value of oxygenates because they will not reduce emissions of nitrogen oxides, the main promoters of smog. "I don't know that it's really going to solve the problem," she says.

New Scientist,
9 July 1994

Figure 5.7.6
US says drivers must use more alcohol

Worldwide, Brazil is the largest producer of liquid biofuels, mainly in the form of ethanol from sugar-cane. In the US ethanol made from surplus maize replaces about 0.5 per cent of national petrol consumption. Here a mixture of 10 per cent ethanol and 90 per cent gasoline is sold as 'gasohol' in several mid-western states. A mixture of state and federal subsidies as well as tax exemptions supports the production of ethanol which increased at a rate of 25 per cent per year during the 1980s, although slowing to 10 per cent per year in the early 1990s. When added to petrol ethanol reduces emissions of carbon monoxide and hydrocarbons (Figure 5.7.6). However, critics point to the pollution caused by the burning of coal within the ethanol distilleries and in the power stations that supply them with electricity. Although the use of ethanol saves on oil imports, it is far more expensive.

In 1993 the Department of Energy announced encouragement for energy crops and energy conversion systems tailored to the resources and needs of specific US regions. Crops to be tested include elephant grass, eucalyptus, bagasse, sorghum and sugar cane. Farmers could be producing for the energy market rather than the food market in future.

In Canada biomass is mainly in the form of liquid waste from pulp and paper processing, wood, straw, vegetable oils, and agricultural and municipal wastes. The industrial sector uses over 90 per cent of such energy, mainly for producing process steam and heating, with a small amount used to generate electricity.

Geothermal energy

Geothermal energy is the natural heat found in the Earth's crust in the form of steam, hot water and hot rock. At present virtually all the geothermal power plants in the world operate on steam resources (Figure 5.7.7).

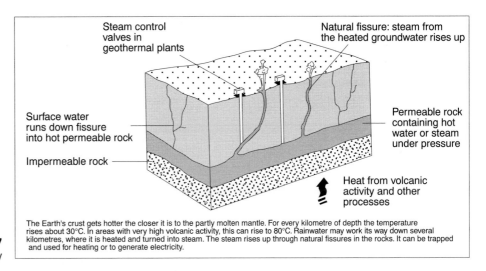

Figure 5.7.7
The production of geothermal energy

Rain water may percolate down several kilometres in permeable rocks where it is heated due to the Earth's geothermal gradient. This is the rate at which temperature rises as depth below the surface increases. The average rise in temperature is about 30 °C per km, but the gradient can reach 80 °C per km near plate boundaries (Section 12.5). The Pacific coast and Western Cordillera is one such region.

The USA is the world leader in the use of geothermal heat to generate electricity. In 1993 this source contributed about 0.3 per cent to the national power supply, enough for three million households. This amounted to over 2800 MW mostly located in California.

The largest facility is at the Geysers, Sonoma County, California, which began operation in 1960. The field has a capacity of 1500 MW, but declining steam pressure in the wells is raising the spectre that the $3.5 billion investment will run dry early in the next century. The underlying problem at the Geysers – and the danger for geothermal development everywhere – is overdevelopment of a poorly understood resource. There is some degree of pollution at the facility because the steam contains impurities such as hydrogen sulphide.

Research into the use of hot-water reservoirs is also in progress. However, technical and environmental problems have become apparent. The hot water below the Imperial Valley in California is very briny. For every litre brought to the surface, up to a quarter may be salts and solid particles that corrode and clog piping. In addition, there is the task of disposing of the waste.

It is more difficult still, at present, to use the heat of hot rocks at depth below the surface. Energy companies have experimented by pumping water several kilometres underground and then bringing it back to the surface after it has been heated. It may be some time before such technology is fully developed.

Another possibility under consideration by the Department of Energy is magma energy extraction. This would involve drilling a hole through the Earth's crust to the molten rock below. A heat exchanger would be installed at the bottom. Energy could be extracted by water or another fluid that is brought to the surface to drive a heat engine. Sites must be found where molten magma is not much more than five km below the surface to keep costs at a reasonable level.

The Interior Department estimates that a staggering 1.2 million quads (quadrillion BTU's) of geothermal energy underlie 1.4 million hectares of US land in the West. However, much of this potential resource is in national parks or other protected areas, or is too deep for current technology.

In Canada, the government of British Columbia has recently given approval to independent power producers to generate geothermal power in the province for long-term contracts to the US.

Figure 5.7.8
Site of Annapolis Royal tidal plant

Tidal power

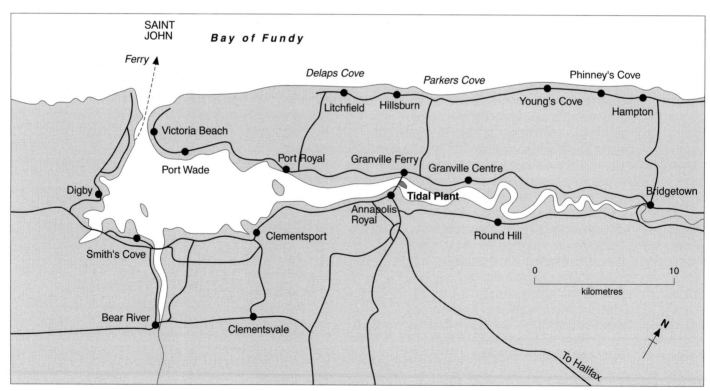

Annapolis Royal is home to the first and only modern tidal plant in North America. The Annapolis Tidal Generating Station, employing the largest straight-flow turbine in the world, is on the coast of Nova Scotia (Figure 5.7.8). This is one of only 40 sites identified worldwide which are suitable for large-scale electricity generation from the tides.

A small island in the mouth of the Annapolis River was selected as the site for the powerhouse (Figure 5.7.9). It is an ideal location not only because the Bay of Fundy has the highest tides in the world, but also because an existing causeway equipped with sluice gates provided a head pond where water levels could be controlled, an essential component in the station's operation. Work began on the $46 million demonstration project in 1980 and it opened in 1984. It generates power from 95 per cent of all tides and each year produces 30 million kW hours, enough electricity to power 4500 homes. At peak output it can generate up to 20 MW.

Figure 5.7.9
The Annapolis Royal tidal plant

The average tide at Annapolis is 7 m. At high tide, sea water passes through the sluice gates and into the head pond behind. At low tide the sluice gates are closed and water flows out through the turbine water passage, as the level of water in the head pond is higher than that in the bay. The turbine generates power when the head of water exceeds 1.6 m. Power is generated during two periods of five to six hours, each beginning two to three hours after high tide.

More ambitious tidal projects have been considered for sites at the head of the Bay of Fundy. The most promising sites are in the Minas Basin and the Cumberland Basin. The former would span 8 km of the basin and be outfitted with 97 sluice gates and a 3 km-long powerhouse with 128 double-effect turbines for a total installed capacity of more than 5000 MW. The proposal for the Cumberland Basin is 42 turbines and an installed capacity of over 1400 MW. But before these projects begin, more information about the possible environmental impact has to be gathered. The main concerns are potential effects on fish populations and the filling in of the head pond with sediment. In addition, because the Minas Basin proposal is so much larger, it may have an impact on tides along the coast.

Ocean Thermal Energy Conversion (OTEC)

OTEC uses the temperature difference between the ocean's warm surface water and the cold deep waters to generate electricity. The system can only work in the tropics where the temperature difference between surface and deep water is at least 20 °C.

The USA has developed the world's first at-sea power plant, located off the coast of Hawaii. Here the warm surface water vaporises ammonia which is used to drive a turbine. The ammonia is then condensed by deep cold water and the cycle begins all over again.

Unlike some other renewable energy sources, OTEC provides 'base load' power (a continuous electricity flow). However, at present the costs of OTEC are high and the amount of electricity produced is very small.

QUESTIONS

1 Use Figure 5.7.2 to assess the proportion of federal R&D energy expenditure allocated to the 'renewables' sector.

2 Write a simple matrix to illustrate the costs and benefits of solar power.

3 (a) What is a 'wind farm'?
  (b) If a community relied totally on wind turbines for its electricity, what problems might it encounter?
  (c) How might technological advance overcome such problems?

4 How does the use of biomass differ between the developed and developing worlds?

5 What are the advantages and disadvantages of ethanol production for use in motor vehicles?

6 Describe the different forms of geothermal energy and identify any environmental problems relevant to each form.

7 (a) Explain how the Annapolis tidal power plant works.
  (b) Why is it unlikely that tidal power will ever be of major importance in North America?

8 To what extent can California be considered the global leader in renewable energy?

6

Agriculture

6.1 Continental Overview

North America is one of only a few world regions that can both feed its own population and export large quantities of food, and in recent decades the agricultural economies of both the USA and Canada have become increasingly geared to foreign demand. Measured by most criteria, North American agriculture is the most productive in the world. Output per agricultural worker is many times greater than the average for western Europe. Although yields per hectare are not generally as high as some more intensively farmed areas around the world, the highly mechanised cultivation of large farms in North America results in huge absolute annual production figures which have steadily risen due to massive direct investment and considerable expenditure on research and development.

Agriculture is but one component of the gargantuan North American economic system. However, the nature of the industry's links with other sectors of the economy has become increasingly complex. The expansion of horizontal and vertical integration has spawned the term 'agribusiness' to describe the nature of modern agricultural operations which have to be highly cost effective to remain competitive. Companies like Safeway and Del Monte own farms, food-processing factories and supermarkets. Another side of agribusiness supplies farms with machinery, fertilisers, pesticides and other needs.

The immense variety of physical environments in North America has allowed a wide variety of crops to be grown and livestock kept. Although technological development and the availability of capital has enabled farmers to expand into increasingly difficult environments, the physical geography of the continent remains a major influence on agricultural practices and patterns. The role of physical factors must however be kept in perspective. Relief, climate and soil fertility set broad limits as to what can be produced, but overall the physical environment provides only a base which presents the farmer with a varying number of choices. The pattern of agricultural production that has emerged is a result of millions of individual decisions since the initial colonisation of the continent, whereby each agricultural unit has had to select the nature of production from the options available. It is in such selection that economic factors and government intervention at various levels have become increasingly important.

While agriculture is in many ways organised in a similar way in the USA and Canada, physical contrasts have resulted in considerable differences in farm production in the two countries.

Agriculture in the United States

Agricultural patterns

Figure 6.1.1 shows the traditional farming belts of the USA. Although the more intense regional specialisation of the pre-war period has been considerably reduced by progressive diversification in virtually every part of the country, broad agricultural regions can, to a certain extent, still be recognised.

In the upper North East and Midwest regions the Hay and Dairying Belt stretches from New England to Minnesota. The cool summers and the long

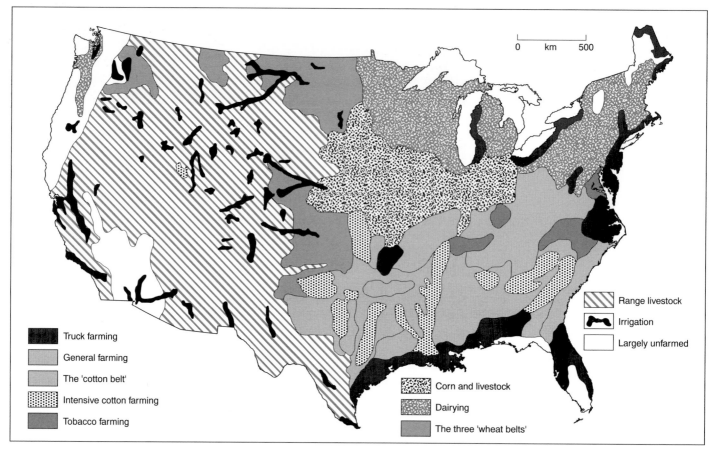

Figure 6.1.1
Agricultural regions in the US

Figure 6.1.2
Principal crops: hectares harvested, (millions) 1994

Corn for grain	29.5
Wheat	25.0
Soybeans	24.6
Hay	23.8
Cotton	5.4
Sorghum	3.6

severe winters present sub-optimum conditions for grain cultivation. Thus the emergence of dairying in this region has been partly in response to physical limitations on other agricultural practices. Other factors which have encouraged this use of the land were the cultural traditions of the German, Swiss, Scandinavian and Dutch farmers who settled in the western part of the belt, and the intense urbanisation of the region which presented a large market for milk. Milk production dominates in the eastern area while butter and cheese increases in importance farther west. However, even here milk remains the main dairy product close to major cities. Wisconsin with 14.6 per cent of US production in 1994 is second only to California in milk.

To the south west lies the Corn and Livestock Belt, one of the most productive agricultural regions in the world. Corn (maize), the most extensively grown crop in this region and in the country as a whole (Figure 6.1.2) is primarily used as a fattening fodder for pigs and cattle. Corn requires a growing season of at least 130 days, and high summer temperatures of 21 °C are necessary along with warm nights. Iowa and Illinois form the nucleus of corn production, accounting for over one-third of the US planted area in 1994. These two states are also the leaders in soybeans. In some states such as Ohio and Missouri the area under soybeans exceeds that planted with corn.

East and south east of the Corn Belt is a sizeable but irregular area between the agricultural systems of the Midwest and the South. This zone was known in earlier years as the Corn and Winter Wheat Belt. While both crops are important to the area indicated in Figure 6.1.1, yields are generally lower than in the major regions for corn and wheat. Within and bordering this zone are important areas of tobacco cultivation. North Carolina, Kentucky, Tennessee and Virginia are the leading tobacco producing states.

Figure 6.1.3
USA: Farm income and principal sources for each state

A   Apples
B   Broilers
Ba   Barley
C   Corn
Ca   Cattle
Co   Cotton
Cr   Cranberries
D   Dairy products
E   Eggs
F   Forest products
Gf   Grapefruits
G   Greenhouse

Gr   Grapes
H   Pigs
Ha   Hay
Ho   Horses
L   Lettuce
M   Mushrooms
O   Oranges
P   Potatoes
Pe   Peanuts

Pi   Pineapples
R   Rice
S   Soyabeans
Sb   Sugar beet
Sc   Sugar cane
Sh   Sheep

So   Sorghum grain
Su   Sunflower
T   Tobacco
To   Tomatoes
Tu   Turkeys
W   Wheat

Truck farming (market gardening) occurs in a number of areas in the USA where a combination of physical and economic factors make this type of farming a profitable concern. Regions most noted for truck farming are the north eastern seaboard, the eastern borders of Lakes Michigan, Erie and Ontario, Florida and the Gulf Coast. The sub-tropical climate of the southern areas allows a range of exotic fruits and vegetables to be grown. Similarly, fruit and vegetables are intensively cultivated under irrigation in southern California and in other more restricted areas of the west. California is the leading state for vegetables and for a range of fruits.

The Cotton Belt of the South is one of the best known of America's farming regions. Cotton production is now highly mechanised and has undergone a major westward shift in location. Southern agriculture, for so long in a distressed state, has been radically transformed in recent decades through rapid mechanisation and considerable diversification. In 1994 cotton ranked as the major source of farm income in just one state, Louisiana (Figure 6.1.3).

In terms of cropland, wheat dominates on the High Plains, with spring wheat concentrated to the north west of the Corn and Livestock Belt and the nucleus of winter wheat to the south west. The westward position of the wheat belts is explained largely by the crop's ability to tolerate drier conditions than the other major grains. North Dakota, the centre of spring wheat production, was the only state where wheat was the major source of income in 1992. Ranching is overwhelmingly dominant from the drier parts of the High Plains, across the Mountain states into the less favourable agricultural areas of the Pacific division. Cattle provide the main source of farm income in 15 states west of Illinois.

The extensive ranching of the drylands is interspersed with elongated zones of irrigation farming, frequently straddling major river valleys. Farming in such areas varies widely but cotton, sorghum and alfalfa are of special importance.

Mountain and arid landscapes present a significant obstacle to agriculture over a large area of the Southwest and Northwest. In contrast the Pacific coastal lowlands and the linked interior valleys are centres of intense production where the climatic regime is a vital factor influencing the type of farming practised.

As Figure 6.1.3 indicates, California is clearly the leading state in terms of total farm income. Only Texas exceeded half of the total farm receipts of the Pacific giant in 1994. The lowest farm incomes occur in the Mountain states and New England.

Fewer but larger farms

Two hundred years ago there were about four million farms in the US with an average size of just over 50 ha. At the time 90 per cent of the total labour force were employed on the land. As agriculture spread into hitherto unfarmed areas both the number of farms and farm size increased. However, in the three decades immediately following World War II, the agricultural landscape underwent a profound change. The rapidly expanding secondary and tertiary sectors of the economy attracted people away from farms and encouraged a considerable rural to urban migration. At the same time technological advance in all areas of farming brought greatly increased productivity. As a result of these two trends the number of farms sharply declined while those remaining grew significantly larger. Between 1945 and 1974, the number of farms fell from 5.9 million to 2.3 million, a decline of more than 100 000 farms a year. At the same time, the average size of farms nearly doubled to 178 ha.

In the last two decades the rate of farm loss has slowed down. Between 1974 and 1992, the number of farms declined from 2.3 million to 1.9 million, while the average size increased to 199 ha. Commercial farming is now basically made up of a few hundred thousand very large and very efficient farms. The gap between large and small producers has widened as the former have benefited from their ability to purchase the latest in agricultural technology. As a consequence farming is now a part-time occupation for most people. In 1991, more than two-thirds of farm households received more income from off the farm activities than from those on the farm. This increase in part-time farming has acted to reduce the rate of decline in the number of Americans engaged in farming (Figure 6.1.4). Labour accounted for half the cost of running a typical farm before 1940. By 1990 it was down to less than 12 per cent of total costs.

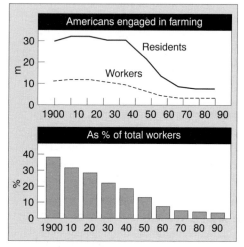

Figure 6.1.4
USA: Agricultural work force

No-till farming

No-till farming means planting crops directly into the previous year's stubble, with minimum soil disturbance. As ploughing is not required, losses of topsoil through wind and water erosion can be considerably reduced. It has been estimated that the average rate of soil erosion from agricultural land in the USA is 17 tonnes per ha per year, whereas new soil is created at the rate of only one tonne per ha each year. The area planted in the USA using no-till methods more than doubled between 1989 and 1993 (Figure 6.1.5). No-till farming can cut soil erosion by as much as 90 per cent, depending on what kind of stubble is holding the topsoil in. On average, crop yields are the same or better than on farms using conventional techniques. No-till farmers use their machinery less intensively, saving on repair and fuel costs and also save in terms of the time and labour needed to plant a crop.

Concerns about no-till farming are centred on the fact that herbicides must replace ploughing to kill weeds. However, after two or three years, the weed seeds in the top few centimetres of the soil come under control and herbicide use declines to below that of the previous tillage system.

No-till farming is an important strategy in the battle against soil erosion. Over 40 per cent of cropland is eroding above tolerance levels that will enable sustained, high level production in the future. As the cultivated area expanded

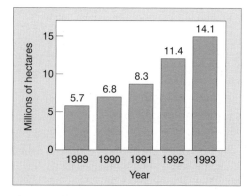

Figure 6.1.5
USA: area under no-till farming (ha)

in the 1970s much marginal land was ploughed up, making it particularly vulnerable to erosion. Most soil experts agree that much more needs to be done to protect this most important resource in the future.

Changing demand patterns

Increasing affluence and greater health awareness have been the major factors behind the changing pattern of domestic demand. For example consumption of chicken in the USA overtook that of beef in 1992 and is now racing ahead (Figure 6.1.6). In 1994, American broiler companies processed some 7 billion chickens. It is not just a desire to eat healthier white meat but also the price factor which had led to such an increase. Chicken is a quarter of its price 30 years ago, after adjusting for inflation. In a few decades the industry has been transformed from a small scale, scattered structure to automated, mass production in a highly concentrated area. About four-fifths of production is in the 'broiler belt' stretching from eastern Texas to northern Florida and up to south-eastern Pennsylvania. The leading producer is Arkansas, where the industry is the biggest private employer.

However, the industry is not without its critics as more people have become aware of its environmental impact. Waste from poultry plants has overwhelmed some municipal sewage-treatment systems. Although newer poultry plants incorporate their own modern water-treatment facilities, environmentalists fear that the growth of the industry offsets the improvements in waste-water treatment and that the mountain of chicken excrement produced on the land remains barely regulated at all.

The export market

In the 1920s and 1930s the USA was a net importer of agricultural products. World War II changed this situation for a while but by 1950 the USA had again moved into deficit in its farm trade. In the 1950s and 1960s the margin between exports and imports was relatively narrow although generally in favour of the former. However, in the 1970s US agricultural exports quadrupled (Figure 6.1.7) and in 1981 broke their record for the eleventh year running, reaching almost $45 billion.

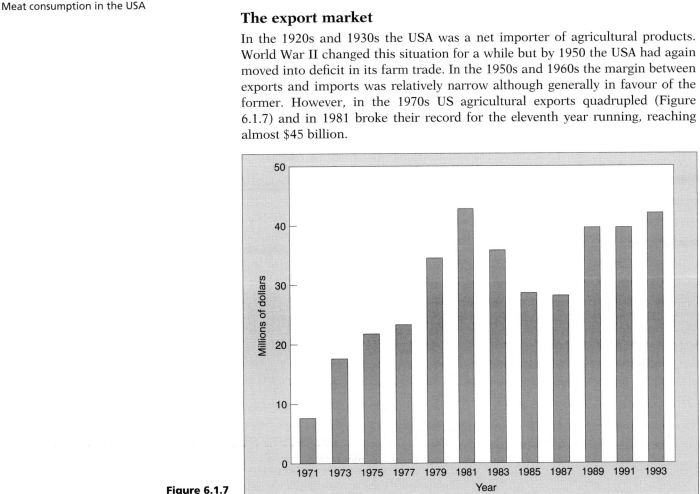

Figure 6.1.6
Meat consumption in the USA

Figure 6.1.7
USA: agricultural exports

In 1994, grains and feeds accounted for 29.6 per cent of exports by value, followed by animal and animal products (20.0 per cent), oilseeds and products (15.8 per cent), vegetables (8.5 per cent), fruits (6.8 per cent) and cotton (5.8 per cent). In the same year the USA continued to dominate the world export market supplying 71.7 per cent of the soybeans, 82.9 per cent of the corn, 33.1 per cent of the wheat and 25.2 per cent of the cotton.

Figure 6.1.8 shows America's leading trading partners in 1993 and the main products involved. In terms of the world as a whole the USA had a healthy surplus in its farm trade balance of $17.6 billion.

Figure 6.1.8
Agricultural trading partners, 1993

| TRADING PARTNER | US$ billion | | | MAJOR COMMODITIES | |
	Exports	Imports	Trade balance	Exported by United States	Imported by United States
Japan	8.7	0.3	8.5	Corn, beef, soybeans	Vegetables and preps., beverages, seeds
European Union	6.8	4.8	2.1	Soybeans, animal feeds, nuts and preps.	Wine and beer, vegetables and preps., cheese
Taiwan	2.0	0.1	1.9	Corn, soybeans, wheat	Vegetables and preps., noodles, biscuits/wafers
Korea, South	1.9	0.1	1.9	Cattle hides, cotton, soybeans	Vegetables and preps., noodles, fruit and preps.
Soviet Union (former)	1.8	(Z)	1.7	Corn, wheat, soymeal	Casein, furskins, cheese
Mexico	3.6	2.7	0.9	Soybeans, grain sorghum, vegetables and preps.	Vegetables and preps., noodles, fruit and preps.
Hong Kong	0.9	0.1	0.8	Fruit and preps., chicken, vegetables and preps.	Vegetables and preps., biscuits/wafers, noodles
Canada	5.3	4.6	0.7	Vegetables and preps., fruit and preps., beef	Cattle, pork, beef
Egypt	0.7	(Z)	0.6	Wheat, corn, wheat flour	Essential oils, vegetables and preps., fibres (ex. cotton)
Turkey	0.4	0.3	(Z)	Tobacco, corn oil, rice	Tobacco, fruits and preps., nuts and preps.
Philippines	0.5	0.4	(Z)	Wheat, soymeal, vegetables and preps.	Coconut oil, fruit and preps., sugar
China	0.4	0.5	−0.1	Wheat, soybeans, chicken	Vegetables and preps., tobacco, leathers
Guatemala	0.2	0.5	−0.3	Wheat, corn, cotton	Coffee, bananas, sugar
Costa Rica	0.1	0.6	−0.4	Corn, soybeans, wheat	Bananas, fruit and preps., beef
Thailand	0.3	0.7	−0.4	Wheat, tobacco, animal feeds	Rubber, fruit and preps., tobacco
Indonesia	0.3	0.8	−0.5	Cotton, soybeans, animal feeds	Rubber, cocoa, coffee
Colombia	0.2	0.8	−0.6	Corn, wheat, soymeal	Coffee, cut flowers, bananas
New Zealand	0.1	0.8	−0.7	Beverages, fruit and preps., soymeal	Beef, casein, fruit and preps.
Australia	0.3	1.1	−0.7	Beverages, vegetables and preps., soymeals	Beef, wool, sugar
Brazil	0.2	1.4	−1.2	Cotton, wheat, vegetables and preps.	Tobacco, coffee, orange juice

Z Less than $100 million.

Agriculture in Canada

Physical geography has imposed far greater limitations on agriculture in Canada than in the USA. In two countries of broadly similar land area the USA possesses more than five times the cropland of its northern neighbour. In addition the USA has a much higher percentage of cropland in 'optimum' climatic conditions. Thus agriculture is restricted to only 7 per cent of Canada and much of this land is only of marginal capability for many field crops. Even so, the products from the 45 million ha of farmland amounted to C$23 632 million in 1992, some 10 per cent of Canada's economy, with more than half destined for export.

Regional variations

70 per cent of Canada lies north of the thermal limit for crop growth and most farms are within 500 km of the US border. Where agriculture is climatically possible adverse relief, soil and drainage conditions impose widespread limitations within three of the major physiographic regions of southern Canada: the Appalachian-Acadian Uplands in the east, the Laurentian Shield and the Western Cordillera. The extent of Canada's agriculturally productive land is illustrated by Figure 6.1.9 which also indicates the dominant farming type in each region and the great contrast between eastern and western Canada.

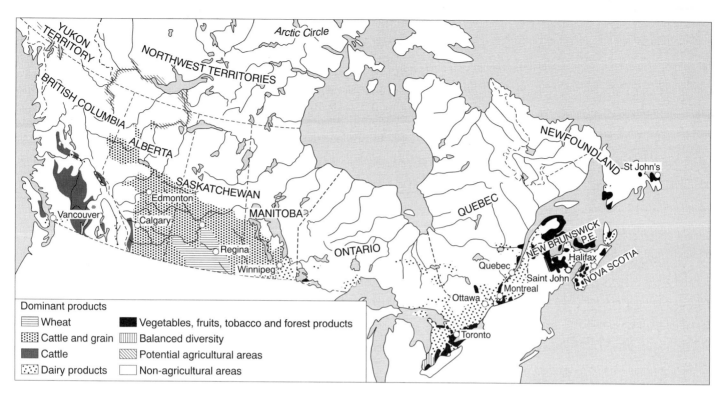

Figure 6.1.9
Agricultural regions in Canada

Dominant products
- ☰ Wheat
- ▦ Cattle and grain
- ■ Cattle
- ⋯ Dairy products
- ■ Vegetables, fruits, tobacco and forest products
- ▥ Balanced diversity
- ▨ Potential agricultural areas
- ☐ Non-agricultural areas

In the Atlantic Provinces agriculture is extremely limited by poor soils and hilly terrain and is characterised by extreme fragmentation of holdings, apart from Prince Edward Island, which is sometimes known as the 'Garden of the Gulf'. Agriculture is practised over most of the island, which specialises in producing seed-potatoes and breeding pedigree cattle. Elsewhere in the Atlantic Provinces dairying is the main source of income, with cattle, hogs, poultry and to a lesser extent fruit and vegetables also of importance. Overall, the Atlantic Provinces contain less than 2 per cent of the total agricultural area of the country.

The southern portions of Quebec and Ontario together comprise the second-largest agricultural region in Canada. Here most of the farms are confined to the shores of the St Lawrence, the Ottawa Valley and southern Ontario, although there are a few isolated agricultural districts farther north. Livestock operations predominate, particularly dairy, beef, pigs and poultry. Corn, mixed grains, winter wheat, oats and barley are cultivated for feed. Among the cash crops of the region are soybeans, potatoes, tobacco, fruit and vegetables. Quebec is the world's largest producer of maple syrup.

Farms in the central region are considerably smaller than in the extensive farming region of the Prairies, being mainly between 30 and 140 ha. However, the nature of production is generally intensive, being close to urban markets; and while the region contributes only 13 per cent of Canadian farmland it was responsible for almost 38 per cent of total farm cash receipts in 1991.

The largest of Canada's agricultural regions, with almost 82 per cent of all farmland, is the Interior Plains in the southern Prairie Provinces, characterised by extensive wheat and cattle farming. Farms average around 300 ha and are highly mechanised. The climate is particularly suited to the production of high-quality hard red spring wheat, but in recent decades oilseed crops have become increasingly important. Saskatchewan produces more than half the Canadian wheat crop and large quantities of other grains. Alberta is Canada's chief producer of feed grains and beef cattle while Manitoba supports more varied farming. The three Prairie Provinces combined account for about half of all Canadian farm sales.

Farming in British Columbia is extremely restricted by the physical environment. Concentrated in river valleys, the southwestern mainland and southern Vancouver Island, farms are mostly small and highly productive. Only 2 per cent of the province is used for agriculture with most of this given over to thriving dairy and livestock farms. The Okanagan Valley is Canada's largest producer of apples and is noted for its peaches, plums, apricots, cherries and grapes. The Peace River Valley in northern Alberta and British Columbia is an anomaly in Canadian agriculture because of its northerly position. The cool climate and short growing season limit agricultural production, but are compensated to some extent by long summer days. Grain and forage are the two main crops, with livestock based on cattle, pigs and poultry.

Development and change

As in the USA, Canadian agriculture has been characterised by rising levels of mechanisation and productivity, increases in farm size and a declining labour force. The farm population, which includes all people living on farms, has fallen from over three million in 1941 to 867 000 in 1991. The farm population now accounts for only 3.2 per cent of Canada's population while in the early 1980s they made up roughly 5 per cent (Figure 6.1.10). Saskatchewan has the highest proportion of farmers to total population at 16 per cent, but even here, the number is falling. It is not only the farm population, but rural and small town populations in general that are declining. Indeed, between 1986 and 1990, one-half of Canadian census divisions lost population.

Of Canada's farmers, 37 per cent supplemented their incomes in 1991 by working off the farm. Men reported an average of 72 days at their off-farm work, while women reported an average of 63 days. Canadian farm families receive over one-half of their income from off-farm sources, and even for families on the largest farms, over 40 per cent have at least one member working off-farm. The increase in off-farm work has stabilised farm incomes in recent years. However, the gap between the incomes of farm and non-farm families is widening. Agriculture and agriculturally-related employment now account for only a small share of rural jobs.

Government intervention and income stabilisation

The Canadian government has intervened in the agricultural markets with the purpose of maintaining a healthy farm sector by ensuring relative stability in farm income and farm prices. The Agricultural Stabilisation Act protects farmers against significant market instability by establishing support prices for many products. The creation of marketing boards, particularly the Canadian Wheat Commission and the Canadian Dairy Commission has done much to develop the efficiency of the nation's agriculture. The federal government's 'Food Strategy for Canada' stresses the aim of ensuring adequate supplies of safe and nutritious food at prices that are reasonable to both producers and consumers. While the export trade forms an important part of the strategy, Canada has maintained an above-average per capita level of food aid to the developing world.

Exports

Exports of farm produce amounted to C$10 856 million in 1991, just under one-tenth of the total value of all exports. In that year 44 per cent of all agricultural production was exported. Traditionally, nearly 90 per cent of Prairie wheat shipments have been exported to the UK and the rest of northwest Europe, but recently an increasing proportion has been going to the Pacific Rim countries and to the former USSR. Global trade has presented tough challenges to Canadian agriculture. All exporting countries are keen to shore up their own producers and protect domestic markets while gaining access to consumers abroad. The protectionist climate and oversupply of goods has made it harder for Canadian producers to get what they perceive to be fair prices for their exports.

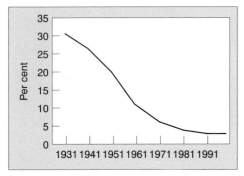

Figure 6.1.10
Percentage of Canadians on farms

Since the signing of the Canadian–United States Trade Agreement (CUSTA) there has been a substantial growth in Canadian grain and oilseed exports: 7 per cent in 1990–91 and 43 per cent in 1991–2. Under CUSTA, agricultural tariffs between Canada and the US are being gradually eliminated over ten years. However, for some commodities there was agreement to reduce tariffs more quickly. CUSTA has been a contributing factor in the significant increase in exports of some of these products. It is also likely that farm exports to Mexico will increase as a result of the North American Free Trade Agreement (NAFTA) (Section 13.2).

Environmental awareness

The 1991 Census of Agriculture showed a reversal in the previously steady upward trend in the use of commercial fertiliser and herbicides. In 1990, 59 per cent of census farms were using commercial fertiliser, compared with 66 per cent in 1985. The percentage using herbicides fell to 49 per cent in 1990 from 59 per cent in 1985. In addition, in 1991, one-quarter of land seeded was prepared using conservation tillage. A number of influences are at work here: an increasing number of consumers is demanding organic produce; more farmers are now concerned about the impact of chemicals on their land; and some conservation techniques are proving to be cost effective.

6.2 Agriculture in the Canadian Prairies

Three provinces

Over 80 per cent of Canada's farmland lies in the Prairie Provinces of Alberta, Saskatchewan and Manitoba (Figure 6.2.1). This is one of the world's major areas of extensive agriculture where farms averaging 300 ha in size stretch for almost 1500 km from the hills of Ontario in the east to the Rockies in the west (Figure 6.2.2). Such enterprises can be run by only two or three workers using an array of modern machinery. Long sunny summer days and generally reliable precipitation allow production of a range of quality grains and oilseeds. In addition native grasslands, cultivated forage crops and feed grains support a large beef cattle industry.

Figure 6.2.1
Canadian prairie provinces: area of farms, 1991

	Hectares	% of Canadian total
Manitoba	7 724 989	11.4
Saskatchewan	26 865 488	39.7
Alberta	20 811 003	30.7
Canada	67 753 701	100.0

Figure 6.2.2
Harvesting wheat in the foothills of the Rockies

Figure 6.2.3 includes all the agricultural products where the three prairie provinces together account for at least 50 per cent of Canadian production. Saskatchewan produces more than half of the Canadian wheat crop and large quantities of other grains. Alberta is Canada's chief producer of feed grains and beef cattle and is second only to Saskatchewan in wheat. Manitoba, with the highest prairie rainfall and over 100 frost-free days, supports more varied farming. The prairies produce all of Canada's peas, flaxseed, sunflower seed, mustard seed, and sugar beet.

Figure 6.2.3
Canadian prairie provinces: agricultural production

	Wheat 1992 '000t	%	Barley 1992 '000t	%	Oats 1992 '000t	%
Manitoba	5241.7	17.4	1393.4	12.9	493.5	16.5
Saskatchewan	16 114.2	53.6	2819.5	26.1	709.4	23.7
Alberta	7242.0	24.1	5181.8	48.0	1210.6	40.4
Canada	30 050.1	100.0	10 796.8	100.0	2998.6	100.0

	Canola/Rapeseed 1992 '000t	%	Sunflower Seed 1991 '000t	%	Mustard Seed 1991 '000t	%
Manitoba	986.6	23.7	124.3	93.8	8.9	7.3
Saskatchewan	1746.3	41.9	8.2	6.2	81.7	67.5
Alberta	1383.5	33.2	–	–	30.5	25.2
Canada	4168.6	100.0	132.5	100.0	121.1	100.0

	Rye 1992 '000t	%	Flaxseed 1992 '000t	%	Peas 1991 '000t	%
Manitoba	57.2	24.5	208.3	50.9	84.4	20.6
Saskatchewan	82.5	35.3	165.1	40.4	160.6	39.2
Alberta	67.3	28.8	35.6	8.7	164.7	40.2
Canada	233.9	100.0	409.0	100.0	409.7	100.0

	Cattle 1991 '000 head	%	Tame Hay 1991 '000t	%	Sugar Beet 1991 '000t	%
Manitoba	1100	8.9	3538.0	12.1	453.6	41.8
Saskatchewan	2200	17.8	3719.5	12.7	–	–
Alberta	4403	35.6	8164.7	28.0	631.4	58.2
Canada	12 369	100.0	29 192.4	100.0	1085.0	100.0

Wheat: still number one

Although first planted in 1604 it was not until 1617 that wheat became a significant crop in Canada. However, the establishment of wheat cultivation on the huge scale that made Canada one of the world's major producers awaited the opening up of the western prairies. The railways were the key to development. As the network extended to the west and north more and more land was put under the plough as the agricultural domain spread. By 1876 wheat was being shipped from the prairies to Ontario for seed. Prairie wheat was recognised early on as unusually hard and of outstanding quality for milling. The first wheat shipment from western Canada to Britain, where the expanding population created an ever-rising demand for bread, went via the United States in 1877. It began to be shipped by an all-Canada route in 1884, the year after the Canadian Pacific railroad reached Calgary.

The major problem encountered by the early farmers was the short growing season, which exceeds 120 days only in the extreme south east where initial prairie cultivation was concentrated. The first really successful wheat strain to be produced was Red Fife, which allowed the westward and northward extension of the wheat growing area from the Manitoba lowland. Shortly after the turn of the century the Marquis strain was developed which ripened a week earlier than Red Fife, allowing wheat cultivation to spread further still. Later developments such as Garnett and Reward combined the characteristics of maturing early with resistance to rust (a fungoid growth).

Innovation has gradually changed the agricultural scene on the prairies. The family-sized farm has steadily grown from a few hectares worked by hand to hundreds of hectares easily managed with machines. Further improved varieties of wheat have reduced the growing season requirement by 20 days compared with the early nineteenth century, while progress in transportation and marketing have combined to make the industry highly efficient. Ploughing is now virtually obsolete as it contributes to soil erosion by making the topsoil susceptible to drifting. Cultivators are designed to cut weed roots a few cm below the surface without turning the soil. The seed drill then plants the wheat in combination with fertiliser in rows 15 cm to 18 cm apart.

Wheat streak mosaic disease, caused by a virus, can be controlled by cultural practices but recent research has produced a new variety with resistance to the disease. New varieties resistant to wheat stem sawfly attack are now cultivated on the prairies. The larvae of the fly develop in the hollow stem of wheat plants and later girdle the stems causing the grain-bearing heads to topple to the ground. The new strains have solid stems and greatly reduce the damage that might occur to part of the more than 6 million ha of wheat grown each year.

Transport to markets

As important as increasing efficiency and reliability on the farm is the task of transporting the grain to market. The railroad network is vital in this movement. The first country grain elevator in western Canada was built in Manitoba in 1881. Today there is a network of some 5000 elevators, spread across about 2000 railway shipping points from which wheat and other grains are picked up. A standard elevator is a tall wooden structure with a capacity of around 2500 m³ divided among perhaps 10 separate storage bins (Figure 6.2.4). The marketing

Figure 6.2.4
Grain elevators at Vulcan, Alberta

of prairie wheat, and for that matter barley and oats as well, is controlled by the Canadian Wheat Board which was established in 1935. So that delivery opportunities are allocated fairly among all farmers, a quota system is operated. Each farmer's quota is determined by the area farmed and the crops grown. The farmer may use his quota for all three crops or apply it totally to one grain. The quota is administered through a system of delivery permit books.

The Wheat Board operates a system of price pooling which helps to stabilise prices, as of course does the price support mechanism. At the time of delivery to the elevator the farmer's wheat is weighed, graded and an initial payment is made by the elevator manager. The full price paid by end-customers will not be known until the crop year is over. Then the price received by the Wheat Board for the type and grade of wheat delivered by the farmer is pooled and the farmer receives a final payment to make up the difference between the pool price and the initial payment.

The movement of grain is a precise operation and is organised on a weekly basis by the Wheat Board following a specified routine:

- six weeks before delivery it is decided what type, quality and quantity of grain will be required during the summer shipping season at the terminals in Thunder Bay, Vancouver, Prince Rupert and Churchill (Figure 6.2.5);
- elevator managers report weekly on the stocks they are holding;
- shipping instructions are issued on the basis of loading zones. The prairie grain-growing area has been divided into 48 blocks for this purpose;
- grain cars load the wheat, barley or oats and deliver to the port where the produce is required, to load on ships or to build up inventories in the huge terminal storage elevators.

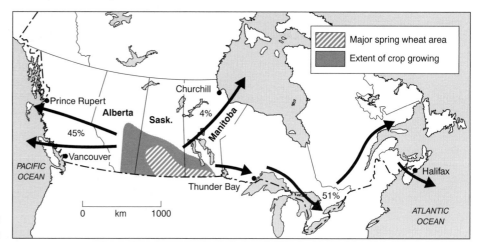

Figure 6.2.5
The prairies: grain export routes

Rural settlement

Rapid change is the hallmark of rural settlement on the Canadian prairies. During the main settlement boom in the early twentieth century, it appeared that the region would become dotted with quarter- and half-section farms from Winnipeg to the edge of the Rocky Mountains and northwards to the heavily forested lands. However, after a relatively short initial period of population increase, which lasted longer among some ethnic groups and in some physical environments than in others, farms became fewer and larger, and rural populations began progressively to decline. Increasingly large-scale machinery, periodic droughts and other natural hazards, a trend to residing off farms, and lack of alternatives to agriculture have all contributed to population decline. Rural population densities have dropped to extremely low levels and it would appear that even lower levels will be reached in the future.

Environmental concerns

In the last 150 years 25 million ha of natural grass cover have been lost to agriculture. As a result, much of the cover necessary for wildlife has disappeared, and as ponds have been drained to make way for farm land, bird nesting sites have vanished too. To make matters worse the use of agricultural chemicals has greatly increased (Figure 6.2.6), contaminating soil, water and air. In recent years many insecticides have been replaced by new 'short-lived' neurotoxic chemicals which affect far more than the intended target.

Figure 6.2.6
Pesticide container dump

It is not just farmers who have suffered the almost inevitable side-effects of such intensive chemical application. Women suffer from higher rates of breast cancer because of the overproduction of oestrogen caused by some compounds. Native tribes such as the Blackfoot, who still hunt and fish extensively, are at an increasing risk of consuming contaminated food. Wildlife has also suffered considerably due to factors such as habitat loss, decline of key prey, and loss of eggs as a result of organochlorine poisoning.

Prairie farmers are now using at least 50 different insecticide and herbicide compounds, each of which can impact all along the food chain. Gulls die from poisoned insects, as do hawks and bald eagles which scavenge birds and mammals killed by pesticides. Products used to limit certain bird populations such as house sparrows and rock doves, poison large birds attracted to their sluggish prey.

The values and attitudes of farmers are only partly responsible for the continuation of this 'poisoning cycle'. Most farmers believe they have to fight pests and weeds with potent chemicals or lose their livelihood, because of intense competition driven by consumer demand for food at the lowest possible price. On the other hand, a small but growing group of organic farmers argue that savings on chemicals compensate for reduced yields. Government policy also plays a big part in maintaining the status quo. Farmers who want to preserve wildlife habitats on their land forfeit tax benefits and government subsidies that encourage the cultivation of land. However, changes in farming methods depend not only on the attitudes of farmers and government but also on the choices made by consumers. People are more conscious than ever before of what they eat and how it is produced, and it is likely that the demand for 'environmentally-friendly' produce will increase considerably in the future.

Soil erosion

In 1986 Alberta classed 3.4 million ha of farmland as being at 'high' or 'very high' risk from wind erosion (Figure 6.2.7) and another 12 million ha as 'vulnerable'. The severe droughts of 1984 and 1985 greatly exacerbated an existing problem of considerable magnitude (Figure 6.2.8). In the worst affected areas of southern Alberta five to six centimetres of topsoil were lost between 1984 and 1986. However, it takes 25 to 30 years for nature to replace 2.5 cm of lost soil.

Figure 6.2.7
Areas in danger of wind erosion in Alberta and Saskatchewan

Figure 6.2.8
Soil erosion in the prairies

Prairie farmers attempt to minimise soil erosion by adopting the following practices:

* crop-fallow rotation – fields are left fallow every other year to replenish soil moisture. The crop stubble is left to further reduce the effect of the wind;
* strip-farming – narrow strips of fallow land, perpendicular to the prevailing wind, are left between seeded fields;
* ripping – caterpillar tractors are used in winter to cut the frozen soil and knurl it into chunks which break the effect of the wind close to the soil surface;
* irrigation – while some of the dry areas of the prairies have good irrigation systems, others are not so fortunate. Dry soil crumbles into fine particles which are easily picked up by the wind.

Exports

Canadian wheat growers have faced some difficult times over the last decade or so. Growing conditions have fluctuated widely, causing considerable variations in production. At the same time the 1980s saw a significant rise in wheat production worldwide, making it harder to sell Canadian wheat. The result was large annual changes in wheat exports which ranged from 12 to 23 million tonnes. In 1991–92 Canada supplied about 16 per cent of the world market, the most important destinations being Asia and Eastern Europe. The USA is Canada's biggest competitor in exporting wheat, causing disputes between the two countries. The subsidy received by US farmers has generally been higher than that accruing to Canadian farmers, allowing the Americans to sell wheat at a cheaper rate to foreign buyers. Canadian farmers see this as deliberate price undercutting.

QUESTIONS

1 Describe the geographical location of the Canadian prairies as an agricultural region.

2 Why has this environment proved so suitable for the cultivation of wheat?

3 Describe and justify the routes used for grain exports (Figure 6.2.5).

4 Discuss the environmental problems related to agriculture in the region.

5 Assess the extent of the soil erosion problem in the prairies. Outline the measures being employed to combat this problem.

6.3 Cotton Production in the Sunbelt

Figure 6.3.1
Cotton growing in Texas

Cotton is the world's major non-food crop, providing approximately half of all textiles. In 1992, 80 countries produced a total of 83 million bales, or approximately 18 billion kg. Most of the 2.25 billion kg that US mills spin and weave into fabric each year, generating revenue of $50 billion, ends up as clothing. Cotton is also used to make bookbindings, fishnets, handbags, coffee filters, tents, curtains and nappies. Oil from the kernels is used in margarine, salad dressings and cooking oils. Meal from the kernels makes fish bait and organic fertiliser and also provides considerable amounts of feed for cattle.

Historical developments

The cultivation of cotton in the USA has undergone a fundamental change in location over the last century. Such a movement illustrates the interaction between physical and economic factors and the increasing role of government intervention.

Cotton has been a cash crop in the USA since the earliest days of settlement. Prior to the Civil War it was the stalwart of the southern economy and because of its heavy labour requirements was a prime factor in the pro-slavery attitudes of the southern states. At this time the south eastern states formed the nucleus of the cotton belt and production was firmly based on the plantation system. However, even in this early period of cotton cultivation, problems began to arise. Cotton, along with the tobacco and maize also grown in the South, was extremely exhausting to the soil. Indiscriminate methods of cultivation led to severe soil erosion and heavily declining yields in a number of areas. Cotton is sown widely apart allowing a relatively free flow of water on sloping land which, in the absence of countervailing measures, can effectively devastate a weakened topsoil in a short period of time.

The period following the Civil War witnessed the wholesale destruction of the traditional system with plantations fragmented into share-crop units. Share-cropping further aggravated the problems of soil erosion as the tenant farmers strove to obtain the maximum output from land which was not theirs, thinking little of the eventual consequences. Declining yields and the abandonment of some areas by cotton growers hastened the movement westward to Texas where large-scale units, on soils which had not suffered from exhaustive cultivation, allowed more profitable farming. The growth of cotton production in Texas, first on the Black Waxy Prairies in the east but later under irrigation on the High Plains, was further encouraged by the ravages of the boll weevil in established cotton growing areas. The boll weevil moved into the US from Mexico in 1862 and wreaked havoc as it spread eastward, reaching the Atlantic coast in 1921. The pest thrives in moist humid conditions and is much less of a problem in the drier west.

Cultivation requirements

The cultivation of cotton, like all crops, is confined by certain physical requirements. Its traditional northern limit broadly follows the 25 °C July isotherm, which approximately corresponds with the extent of the 200 day growing season (Figure 6.3.2). The southern limit of cotton growing is generally governed by autumn rainfall in excess of 250 mm, thus excluding the middle and eastern sections of the Gulf Coast Plain. As cotton tolerates a wide variety of soils, and rainfall is nowhere too great within the limits noted above, its longitudinal extent is ruled primarily by lack of water; this problem has been overcome by extensive irrigation. By the 1920s production on the High Plains slightly exceeded that of the Black Waxy Prairies and was not far behind that of the Mississippi Valley. At this time areas to the west of Texas accounted for less than 1 per cent of total US cotton production. However, the following decades were characterised by a decline in the area planted, relatively stable production and a significant shift in location westwards into California, Arizona and New Mexico.

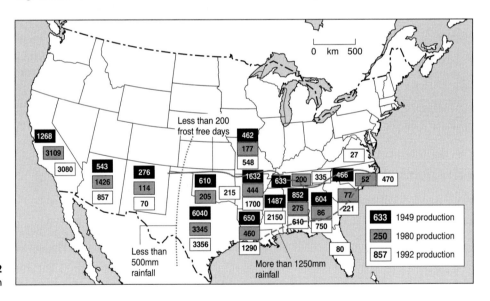

Figure 6.3.2
The physical limits of cotton production

The decline in the cotton producing area can be traced to the years following the First World War when European customers, trying not to increase their existing debt to the USA, sought supply elsewhere. The subsequent fall in prices encouraged US farmers to attempt to safeguard their incomes by increasing production, which had the effect of depressing prices even further. In an attempt to stabilise prices the government restricted the cultivation of cotton to a set area, reviewed each year, which was linked to estimated demand. The total cotton area harvested declined from 15.9 million ha in 1924 to 6.1 million in 1959 and 5.2 million in 1992. However, because technological developments have produced higher-yielding cotton, the level of production remained relatively stable (Figure 6.3.3).

Figure 6.3.3
US cotton production and area harvested, 1949–1992 (1 bale = 225 kg)

	1949	1959	1969	1980	1992
Production (millions of bales)	16.1	14.6	10.0	11.1	16.3
Area harvested (millions of hectares)	11.0	6.1	4.5	5.3	5.2

Along with cheaper production in other countries, intense competition from artificial fibres gave impetus to efforts to achieve greater efficiency in cotton production using large units in the most fertile regions. The areas worst affected by soil erosion, and those where cotton production was no longer economically viable, rapidly abandoned the crop.

Locational change

Production west of Texas, confined mainly to California and Arizona, grew rapidly in the post-war period (Figure 6.3.4). Undoubtedly this western shift of cotton production would have been on a larger scale but for government intervention. Federal price supports allowed many inefficient growers in the south east to remain in business while western growers have had to await the abandonment of acreage allotments held in the south east before expanding their own units. Price supports were fixed in relation to the costs of the smaller-scale units of the south east and thus proved a great incentive to western farmers. However, by 1992 production west of Texas had fallen back to less than 25 per cent of the total, due mainly to significantly increased production in a number of 'traditional' cotton states.

In California, the southern part of the San Joaquin valley is the main cotton growing area and here cotton is the dominant crop. Cultivation depends on irrigation, requiring a large amount of water in addition to normal precipitation (Figure 6.3.5). The low cost of irrigation, an increasingly controversial issue, has been a major impetus to this trend. Cotton is also grown in the Imperial Valley. In Arizona production is concentrated along the Salt River downstream of Roosevelt Dam and along the Gila River downstream of Coolidge Dam. Cotton is also cultivated along the Colorado River downstream of Parker Dam. Yields in the western region are extremely high due to a combination of good soils, ample irrigation water and the intensive methods of farming.

Figure 6.3.4
US: percentage of the cotton crop grown to the west of Texas, 1924–1992

	Percentage
1924	0.8
1934	3.5
1949	12.9
1959	20.3
1969	21.0
1980	41.8
1992	24.6

Figure 6.3.5
Cotton fields being irrigated by sprinklers

Environmental concerns

US cotton farmers are currently spending $500 million a year on pesticides. However, as the use of pesticides has increased, the pests have built up resistance, and farmers have had to spray more often or use stronger chemicals. In California's Imperial Valley, the pink bollworm, a caterpillar that rapidly eats its way through cotton bolls, has caused the area of land planted with cotton to drop from 56 600 ha to only 2800 ha between 1977 and 1994. Numerous chemical sprays proved to be ineffective. To compound the problem, many beneficial insects that otherwise would have helped control harmful bugs were destroyed, leaving crops vulnerable to a whole series of pests. The pink bollworm was finally controlled in the San Joaquin Valley by imposing a 90-day cotton-free period. At the end of each growing season, cotton crops must be completely mowed and ploughed in, leaving no place for the pests to hide.

In California as a whole, some 6000 tonnes of pesticides and defoliants are used on cotton each year. Each autumn during the defoliation period in the cotton fields, people in the San Joaquin Valley complain of nausea, diarrhoea and throat irritation. In neighbouring Arizona the use of pesticides, fertilisers and saline irrigation water has caused a build up of salts in the desert soil, disturbing the natural drainage and destroying the native vegetation.

King once more

COLUMBIA, SOUTH CAROLINA

THUMB through the scores of commercial catalogues stuffed regularly into millions of American mailboxes, and you will see that cotton has made a comeback. The fibre is now highly fashionable, much to the delight of American cotton farmers. Moreover, bad weather and insect infestations have damaged production of the crop overseas. Between them, these factors have bumped cotton prices up to over $1.10 a pound [50 cents a kg], a level not seen since the Civil War.

Such prices are especially welcome in the South, the "land of cotton" until the mid-19th century. After the Civil War, however, yields plummeted as the slaves who had worked the cotton fields were set free. A century later, in the 1960s and 1970s, burrowing boll weevils and competition from synthetic fibres nearly wiped cotton out.

The 1990s, though, have been kinder. Last year, the biggest problem faced by southern cotton farmers was finding enough extra land on which to plant their crops and sufficient equipment to harvest it. Surging worldwide demand in 1994 resulted in an American harvest of 19.7m bales (4.3m tonnes) of cotton, compared with 16.1m bales the previous year. Economists with the National Cotton Council in Memphis suggest several reasons why the market has improved so radically. Heavy advertising over the past few years seems to have paid off. Technology has helped, producing wrinkle-free trousers and easy-care shirts. The trend towards more comfortable leisure wear, the popular explosion of coloured cotton T-shirts, and "casual Fridays" have all made an impact.

Abroad, cotton has always been in demand. Poor harvests in China, India, and Pakistan last year (for reasons ranging from boll-worm infestations and leaf-curl virus to poor weather) meant that American growers had to take up the slack. Eight southern and south-western states set cotton-production records in 1994. Georgia (1.5m bales) and Virginia (79,000 bales) more than doubled their yields from the previous year, while South Carolina (380,000 bales) and North Carolina (820,000 bales) nearly did so.

Record yields have been helped by better technology. South Carolina's previous high production mark was set 77 years ago on 2.8m acres (1.1m hectares). Last year's bumper crop was grown on 223,000 acres. Nonetheless, most cotton-producing states plan to increase their cotton plantings this year.

Southern states have also made an effort to eradicate the boll weevil. The programme, started in North Carolina in the late 1970s and later adopted elsewhere, features chemical sprays and pheromone traps, scent-filled containers that attract the insects and then ensnare them.

Exports are also rising. Although the United States is the world's largest consumer of cotton per head, the Department of Agriculture estimates that 10m bales—more than half the current American crop—will go overseas. This, the highest export level for American cotton since 1980, is yet another reason why southern cotton farmers are smiling.

The Economist, 29 April 1995

Figure 6.3.6
King once more

QUESTIONS

1 Trace the changing location of cotton production since early settlement and explain the reasons for such a shift.

2 What is the main hazard faced by cotton farmers? Why has it proved so difficult to control?

3 How can cotton cultivation and manufacturing damage the environment?

4 To what extent and why has cotton made a comeback in the 1990s?

One response to such problems has been the growth of organic cotton farming, now covering 4000 ha in California. However, because of lower initial yields and higher labour costs, going organic costs more than conventional farming, although this is partially offset by the absence of the costs of chemicals.

Another environmental concern is cotton's manufacturing process, especially dyeing, which is one of the most polluting aspects of the clothing industry. Every year large quantities of dyes, bleaches and heavy metals that are used to fix dye to cotton fabric end up as toxic waste. However, in Texas varieties of naturally coloured cotton have been bred that can be machine spun and woven into fabric. An added bonus is that the colours resist fading. Such cottons save mill owners money while eliminating much of the pollution caused by dyeing. These plants, which now grow on 4000 ha, including farms in Arizona and New Mexico, may also help organic farmers because they derive from seeds with a higher resistance to pests than other strains of cotton.

New markets

In recent years US cotton has undergone somewhat of a renaissance (Figure 6.3.6). Natural fabrics have largely recovered their fashion status in the US while demand from overseas has increased considerably. Rising exports have been accompanied by record yields and new measures to combat the dreaded boll weevil.

6.4   California: an Agricultural Giant

If California were a country in its own right it would be the world's sixth major food producer. It has been the nation's largest agricultural state for nearly 50 years. In 1992 farm income totalled a staggering $18.2 billion. This was $6.6 billion more than Texas, the next biggest farming state in the USA. In that year California's leading farm sectors in order of importance were dairy products, greenhouse products, grapes and cattle. The state's farm assets were valued at almost $70 billion.

A multi-product industry

California produces about half of the USA's fruit and vegetables and almost 15 per cent of its milk. It is the main producer of about 30 crops and the only state producing some of them. Over 200 different crops are cultivated. However, agriculture accounts for only 2 per cent of California's gross state product, an indication of the massive contribution of other sectors of the economy. The 83 000 farms in the state average 149 ha in size. As Figure 6.4.1 shows, agriculture is concentrated in a number of specific areas.

Livestock

Livestock and livestock products make up about 30 per cent of the state's farm income. Milk is the state's leading farm product and only Wisconsin produces more. The greatest concentration of production is in San Bernardino County in the south east. Beef cattle rank second in order of importance, the focal points being Imperial County in the south east corner of the state and the area around Fresno in the Central Valley.

Fruit and nuts

Fruit and nuts account for about a quarter of all farm sales. Grapes rank first in importance, followed by almonds, oranges, and strawberries. The grape crop includes table grapes, wine grapes, and raisin grapes. Ernest Gallo, based at Modesto in the Central Valley, is the world's biggest wine producer with an out-

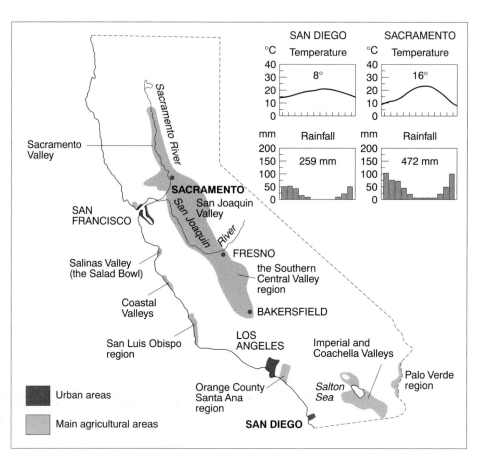

Figure 6.4.1
The main agricultural areas of California

put of more than 2 million bottles a day. California produces almost all of America's almonds, apricots, dates, kiwi fruit, nectarines, olives, pistachios, prunes, and walnuts. California is also the leading state in avocados, grapes, lemons, melons, peaches, pears, plums, and strawberries. Only Florida produces more oranges.

Variations in physical conditions have resulted in considerable regional specialisation within California. For example, the San Joaquin Valley grows almonds, apricots, cantaloupes, grapes, kiwi fruit, nectarines, olives, oranges, peaches, pistachios, plums and walnuts. The Sacramento Valley specialises in honeydew melons, prunes and pears. Southern coastal counties lead in the production of avocados, lemons, oranges and strawberries while the southeastern areas raise dates, grapefruits and melons.

Vegetables

Vegetables contribute around 20 per cent to the state's farm income. Tomatoes and lettuce are the leading vegetable crops and California ranks first for both in the US. The main tomato growing areas are the northern part of the San Joaquin Valley and the southern part of the Sacramento Valley. Lettuce is grown mainly in the southeastern corner of the state and in the regions that lie west of Fresno and southeast of Monterey. Broccoli and potatoes rank next in importance and considerable quantities of asparagus, carrots, cauliflower, celery, mushrooms, onions, and peppers are also cultivated.

Field crops

Field crops account for approximately 15 per cent of farm revenue, with cotton in the leading position. The San Joaquin Valley is one of the world's major cotton-growing regions and in the USA only Texas grows more. The other leading field crops are hay, rice, corn, sugar beet and wheat.

Greenhouse products

Greenhouse and nursery products contribute about 10 per cent to farm income. California is the largest US producer of cut flowers, potted flowering plants, ornamental shrubs and trees, and flower seeds and bulbs as well as indoor foliage plants and seedlings. Most greenhouse and nursery products are grown in coastal areas from San Francisco to San Diego.

Locational advantages

California's farm wealth is due to:

- large areas of fertile soils, particularly in the Central, Imperial and Coachella valleys and the coastal valleys;
- a sophisticated irrigation system, which brings water to the drier southern parts of the state from northern California, the Sierra Nevada Mountains, and the Colorado River. Irrigation is heavily subsidised by the federal government;
- the Mediterranean-type climate which favours a wide variety of crops: sub-tropical fruits can be grown if frosts are guarded against and temperate crops can be cultivated if water supply is ensured. This large, physically varied state has many different sub-climates which have helped to form distinct farming areas;
- a large 'home' market. With over 31 million people California is the most populous state in the USA;
- excellent processing and distribution systems, promoting rapid marketing and export of farm produce;
- an extremely high level of investment. Large corporations own many farms, operating with state-of-the-art technology;
- advanced agricultural research by state universities and farm companies has steadily raised output.

America's Salad Bowl

One of California's many areas of specialised farming is known as America's 'Salad Bowl'. It is the coastal valley area centred on the town of Salinas and extending into the counties of Monterey, San Mateo and Santa Cruz. Lettuce and strawberries are the main crops but others include broccoli, sprouts, cauliflowers, celery, spinach, radishes, tomatoes and artichokes. The Salad Bowl has rich alluvial soils and its climate is ideal for its main crops. It is protected by coastal mountains and is close to the Pacific Ocean. These two factors give rise to mild-to-warm equable temperatures all year round. Rainfall is seasonable with 300–500 mm between November and April and very little in the summer. Wells, some more than 200 m deep, provide an alternative water supply.

Strawberries are grown on absolutely flat furrow-irrigated fields which must be fumigated to give a good quality crop. Other vegetables grown in rotation benefit greatly from this. Much of the production is on a share-cropping basis with small growers tending units of 1–1.5 ha. The owner (usually a large corporation) provides housing, plants, fertiliser and other inputs. The share-cropper provides the labour and profits are divided 60 : 40 in favour of the owner. Within half-an-hour of picking, the crop is in cooled stores, remaining cool until reaching retail outlets, maybe in Britain three days later.

Lettuce occupies the largest area in the Salad Bowl with the most modern harvesters spanning 30 rows. The crop is cut, film-wrapped and boxed in the field. It is then sped quickly to markets across America and beyond. Much of the labour for lettuces and other crops comes from Mexican and Filipino workers.

The highly-intensive farming of the Salad Bowl requires an extremely large investment per hectare but the profits achieved make it worthwhile. Unlike some of the USA's major crops the products of this region are not in surplus, but enjoy a ready demand both in the domestic and foreign markets.

Exports

In 1994 California's agricultural exports reached an all time high of $11.8 billion. Only five nations actually achieved a higher value of exports than this. With the implementation of the North American Free Trade Agreement in January 1994, exports to Mexico jumped $1.03 billion, an unprecedented 50 per cent increase over the prior year. Exports to Japan were down 5 per cent from 1993; however, Japan continues to be by far the largest single market for California's agricultural products. California exports 32 per cent of its total farm production to overseas markets. It is estimated that for every $1 billion in agricultural export sales, 27 000 jobs are created. Thus the sector is a vital part of the state's economy and the mainstay in many rural areas. The leading four export products in 1994 were beef products ($988.8 million), cotton lint ($978.9 million), almonds ($717.9 million), and grapes ($589.9 million).

Leading issues in Californian agriculture

Water supply and prices

The fall in the water table, on which at least half of the state's farmland depends for irrigation, is almost certain to increase irrigation costs in the future. The state government is under pressure from other interest groups to reduce water subsidies to farmers. Most of California's crops, including pasture and hay, are irrigated. Without its water infrastructure the state's agricultural system would be dominated by ranching. California has by far the largest irrigated area of any state, accounting for about one-sixth of the national total.

Though California's farmers are proud of receiving a relatively small share of direct crop subsidies compared with the grain producers in the Midwestern states, they are still the biggest beneficiaries of federally-supplied water, which is not included in the calculation of agricultural subsidies. California's 83 000 farms consume 85 per cent of the state's water even though they produce only 2 per cent of the state's gross output. And they still get most of this water at government-subsidised rates from federal and state irrigation projects; in the case of federally supplied water, for as little as 10 per cent of the price most urban and industrial users pay. Growing the state's $550 million alfalfa crop alone consumes enough water for 41 million city-dwellers a year. California's rice and alfalfa fields have at times been cited as one of the world's great agricultural absurdities because of the huge amounts of irrigation water required.

In early 1992 the irrigation waters of the Central Valley Project (CVP) were turned off for the first time in its 40 year history to save the chinook salmon. The fish spawn just below the Shasta Dam at the top of the Sacramento River. However, their young die if the water gets too warm, which will happen if the water level behind the dam drops too low during summer. The Federal Bureau of Reclamation which runs the CVP had been threatened with lawsuits under the Endangered Species Act if it allowed the water to drop to a critical level.

However, some farmers are now making significant efforts to reduce consumption. PureHarvest Corporation, a privately owned firm based in Napa, has devised a way of growing rice in the Sacramento Valley that uses 25 per cent less water, no herbicides and pesticides, and some innovative machinery, which it licenses out to farmers. Yields are 20 per cent lower than those of conventional methods but this is balanced by lower input costs. The trend in California must increasingly be to higher value crops which are less thirsty.

Urban sprawl

Suburban expansion has consistently eaten into farmland with losses totalling thousands of hectares a year. For example, before the Second World War agriculture was the biggest industry in Los Angeles County but today it hardly exists. The citrus grove has almost disappeared from the southern Californian landscape and it is now hard to see how Orange County came by its name. In an attempt to limit the impact of this problem the California Land Conservation Act was passed in 1965. Under the Act landowners and city or county governments join in a voluntary contract under which both parties give up certain benefits in return for other advantages. Landowners forgo the possibility of development on their land for the duration of the contract. In return their property tax assessment is related to the income-producing ability of the land, leading to a considerably lower tax bill than otherwise. One survey concluded that the Act offered farmers an average 83 per cent per hectare reduction in property taxes compared with taxes based on the current fair market value. In 1990 the agricultural area covered by the terms of the Act had stabilised at approximately six million hectares.

In recent years the transfer of land from agricultural to urban has been aided by the 1993 Millar-Bradley law which allows farmers to sell water rights to cities. As urbanisation increases the farmers' political base declines. Central Valley cities such as Stockton, Fresno, Sacramento and Bakersfield are growing quickly; by 2000 Fresno is expected to be bigger than San Francisco.

Labour

Most of California's speciality crops were originally cultivated and harvested by hand, using a large migrant labour force. The workers start in the south and move north to harvest the temperate crops which ripen later, then return to the far south for winter-harvesting work. After long enduring poor pay and conditions migrant workers became increasingly unionised and militant in the 1960s and 1970s. At its peak in the mid-1970s, the United Farm Workers (UFW) Union had 100 000 members in California. Effective action included campaigns to persuade the public not to buy particular Californian farm products. But as the state closed farm labour camps and imposed licensing on employers the UFW's power eroded. Ground down by hostile growers and unsympathetic Republican governors, the union also lost support by ostracising illegal farm workers as strike-breakers. Today its membership has shrunk to a few thousand in the state and its influence to marginal campaigns. At the same time, tighter regulations have persuaded growers to shift employee responsibilities to labour contractors, thus avoiding liability. In 1992, contractors employed over half the migrant workforce. In an intensely competitive labour market, wages have been forced down to the state minimum in recent years. In 1993 it was estimated that half of California's one million agricultural workers were illegal immigrants, mainly from central and northern Mexico. Conditions for such workers are getting tougher as employers look for more and more cost reductions.

Natural hazards, disease and pests

The state's farm sector is subject to serious disruption from a range of physical hazards. Earthquakes, fire-storms, floods and drought are all quite frequent occurrences. Between late-1989 and early 1995 natural and human disasters cost California more than $32 billion. 'We've conquered every challenge that man or Mother Nature could throw our way', Governor Pete Wilson said in January 1995, in his State of the State address, citing the Loma Prieta earthquake in 1989, the Los Angeles riots in 1992, wildfires in 1993 and, the Los Angeles earthquake in January 1994. However, more was to come, with major storms following later that month and in March. The latter was one of the most devastating storms to hit California this century, with farm losses estimated to be in excess of $300 million.

In spite of high inputs of herbicides and pesticides, disease and pests can still have a devastating effect on California's hi-tech farming system. In recent years the state's vineyards have been particularly hard hit (Figure 6.4.2). Although the industry has suffered other hazards such as drought and soil erosion, the phylloxera epidemic has been by far the greatest setback to the industry since the Prohibition era. The necessity of replanting vineyards ridden with the deadly phylloxera has, however, provided growers with the opportunity to switch to new popular varieties.

Vintage wine bug that spoiled the California dream

TONY Truchard, owner of 170 acres of vineyards in Napa Valley, likens phylloxera, a microscopic yellow insect wreaking havoc on northern California's premium wine country, to a cancer.

Standing under a blazing hot sun among his Chardonnay crop, Mr Truchard points out the affected vines. To the untrained eye they do not seem so different from their healthy counterparts. On closer examination, the bushes are not so tall or luxuriant, and they bear fewer grapes – the tell-tale signs that something is amiss. Eventually these vines will turn into stunted shoots and shrivelled leaves as the louse-like aphid devours the roots.

"The key is how fast the phylloxera will spread," said Mr Truchard, who detected two "weak spots" while inspecting vines from his pick-up truck a few weeks ago.

As the wine harvest begins in Napa Valley about now, the full impact of the phylloxera infestation is beginning to dawn on the area.

Two years ago, experts were predicting that replanting diseased vines in the Napa Valley alone could cost $250 million (£130 million). Now they have doubled that figure for the Napa region and are talking about replacing 150,000 acres of vines in Napa and Sonoma counties, about 60 miles north of San Francisco, at a total cost of $1 billion.

The phylloxera epidemic could hardly have come at a worse time for California's wine industry, already reeling from the recession that has turned the California dream into a nightmare for more and more people.

Patches of dying vines were first noticed in the Napa Valley, and to a lesser degree in Sonoma County, in the 1980s. Now phylloxera has become the biggest threat to California's vineyards since Prohibition. Should the infestation spread elsewhere the cost could shoot higher; already traces of phylloxera are in Washington State's vineyards to the north.

The Guardian
17 August 1992

Figure 6.4.2
Vintage wine bug that spoiled the California dream

The Mediterranean fruit fly has also caused increasing concern in recent years. Back in 1989, the state tried spraying malathion from the air, but this caused outrage among people who did not like the idea of being hit by a chemical whose effects on humans were insufficiently known. Now target-specific ground action only is employed, with sterilisation one of the main methods employed. It is yet to be seen how effective this campaign will be in the long term. Even in the world's most high-tech agricultural system the fight against crop and livestock disease is an ongoing one.

Biotechnology

Bigger, more efficient farmers have been steadily buying out their smaller neighbours. These larger enterprises are increasingly turning to biotechnology to boost yields and so reduce disease among crops. This promises to help them cut production costs and so to sharpen their competitive edge. Biotechnology is tipped to replace microelectronics as the key industry of the next century in the state. California's farmers will increasingly rely on it to increase the production, quality, range and value of their operations.

QUESTIONS

1 (a) Identify the main agricultural regions in California.
(b) With the help of an atlas and any other relevant resources explain why large parts of the state are unproductive agriculturally.

2 Why is the value and range of farm products grown in California greater than that of any other state in the USA?

3 Assess the importance of the foreign export market to the state's farmers?

4 Explain the reasons for the broad rural/urban divide that often occurs over environmental issues.

Environmental problems

Increasing concern about the impact of farming on the environment has affected almost every type of agricultural enterprise in the state. For example rice growers are worried about a number of issues under debate which are critical to production and costs (Figure 6.4.3). As export opportunities show increasing promise rice growers fear that more stringent restrictions will prevent them maximising their potential.

The environmental debate is to a considerable extent a rural/urban conflict. Many farmers see popular environmentalism as mainly an urban phenomenon fuelled by a very limited understanding of the realities of agriculture as an industry.

Stars-and-stripes sushi

WHILE President Clinton in Seattle was talking of improvement in trade with Asia, 14,700 tonnes of rice were being loaded into a Japanese ship docked in the channel at Sacramento in California. Four more shiploads, another 53,500 tonnes, are scheduled for December. For Japan is short of rice, and the result is the first large export of American rice to Japan (except for one brief emergency in 1984) since the 1960s.

Japanese officials insist that only a temporary crisis caused by bad weather led to these shipments. But most American growers and millers think they have opened the door, and mean to make sure it does not shut again. Mike Espy, the secretary of agriculture, says shipments to Japan next year may run as high as 500,000 tonnes.

Only 3–4% of world rice production—roughly 14m tonnes—goes for export. America's share of world production is barely 2%, yet it ships out about 18% of all the rice in international trade, a proportion second only to Thailand's. American rice goes to many Asian countries, the Middle East, Africa and South America.

California produces just the sort of rice—a small or medium-sized glutinous grain, called japonica—that is favoured in Japan. The second most common form of rice, the drier long-grain or indica, is found in the tropics and in the American south.

California, which is mostly desert, seems an odd place to grow rice. At times of severe drought—like last year—rice growers come in for criticism, since their crop appears to rely on a lot of water. They reply that the soil in most rice-growing areas in California is hard clay, which could not be used for other crops. The rice industry also insists that its run-off water is profitably used later on. And yield per acre is high.

Environmental disputes dog the growers. Restrictions have already been placed on the annual post-season burning of rice-straw. Now there is talk of trying to save salmon stocks by increasing water flow in the rivers that drain into the San Francisco delta, thus diminishing their use for irrigation. Growers are also worried about the transfer of some agricultural water to the cities. The biggest rice-producing area, in the Sacramento valley, gets free-flowing river water at a very low price. The same water might be sold to southern Californian cities for three or four times as much.

On the MidAmerica Commodity Exchange in Chicago the rice-futures market has erupted in recent days, with a record number of traders in the market. On one day last week the price rose about 80% above last spring's futures. Much of this excitement is based on the supposition that a rice trade is beginning between America and Japan. It is still too early to be sure of that.

The Economist
27 November 1993

Figure 6.4.3
Stars-and-stripes sushi

7

Water Resources: Supply and Distribution Issues

7.1 Continental Overview

Regional supply imbalances

Water – that most basic necessity of life – has long been a dominant factor in establishing patterns of settlement and economic activity in North America. While increasing concern has been expressed about water quality in recent decades, an even greater worry over the sheer availability of water has also developed.

While there is no general water shortage in the continent as a whole, imbalances between supply and demand have long existed within both Canada and the United States. In the former, the major hydrological dichotomy is north–south: 90 per cent of the population is concentrated within 200 km of the southern boundary with the USA, but 60 per cent of the total available runoff is carried northward by rivers draining into the Arctic Ocean (Figure 7.1.1). In addition, over a third of Canada's mean annual precipitation occurs as snow, most of which accumulates over several months before the spring melt. However, except in specific localities such as the western Prairie region where low precipitation combines with relatively high evaporation rates, there is little likelihood of widespread water shortages in Canada. After all, the average annual discharge of all Canadian rivers is about nine per cent of the world's renewable water supply while Canada has less than 0.5 per cent of the world's population.

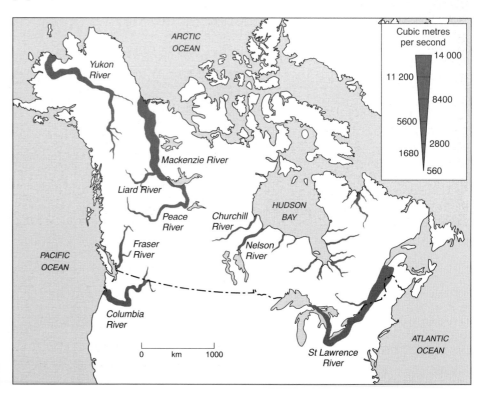

Figure 7.1.1
River discharge in Canada

In the United States the major dichotomy is east–west: the western states contain nearly 40 per cent of the total population and 60 per cent of the national land area but receive only 25 per cent of mean annual precipitation (Figure 7.1.3). Indeed, almost all the land lying west of 98° West comprises a dry zone (less than 500 mm a year) extending westward almost to the mountains along the Pacific. Undoubtedly, the most seriously affected region is the Southwest, and in particular southern California, where Los Angeles, the USA's second largest city, has mushroomed in a desert area.

Figure 7.1.2
Oroville dam, California

As the western states expanded their demands for water – first for agriculture and more recently for rapidly growing cities and industry – so it became increasingly clear that local supplies alone would be inadequate to meet predicted shortfalls. In California a network of canals and aqueducts has been developed during the twentieth century to carry water from the water-abundant north of the state to the water-deficient south. These north–south transfers have been supplemented by supplies drawn from the Colorado River (Figure 7.1.4), but as yet no artificial transfer of water occurs across any state boundary in the US.

Figure 7.1.3
USA: regional contrasts in water supply and usage

% US total*	WEST†	Water Supply (million litres)	EAST	%US total*
51.2	777 799	Total fresh water	738 053	48.8
45.0	546 932	Surface water (rivers, lakes)	667 510	55.0
76.6	230 867	Ground water (aquifers)	70 543	23.4
		Water Usage (million litres)		
94.4	485 556	Irrigation agriculture	28 917	5.6
44.4	68 974	Domestic/Commercial	86 512	55.6
35.6	36 311	Manufacturing industry	65 552	64.4
23.6	170 108	Energy (thermoelectric)	550 912	76.4
73.2	16 850	Other uses	6160	26.8
59.6	5462	Land area (1000 km²)	3702	40.4
39.9	104 773	Population (1000) (1995)	157 982	60.1

*Excluding Alaska and Hawaii.   †States west of Mississippi River (excluding Alaska and Hawaii)

QUESTIONS

1 (a) Explain why, in much of Canada, the supply of water may be extremely variable through the year.
  (b) In which season is the western Prairie region most likely to experience drought conditions, and why?

2 With reference to Figure 7.1.3, write an explanatory account of the principal regional contrasts in the USA's (i) water supply; (ii) water usage.

Figure 7.1.4
California: major water transfer schemes

The NAWAPA scheme

Since the 1960s, several major proposals have been put forward to transfer water from regions of excess capacity to southern California and the water-deficient areas of neighbouring states. Some of these inter-state and international schemes are indicated on Figure 7.1.5. The most ambitious of these proposals is undoubtedly the North America Water and Power Alliance (NAWAPA), which emanated from engineering consultants based, unsurprisingly, in Los Angeles.

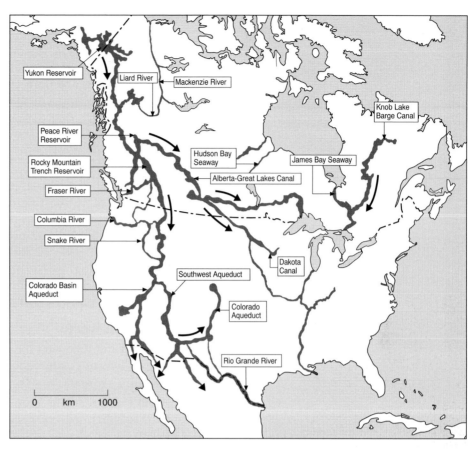

Figure 7.1.5
NAWAPA diversion scheme

Figure 7.1.6
The Rocky Mountain Trench

NAWAPA proposed a series of inter-basin diversions extending from the Canadian Arctic to Mexico in the south and to the Great Lakes in the east. A complicated system of dams, tunnels, canals and reservoirs (including the flooding of the Rocky Mountain Trench (Figure 7.1.6) to store more than 600 trillion m³ of water – enough to supply a city the size of London for 300 years) would take 30 years to build but would deliver the following benefits:

*   14 million ha of additional irrigated land;
*   70 million kW of additional power;
*   45 trillion m³ of additional water for municipal and industrial uses;
*   water to flush out the heavily polluted Great Lakes;
*   a navigable waterway linking the Great Lakes with Vancouver;
*   jobs for more than four million construction workers;
*   increases in agricultural and industrial production amounting to perhaps $100 trillion a year.

Despite these potential benefits, a number of factors have combined to ensure that the NAWAPA scheme remains on the drawing board:

*   political objections by Canadians, who regard water – though renewable – as different from other resources. It is seen as part of Canada's national heritage and therefore not for sale to the USA;
*   fear of an environmental disaster if any of the proposed half-dozen dams impounding the huge Rocky Mountain Trench reservoir collapsed (the Trench is located in an active earthquake zone);
*   large-scale diversions of fresh water away from the Arctic Ocean could cause marine ecological imbalances and trigger a major increase in snowfall across North America;
*   other climatic and ecological changes would result from the increased evapotranspiration and precipitation around the major reservoirs;

QUESTIONS

1 The completed NAWAPA scheme would produce 70 million kW of additional electrical power. How would this be generated?

2 Explain the environmental concerns that have arisen over the NAWAPA proposals.

3 Referring to examples of completed dam-building and river-diversion schemes elsewhere in the world, explain why the impact of NAWAPA on local communities is regarded as likely to be adverse.

4 (a) Discuss the pros and cons of the NAWAPA proposals from the viewpoint of (i) a southern Californian; (ii) a western Canadian.
  (b) Outline and justify your own assessment of NAWAPA.

- the impact of NAWAPA on human activities in the areas affected by dam-building and river diversion would almost certainly be adverse;
- the financial costs of NAWAPA would be astronomical, especially for Canada, while the economic benefits would be very long-term, rather uncertain, and would accrue mainly to the USA;
- NAWAPA's successful implementation would require the co-operation of three countries, seven provinces and 33 states. The chances of reaching agreement across such a wide and varied spectrum of government interests seem remote to say the least.

Since Canadians appear adamantly opposed to NAWAPA, and the US government has never officially supported it, the huge pan-continental scheme seems unlikely to be adopted in the foreseeable future. Yet the demand for water exports from Canada's rivers seems almost certain to become more vociferous as shortages in the south western United States and Mexico become more acute. Agreed transfers – if any – will probably be on a much smaller scale than those called for in NAWAPA.

Alternative sources and strategies

Reducing present waste

Implementation of the following measures would conserve huge quantities of water in those regions of greatest shortage and would obviate, or at least defer, the need for costly new water transfer schemes like NAWAPA.

- Plugging leaks and reducing evaporation losses from the canals and aqueducts of the existing delivery systems would save up to 25 per cent of all water moved.
- Recycling water in industry (where it takes 225 000 litres to make one tonne of steel, for example) could represent huge savings.
- Similarly, recycling of municipal 'waste stream' (sewage) for watering lawns, gardens and golf courses could be implemented or extended, as Los Angeles has already shown. Technically, waste water can be further purified until it is suitable for human consumption – though an obvious psychological barrier remains to be overcome!
- Introducing more efficient toilet systems, which use only 6.5 litres of water for each flush instead of the conventional 26 litres, can make welcome savings, especially in large urban areas where water is most expensive.
- Charging more realistic prices for irrigation water. Many farmers pay only one-tenth of the true cost of water pumped to them; the rest is subsidised by the federal government. When long-term water contracts are eventually renewed, prices could be raised to more economic levels. After all, farmers usually take the lion's share of all water available in arid areas but generate relatively little local wealth: in California the figures are 80 per cent and three per cent respectively.
- Adopting drip irrigation systems, which allocate specific quantities of water to individual plants, and which are 100 times more efficient than the open-ditch system still used by many farmers; or sprinkler systems, which are up to ten times more efficient than open-ditch irrigation.
- Changing from highly water-dependent crops such as rice and alfalfa (for growing which farmers often receive large direct subsidies from the federal government) to those needing less water. Calculating the water needed to irrigate the crops fed to cattle, one steak can represent over 13 000 litres, while one bushel (37.8 litres) of wheat takes over 50 000 litres; and it takes around 450 litres of water to bring a single egg to the American table.
- Changing the law to permit farmers to sell surplus water to the highest bidders. Since 1992, this has been allowed in California, where an emerging net-

work of specialist brokers sells 'agricultural' water to cities for less than they already pay but at a profit for the farmers.

- Requiring both cities and rural areas to identify the source of water to be used before new developments can commence. This sensible proposal, first mooted in southern California in 1994, has been shelved as politically unacceptable.

Exploring future options

Such options include developing ground water resources. In some areas these remain virtually untapped because, historically, supplies of surface water from rivers and lakes have been more easily developed.

Secondly, it has been claimed that various techniques of weather modification – especially cloud-seeding – can provide water at a unit cost of little more than two per cent of that from some of the water transfer schemes.

A third option is the desalination of sea water. Although extremely expensive, providing additional fresh water supplies from the ocean is not only technically feasible but has become a practical reality. In 1991, after several years of drought, the city of Santa Barbara in southern California approved the construction of a $37.4 million desalination plant scheduled to provide 9.1 billion litres of water for Santa Barbara and neighbouring communities, essentially for domestic and culinary purposes. The high costs of desalination will, however, make it prohibitively expensive for irrigation well into the foreseeable future.

A fourth option could be to exploit the frozen reserves of Antarctic water. Serious proposals have been made to find a 100 million-tonne iceberg off Antarctica (Antarctic bergs are broad, flat and better suited to towing than Arctic ones which tend to be pyramidal and irregular), wrap it in sailcloth or thick plastic, and haul it to southern California (Figure 7.1.7). There, Los Angeles could derive about 75 billion litres of water (over five per cent of the city's annual consumption) from the iceberg. Potential problems include not only the huge cost (perhaps $100 million) but the loss through melting during the year-long journey. Optimistic estimates range from a mere five per cent loss up to 20 per cent, but one cold-regions research scientist has warned the iceberg-moving enthusiasts 'Once you get north of the equator, you'll have nothing but a rope at the end of your tow'. Even if the scheme proves practicable,

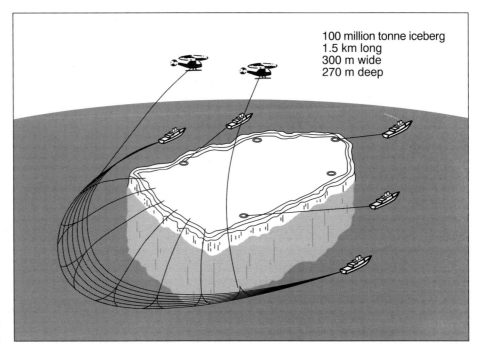

100 million tonne iceberg
1.5 km long
300 m wide
270 m deep

Figure 7.1.7
How an iceberg might be moved

another question would arise: what would be the environmental effects of anchoring such a huge block of ice off an arid coast? So far, precise answers are lacking.

There is now general agreement that planning for the future water supply of the Southwest should embrace all practicable options. Some proposals may still seem rather far-fetched, but sensible management of this priceless resource should rule out no feasible strategy if this important region is to sustain its economic viability and growing population.

QUESTIONS

1 (a) Why is evaporation a major cause of losses from existing water delivery systems in the Southwest of the USA?
   (b) What could be done to reduce evaporation loss?

2 Why is cloud-seeding, to increase precipitation in water-deficient areas, not universally regarded as beneficial?

3 The introduction of water meters could reduce consumption by up to 55 per cent. Argue the case for and against metering.

4 Discuss the contention that the South West region of the United States is not so much short of water, only of cheap water.

5 With the aid of a diagram, indicate the possible environmental effects of anchoring an Antarctic iceberg off the coast of southern California.

7.2 The Colorado: Western River Drained Dry

In 1912 Joseph B Lippincott, an official seeking future water supplies for the growing city of Los Angeles, described the Colorado as 'an American Nile awaiting regulation'. Since then the Colorado has become one of the world's most extensively controlled rivers. Scores of dams, reservoirs, pumping stations and HEP plants together with thousands of kilometres of canals, aqueducts and tunnels have diverted so much of its flow that the river reaches the sea only as a small trickle. The Colorado has been 'regulated' almost out of existence.

In its original state, the Colorado was a formidable river. Rising 4250 metres up in the Rocky Mountains of Colorado, it flowed some 2300 km to the Gulf of California. Its drainage basin is larger than France and includes parts of seven states in the most arid region of North America. The Colorado is unique in that no other river in the world has cut so many deeply-incised trenches, of which the Grand Canyon is the largest and most spectacular. Downstream the river forms the international boundary for about 30 km, then its remaining waters struggle across 130 km of Mexico to the Gulf.

The Colorado was the first river system in which the concept of multiple use of water was attempted by the Bureau of Reclamation. In 1922 the Colorado River Compact divided the seven states of the basin into two groups: Upper Basin and Lower Basin (Figure 7.2.1). Each group of states was allocated 7.5 million acre-feet (9.25 trillion litres) of water annually, while a 1944 treaty guaranteed a further 1.5 million acre-feet (1.85 trillion litres) to Mexico. Completed in 1936, the 221-metre high Hoover Dam and Lake Mead, which was created from impounded water (Figure 7.2.2), marked the beginning of the era of artificial control of the Colorado. Glen Canyon Dam and Lake Powell, Parker Dam and Lake Havasu, and Flaming Gorge Dam and Lake are among the most notable of several major units replicating the Hoover Dam's functions of flood and silt control, power, irrigation, domestic and industrial water supplies, and recreation. All these schemes together comprise the Bureau of Reclamation's Colorado River Storage Project.

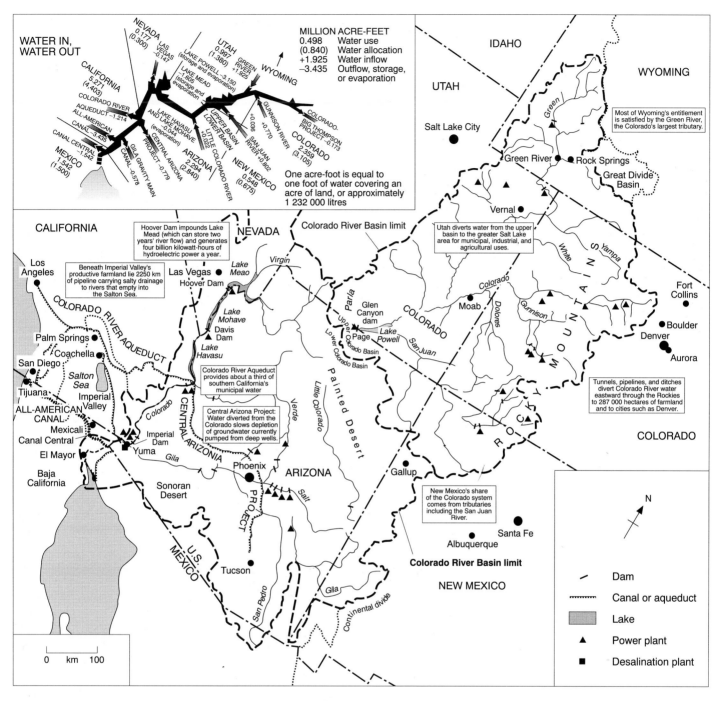

WATER IN, WATER OUT

MILLION ACRE-FEET	
0.498	Water use
(0.840)	Water allocation
+1.925	Water inflow
−3.435	Outflow, storage, or evaporation

One acre-foot is equal to one foot of water covering an acre of land, or approximately 1 232 000 litres

Hoover Dam impounds Lake Mead (which can store two years' river flow) and generates four billion kilowatt-hours of hydroelectric power a year.

Beneath Imperial Valley's productive farmland lie 2250 km of pipeline carrying salty drainage to rivers that empty into the Salton Sea.

Colorado River Aqueduct provides about a third of southern California's municipal water

Central Arizona Project: Water diverted from the Colorado slows depletion of groundwater currently pumped from deep wells.

Colorado River Basin limit

Utah diverts water from the upper basin to the greater Salt Lake area for municipal, industrial, and agricultural uses.

Most of Wyoming's entitlement is satisfied by the Green River, the Colorado's largest tributary.

Tunnels, pipelines, and ditches divert Colorado River water eastward through the Rockies to 287 000 hectares of farmland and to cities such as Denver.

New Mexico's share of the Colorado system comes from tributaries including the San Juan River.

Colorado River Basin limit

Legend	
Dam	
Canal or aqueduct	
Lake	
Power plant	
Desalination plant	

0   km   100

Figure 7.2.1
The Colorado River Storage Project

Despite the inter-state and international agreements, the Colorado has become the most contested and debated river in the world. One major natural problem is that while the river was committed to deliver 16.5 million acre-feet (20.35 trillion litres) every year, its annual flow has averaged only 14 million acre-feet (17.25 trillion litres) since 1930, evaporation from artificial lakes and reservoirs removing another 2 million acre-feet (2.45 trillion litres); in the drought years of the late 1980s–early 1990s, this shortfall was accentuated.

A second, human, problem is that the population and urban/industrial development of the seven Compact states has grown extremely rapidly in recent decades. In 1970–1990 for example, the population of these states expanded from 22.8 million to 36.1 million, an increase of 58 per cent compared with 18 per cent for the rest of the USA. Consequently the demands placed on the Colorado are beginning to exceed anything envisaged in the 1922 Compact. The

Figure 7.2.2
The Hoover Dam and Lake Mead

river now sustains around 25 million people and 820 000 ha of irrigated farmland in the USA and Mexico.

Obviously, the supply of Colorado water is limited but demands for it seem set to continue to grow. As long as some states use less than their allocation, others can continue to take more. But as populations rise and states in both Upper and Lower basins complete irrigation and other water projects, conflict among users is bound to result. For example, until 1990, a drought year, the Metropolitan Water District (MWD) of Southern California's Colorado River Aqueduct abstracted nearly four billion litres a day from Lake Havasu for use in the Los Angeles region – twice the MWD's official allocation. But when the MWD requested extra supplies, the Bureau of Reclamation refused.

Figure 7.2.3
The Central Arizona Project, north west of
Phoenix

The reason was that Arizona had opened the first phase of the Central Arizona Project (CAP) (Figure 7.2.3), causing the Lower Basin to use up its full share of Colorado water for the first time. Since the CAP was completed in 1992, 1.5 million acre-feet (1.85 trillion litres) of water a year has been distributed to farms, Indian reservations, industries and fast-growing towns and cities along its 570 km route between Lake Havasu and Tucson. However, in this hydrological equation, providing more water for Arizona will mean that less is available for California.

Thus the Colorado River Compact no longer seems suited to modern circumstances. A new strategy needs to be implemented – in demand as much as in supply – before the Colorado dries up completely.

QUESTIONS

1 With reference to Figure 7.2.1, exemplify the benefits of the Bureau of Reclamation's multi-purpose dams along the Colorado River and its tributaries.

2 Identify and discuss at least two physical problems that result from disrupting the river's natural processes by building dams.

3 In order to pump water along the Central Arizona Project to Tucson and subsidise its cost for irrigation farmers, the Bureau of Reclamation helped build a coal-fired power station near Page, Arizona (Figure 7.2.1). Suggest why the construction of the CAP in general and the power station in particular have been criticised by environmentalists as 'the ultimate in desert folly'.

4 (a) Tabulate the state and Mexican data in the 'Water in, water out 1990' diagram (Figure 7.2.1), using the headings Water allocation, Water use and Surplus (+)/Deficit (−).

   (b) Calculate the net surplus/deficit for the Upper and Lower basins, the seven states combined, Mexico, and the whole Colorado Basin.

   (c) Explain (i) the regional variations in surplus and deficit; (ii) why the apparent overall surplus is actually a deficit.

5 Suggest how a new 'demand-side' strategy might solve the problem of allocating the finite supply of Colorado water.

7.3 Lake Erie: Reviving a Dead Lake

Figure 7.3.1
Lake Erie industrial pollution before the clean-up campaign

Although the Great Lakes comprise the world's largest body of fresh water, their drainage basin is disproportionately small. A glance at an atlas map will show that none of the major rivers of the continental interior flows into the Great Lakes, though several watersheds lie close to their southern shores – a result of the glacial action that shaped the lakes and the surrounding terrain (see Section 1.3). In fact, the Great Lakes themselves contain nearly all the water in the drainage basin, the tributary rivers being comparatively tiny. Thus, the lakes' natural flushing system is not very efficient.

This fact would be of little consequence if the Great Lakes basin did not contain almost one-fifth of the entire US population and one-half that of Canada, with especially heavy concentrations of people and industry along the lakes' southern shores. Until the 1970s it was standard practice to discharge sewage and domestic waste directly into the lakes, together with waste-water from steel and paper mills, and from chemical and metal-refining works (Figure 7.3.1). To all this were added the outwashings of artificial fertilisers, insecticides and herbicides from streams draining Corn Belt farms, and the discharge of engine oils, bilge-water and other wastes from vessels navigating the St Lawrence Seaway–Great Lakes system. Small wonder that water pollution became a major problem throughout much of the Great Lakes basin, seriously affecting the water supply of millions of people, their recreational facilities, and once-prosperous commercial fisheries.

The worst situation occurred in and around Lake Erie, which is unique among the Great Lakes for a number of reasons:

- it is the shallowest lake and relatively warmer than the others; consequently it is (without pollution) one of the world's most productive freshwater ecosystems;
- it is smaller by volume than all the other Great Lakes (though Lake Ontario is smaller in area);
- population density in the Lake Erie basin is easily the highest among the Great Lakes;
- the use of Lake Erie water for power generation, and domestic, manufacturing and commercial purposes exceeds that of any other lake; only Lake Michigan comes close.

By the 1970s Lake Erie pollution was so critical that the bottom of a sampling bucket, left too long in the water, became completely corroded and dropped out! This and similar dramatic incidents brought matters to a head, triggering in 1972 the US government's Clean Water Act and the US–Canadian Great Lakes Water Quality Agreement. In the following account, the problems of Lake Erie and the effects of the 1972 legislation are reviewed from the perspective of Ohio, the state occupying most of the lake's southern shore and most affected by pollution problems.

Figure 7.3.2

Lake Erie: Ohio water usage, 1994 (million litres/day)

Power stations (fossil fuel)	9000
Power stations (nuclear)	550
Domestic water supply	1790
Private industrial	194
Other	4
Total	11 538

QUESTIONS

1 Explain why, by 1972, Lake Erie had become the most polluted of all the Great lakes.

2 (a) Convert the statistics in Figure 7.3.2 to an appropriate diagram.
   (b) Account for the preponderant use of Lake Erie's water by power stations (both nuclear and fossil fuel).
   (c) Which category of user has the greatest need for good-quality water, and which the least need? Explain why.

The impact of the Clean Water Act

In 1970 Lake Erie was declared dead from serious pollution problems that threatened its productivity and value as a natural resource. The Lake's critical condition was, to a great extent, responsible for the public outcry that resulted in the passage of the Clean Water Act in 1972. The Act provided for the development of clean-up objectives, prompt action, treatment technologies, and, most importantly, the funding to implement the proposed programme.

The four major categories of pollutants that contributed to the Lake's failing health, their impacts, and the improvements brought about by the Clean Water Act may be summarised as follows:

1. *Nutrients*

Nutrients including phosphorus and various forms of nitrogen, support the phytoplankton community, the primary link in the Lake Erie food chain, but excessive amounts can wreak havoc on the whole system. Phosphorus was identified as the nutrient of greatest concern in Lake Erie in 1970.

Before 1972, most sewage treatment plants and septic systems did not remove phosphorus which is found in detergents. Consequently effluent concentrations of phosphorus averaged a very high 7 mg/l (milligrams per litre). Runoff from farms was another major source of phosphorus. The Detroit River, which supplies approximately 90 per cent of the flow to Lake Erie, was the largest source of phosphorus, followed closely by the Maumee River. Total external phosphorus loads to the lake peaked at 28 000 tonnes in 1968.

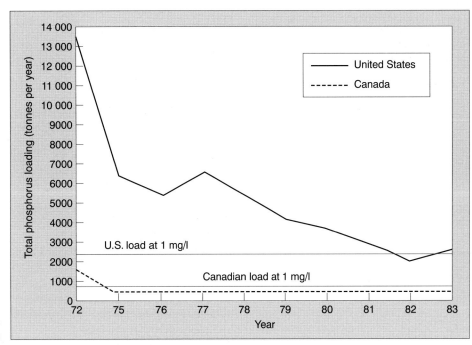

Figure 7.3.3
Lake Erie: municipal phosphorus loadings

The effects of phosphorus on Lake Erie were visible and very disturbing. Massive algal blooms, composed mostly of nuisance blue-green algae, became commonplace. During the summer months, waters in sheltered bays and the near shore often resembled pea soup. Shorelines were rimmed with what appeared to be aqua paint.

Floating rafts of scum were found even in the open lake. Long filamentous algae called *Cladophora* attached to nearly every available hard surface and washed ashore in thick mats along with numerous dead fish. The visual impact and the smell of this decaying mass kept many beaches deserted. Taste and odour problems plagued shoreline communities who drew their water supplies from the lake.

Tonnes of decaying algae raining down on the lake floor totally depleted the bottom waters of oxygen, killing aerobic macroinvertebrates and driving desirable fish species from the lake. Under such anoxic conditions, certain pollutants in the lakefloor sediment are regenerated back into the water above; there was now an internal, largely uncontrollable source of phosphorus within the lake itself (Figures 7.3.3 and 7.3.4).

In 1972, the USA and Canada signed the Great Lakes Water Quality Agreement, which was the impetus for reducing Lake Erie phosphorus levels.

The Clean Water Act provided funding for constructing phosphorus removal facilities at municipal wastewater treatment plants and implementing programmes to reduce runoff from agricultural land. In 1988, Ohio passed a law banning the use of phosphorus in laundry detergents throughout the Lake Erie basin.

Today, phosphorus loading from Ohio's major wastewater treatment plants is 0.5 mg/l, the level established by the Great Lakes Water Quality Agreement. Throughout the Lake Erie Basin, county committees are working to meet the phosphorus reduction goals set by the state government and farmers, voluntarily participating in phosphorus reduction activities.

It is now rare to see an algal bloom on Lake Erie. The area of anoxia occurring on the lake bottom has been reduced, and total annual phosphorus loading is close to the values established in the Great Lakes Water Quality Agreement.

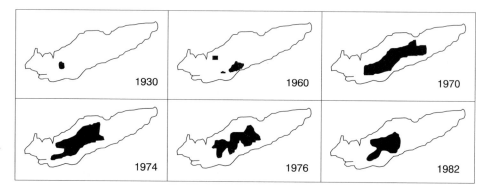

Figure 7.3.4
Lake Erie: anoxic bottom waters in the Central Basin, 1930–1982

2. Sediment

As the Lake Erie Basin became increasingly populated, wetlands were drained and cultivated, thus eliminating a natural filter which prevented sediment from entering the lake. Soil erosion increased as fields were ploughed right to the edge of streams and rivers, many of which were dredged and channelised to prevent flooding. This, in turn, provided faster flows and a more direct route for sediment to reach the lake.

Increased sedimentation covered invertebrate communities in rivers and the lake, suffocating fish eggs and choking spawning grounds. The increased turbidity destroyed beds of aquatic plants by restricting light penetration, and eliminated less tolerant, clean-water fish species.

Much of the sediment washing into the lake consisted of fine silts and clays, the type of particles to which pollutants adhere. Thus, in addition to the obvious visual impacts, the increased sediment loading also meant increased loads of nutrients and toxic chemicals.

3. Bacteria

The original sewage systems of the cities expanding along the Lake Erie shoreline, were often little more than a series of sewers draining wastes away from populated areas and converging to empty into the rivers and lake at one common point. Sewers were generally constructed to contain both sewage and stormwater runoff. Up until the early 1970s, most sewage treatment plants had only primary treatment, which did little more than remove solids before discharging wastes to the nearest body of water. The discharge contained extremely high concentrations of bacteria, with potential for spreading disease and infection. Overloaded sewer systems designed to discharge only during heavy rain, often overflowed continuously, even in dry weather. Raw sewage floating in rivers and along the lake shore was a common

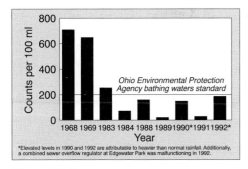

Figure 7.3.5
Edgewater Beach recreation area, Lake Erie: bacteria counts, 1968–1992

sight. Most of Lake Erie's beaches were closed as health hazards during the 1960s (Figure 7.3.5).

The Clean Water Act required all sewage treatment plants to upgrade to at least secondary treatment. The construction of retention basins and interceptors and the separation of sewers from stormwater outlets have reduced the discharge from combined sewer overflows. However, many areas still experience high bacteria counts following heavy rainfall.

4. Toxics

The presence of toxic chemicals in Lake Erie was less obvious than the visual impacts resulting from phosphorus enrichment and sediment loading, but their effects were farther reaching, longer lasting, and more devastating to the ecosystem. The unrestricted point source discharge of metals and organic chemicals, and the non-point source runoff of pesticides allowed these toxics to contaminate sediment and accumulate in fish and wildlife. Some of these pollutants biomagnify as they move up the food chain; e.g. a seagull feeding on many slightly contaminated fish eventually will become highly contaminated.

High loads of mercury from chlor-alkali plants on Lake St Clair and the Detroit River resulted in a ban on commercial fishing for walleye and white bass in Lake Erie in the early 1970s. High concentrations of the pesticides DDT, its by-products (e.g. DDE) and dieldrin caused eggshell thinning and reproductive problems for herring gulls and bald eagles, resulting in declining populations.

All of the harbours at the mouths of the major Ohio tributaries contain heavily polluted sediments. The deep, dredged channels became traps for pollutants moving down the rivers. Discharges from oil refineries, steel mills, and chemical works caused foul-smelling waters and floating patches of oil and grease. Oil-soaked debris, accumulating along the bends of the Cuyahoga River, ignited several times. A large fire in 1969 branded the river as a national symbol of environmental neglect and helped trigger the first serious programmes to clean up the Great Lakes.

The Clean Water Act required industrial companies and others to obtain a permit to discharge into US waters. The act also required pre-treatment of industrial waste to restrict the flow of toxics to municipal wastewater treatment plants. Together, these measures have helped considerably to reduce levels of toxicity in Lake Erie.

In addition the Great Lakes Critical Programme Act of 1990, established the Great Lakes Initiative to develop water quality criteria common to all the Great Lakes to protect human health, aquatic life, and wildlife. As part of the initiative in Ohio, Remedial Action Plan groups are developing programmes to restore water quality in the state's four Areas of Concern (Figure 7.3.6).

All of the above efforts have helped to improve the Lake Erie ecosystem. The Cuyahoga River has been dramatically transformed from a debris-choked sink to a 'public playground'. Aquatic life has returned to all of the harbours, and the nearly exterminated bald eagle population has made a progressively successful comeback.

However, the lake still has many problems: contaminated sediments remain a source of pollution; anoxia persists in the central basin; nitrate concentrations are increasing; some fish remain on the threatened species list; and many shoreline habitats and buffering wetlands have been destroyed forever.

It is clear that the Clean Water Act alone is not sufficient to restore the water quality in Lake Erie; all the environmental laws passed at the state and federal levels must be brought to bear. Moreover, pressure can no longer be directed at individual facilities, individual actions, and individual sources: the focus must shift to a holistic approach. It remains to be seen whether the sharing of scientific data and coordination across a broad spectrum of both public and private sectors will ensure the continuation of improvements to Lake Erie.

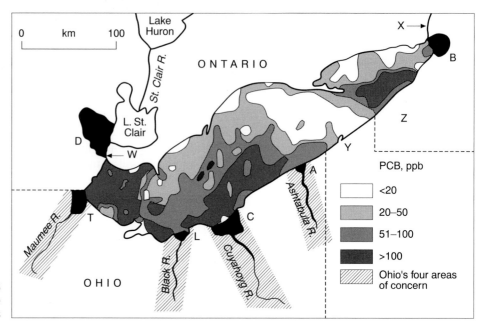

Figure 7.3.6
Lake Erie: surface toxic chemical levels in the early 1970s

1 (a) With reference to Figure 7.3.3, account for the different US and Canadian loadings of municipal phosphorus.
   (b) What is the other main source of phosphorus effluent?

2 (a) What do you understand by the term 'anoxia'?
   (b) What were its causes and consequences in Lake Erie?

3 How were the improvements shown in Figures 7.3.3 and 7.3.4 achieved?

4 (a) Explain in your own words the causes and effects of increased sediment flow in Lake Erie.
   (b) Why are many of Ohio's phosphorus control programmes aimed at decreasing the runoff of sediment?

5 Distinguish between and exemplify a 'point source' and a 'non-point source' of toxic pollution.

6 (a) With reference to Figure 7.3.6, identify the rivers W and X, the states Y and Z, and the industrial towns and cities A, C, D, L and T.
   (b) Explain the pattern of heaviest toxic chemical pollution in Lake Erie in the early 1970s.

8

Other Renewable Resources and their Management

8.1 Fishing: Overproduction and Resource Management

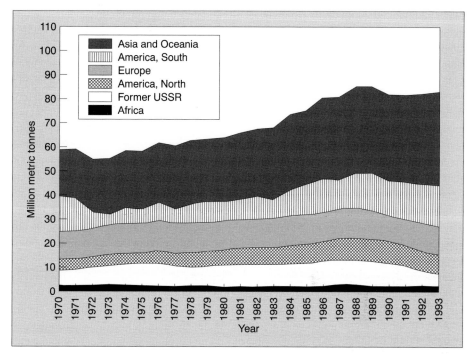

Figure 8.1.1
World fish catches in marine waters

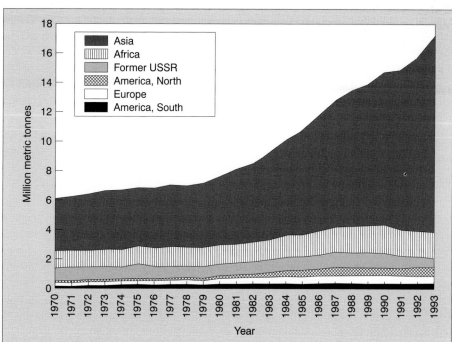

Figure 8.1.2
World fish catches in inland waters

By the mid-1970s the North American fishing industry was in trouble, mainly because of over-fishing. Over the years more and more foreign fishing fleets had come to the rich fishing grounds close to the Atlantic and Pacific coasts. More fish were caught, but there is an upper limit to the exploitation of this valuable resource. Not enough fish were being left to reproduce and eventually stocks began to fall. However, declining stocks and falling catches were trends not confined to North American waters but were characteristic of a much wider global fishing problem.

North American fishing grounds

US commercial fishing is concentrated in three fishing areas (Figure 8.1.3) but the Pacific Northeast is by far the most important, accounting for 50 per cent of total US catch. The only other fishing grounds of real significance for US fishing are the Atlantic Northwest and the Atlantic Western Central. However Canadian fishing is even more concentrated, with the Atlantic Northwest accounting for 71 per cent of the total catch. Inland waters, whilst locally significant, are of limited general importance, supplying 5.8 per cent of US and 3.1 per cent of Canadian production. Overall the USA and Canada rank 5th and 18th in the world fishing league table.

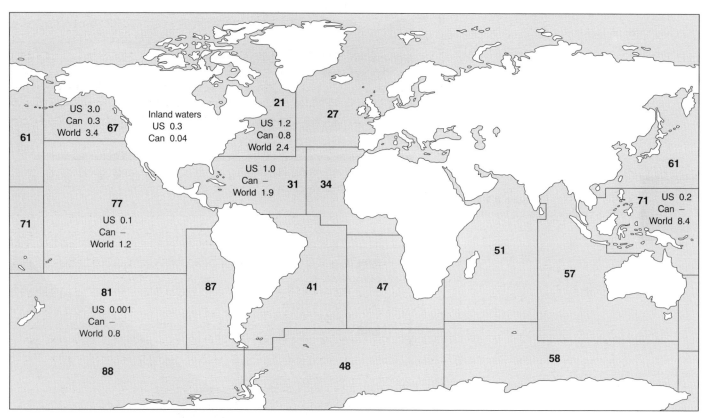

Figure 8.1.3
US and Canadian catches in tonnes in international fishing zones and world totals for those zones, 1993

200-mile limits

Until the 1970s, most of the sea outside a narrow coastal strip was regarded as international waters. In 1972, Iceland became the first country to claim an extended fisheries limit of 50 miles (93 km), which was increased to 200 miles (372 km) in 1975. In the latter part of the 1970s the 200-mile fishing limit became accepted in global practice and thus in 1982 the UN Convention on the Law of the Sea established a 200-mile limit as the norm for all coastal countries.

The contention that open-access exploitation of a common property fish stock attracts excessive effort leading to depletion of stocks became widely accepted. The objective of, and the main justification for, increased jurisdiction was that by limiting access and managing effort wisely, coastal states could halt over-exploitation and regenerate net economic benefits. The 200-mile limit encompasses most of the world's continental shelves and slopes so that the important demersal stocks that are tied to these shallower areas mostly come under coastal state management. Areas within 200-mile limits are often fished by foreign fleets which pay substantial licensing fees for the privilege. Nearly 90 per cent of the global marine catch takes place within these limits.

The Canadian government in 1976 developed a comprehensive policy document concerning resource management and in the following year declared its own 200-mile limit. The country's most extensive fisheries, in the north west Atlantic, had long been in a generally depressed state because of resource depletion resulting from a considerable increase in foreign fishing. Restrictions on foreign operations were partially balanced by quota concessions.

The USA adopted similar management policies in its Fishery Conservation Zone (FCZ), although it must be noted that the USA and Canada have come into conflict in the delimitation of boundaries. The main dispute concerned where the line should be drawn off Maine, New Brunswick and Nova Scotia, with the rich Georges Bank at issue.

American fishery managers estimate that the US catch is almost half as valuable as it could be if fish stocks were allowed to recover. In 1975 the Alaskan fleet enjoyed a season for Pacific halibut lasting 120 days. The fleet can now take the year's entire catch in one or two 24-hour 'derbys'. If fishing went on longer there would be too few halibut to spawn future catches. The Alaskan herring-roe fishery is open for a mere 40 minutes a year. Boats queue to sell their catch to processors, who gut and freeze the year's supply as quickly as they can. Overall more than 40 per cent of the fish populations in US waters are considered overfished.

Conflict in the north west Atlantic

Since 200-mile limits were recognised in 1982, every valuable fish that straddles or migrates across these borders has been hit hard by foreign fleets. Not surprisingly the once-rich fishing grounds off Canada's Atlantic coast have become one of the focal points of such intense activity. Fierce competition has led to violent confrontations between fishermen in many parts of the world.

Newfoundlanders have depended on the annual influx of cod for their livelihood for 400 years but recent decades have proved to be troublesome times for the numerous small fishing communities in the province. The economic decline of Newfoundland cannot be attributed to fishing alone but the problems of the Atlantic ground fisheries have produced cumulative detrimental effects. Until the mid 1950s only Canada, the USA and five or six west European countries fished off Canada's Atlantic coast. However, fishing intensified continuously from the late 1950s onwards. The groundfish catch more than doubled from 1.26 million tonnes in 1951 to a peak of 2.83 million tonnes in 1965, but by 1980 it was down to 1.21 million tonnes. In the latter year the cod catch alone, which accounted for 50 per cent by value of all Newfoundland landing, was down to less than a quarter of what it was in the early 1960s.

The decline was reversed for a while by the introduction of the 200-mile fishing limit, but increased activity focusing on stocks straddling the limit has brought the industry to the edge of catastrophe in recent years. The cod stock declined to such an extent that in July 1992 Canada's Department of Fisheries and Oceans declared a two-year moratorium on its Newfoundland and Labrador cod industry, with the loss of 40 000 jobs, to prevent stocks of Atlantic cod from being totally fished out. EU fishermen, operating just beyond the 200-mile limit, had contributed to a total haul of cod, redfish, flounder and plaice which was 16 times the quota recommended by the regulatory authorities. In September 1993, the ban was extended to the whole of Eastern Canada, and in spring 1994 all recreational fishing for cod was forbidden.

It is clear that the moratorium will have to stay in place for the forseeable future. No one knows for sure what caused the collapse but constant overfishing appears to be the most obvious factor. Small-time inshore fishermen, who often use hook and line, blame the offshore trawlers whose nets drag along the bottom and pick up fish they don't want as well as those they do. The offshore fishermen counter that they often have observers aboard to make sure that they do not exceed their quotas, unlike the thousands of little boats fishing from tiny, isolated harbours. For years Canada has accused European boats, especially Spanish and Portuguese, of catching too many fish. The Europeans argue that Canada has also been guilty of overfishing. After declaring its 200-mile limit in 1977, the Canadian government subsidised new boats and fish-processing plants against scientific advice. Other factors put forward as exacerbating the situation are the increasing seal population, colder water than usual and higher salinity.

The result has been extremely costly both for Newfoundland and Canada as a whole. Under a special scheme, 20 000 Newfoundlanders made jobless by the cod moratorium are drawing an average of C$335 a week from the government in compensation for their lost earnings. This is in the province with the lowest per capita income and the highest rate of unemployment in Canada. In 1994 a former Premier of the province declared that Newfoundland had 250 000 too many people to support itself. In some fishing communities unemployment has reached 80 per cent. Many villages may eventually die out, virtually eliminating the island's traditional culture.

Concerned that the situation was continuing to deteriorate, the federal parliament unanimously approved the introduction of legislation in May 1994, enabling Canada to take action to protect important fish stocks on the high seas that straddle Canada's 200-mile limit. However, it was clearly only a matter of time before this action would bring Canada into dispute with other fishing nations (Figure 8.1.4).

Canada takes firm line on EU overfishing

A Spanish trawler was arrested last week and taken to St John's, Newfoundland, where the captain will face charges under Canadian fisheries conservation laws. When the holds of the Estai were inspected 80% of the catch was found to contain under-sized fish.

A proposal put forward by Prime Minister Chrétien for a 60-day moratorium on turbot fishing has been ignored by the European Union. Their ships, mainly Spanish and Portuguese, continue to reject turbot quotas set by the North-west Atlantic Fisheries Organization and, after initially pulling back, some trawlers returned to the Grand Banks.

Fisheries Minister Brian Tobin says everytime Europe ignores a quota, the type of fish being caught has been decimated. Mr Tobin says it's time for that to stop.

Mr Tobin also noted that 39 Spanish vessels just outside the 200-mile zone have already taken about seven-thousand tonnes of turbot. That's almost double the quotas set earlier for the 15-country European Union.

Portugal condemned Canada's ultimatum to stop fishing for turbot just outside the 200-mile limit. Portugal's maritime affairs minister called Canada's actions 'deplorable' and a violation of international law.

Mr Tobin said patrol vessels are ready to make more arrests if necessary. Canada's chief negotiator at the United Nations talks to regulate international fishing, Paul Lapointe, said the dispute need never have reached a boiling point but, under the circumstances, most coastal nations would accept emergency action.

Canada Focus
17 March 1995

Figure 8.1.4
Canada takes firm line on EU overfishing

After tough negotiations Canada and the European Union reached a conservation agreement in April 1995 that was hailed in Ottawa as a model for saving endangered fish stocks around the world. The accord ended one of the most serious confrontations between North Atlantic Treaty allies seen in recent years. It commits the 15 EU-member countries to Canadian demands for tougher enforcement as well as strong conservation measures. In return, Canada gave up 5000 tonnes of its quota of turbot (also known as Greenland halibut). The agreement was not just about providing immediate protection for turbot stocks, but also about rebuilding cod and flatfish stocks already under moratoria. Major components of the accord include:

- independent, full-time observers on board vessels at all times;
- enhanced surveillance via satellite tracking;
- increased inspections and quick reporting of infractions;
- verification of gear and catch records;
- timely and significant penalties to deter violations;
- new minimum fish-size limits;
- improved dockside monitoring.

Fishing law is complex, confused and generally inadequate to meet the needs of world fishing in the 1990s and beyond. Hopefully agreements such as that between Canada and the EU will set a model of 'best practice' to be followed on a global basis in the future.

Development and pollution

Overfishing is not the only threat to the world's fisheries, although it is the most severe. Development and pollution are also reducing stocks. According to Paul Brouha, director of the American Fisheries Society, 11–15 million salmon once spawned in the Columbia River system. Now there are only 3 million, of which 2.75 million come from hatcheries. So much of the river system has been dammed that only 250 000 salmon can find their way back to old spawning grounds (Figure 8.1.5).

According to a recent study, three-quarters of the entire American fish catch comprises species that depend upon estuaries (often as a habitat for juveniles, which can safely feed in the shallows). But estuaries have attracted heavy settlement and considerable industrial location. Agricultural chemicals also pour into them, and the ecosystems of some lagoons, wetlands and mangrove swamps have been altered beyond recognition.

Figure 8.1.5
West Coast salmon: belly up

West Coast salmon Belly up

SEATTLE

EACH year, millions of salmon from the Columbia river in Washington state make a long, strange trip. Newly hatched fish, called smolts, are loaded on huge barges and sent downriver to the ocean. In this way, they bypass a series of huge hydro-electric dams where spinning turbines kill 80% or more of the fish trying to pass through them.

This elaborate piscine transport system has only one flaw. It doesn't work. Despite the barges, and despite a string of fish hatcheries along the Columbia, the salmon that once swarmed up and down the river are fading fast. A century ago, perhaps as many as 16m salmon migrated upriver to spawn each year. Photographs from that era show boatmen heaving salmon the size of men's legs into huge, shimmering piles outside the canneries that dotted the coast. Today, barely 2m fish make the same journey. Worse, a scant quarter-million or so of those are descendants of the once-mighty native runs; the rest are hatchery-raised fish.

In recent years, fish

1 Describe the changes illustrated by Figures 8.1.1 and 8.1.2.

2 Summarise and explain the spatial location of Canadian and US fishing.

3 Briefly review the history of 200-mile limits.

4 (a) Why did the dispute between Canada and the EU arise?
   (b) To what extent can the agreement reached be regarded as a model for future resource management?

5 Exemplify the impact of development and pollution on fish stocks in North American waters.

stocks up and down the coast have plunged to 5% or less of past numbers. One study completed during the past year suggests that as many as 214 west coast salmon runs are in serious trouble. The figures are causing panic. Salmon, even more than timber, was the first source of wealth in the north-west. It was a vital part of native American culture, helped to draw in the first white settlers, and remains both a big industry and source of keen recreational pleasure.

There are, alas, many causes to point fingers at. On the Columbia and several other rivers, dams have done their worst. Their turbines chew up young fish because some dams were built with no provision at all for allowing spawning salmon past. The day the Hell's Canyon Dam on the Snake river in Idaho was completed, for instance, salmon simply ceased to exist upstream.

The barging operations, started in the 1970s, are thought to be failing because crowded fish are more vulnerable to disease, or because their unusual mode of reaching the ocean interferes with the homing instincts they need to find their way back. Hatchery-raised fish, besides, are almost universally held in contempt. They are too stupid to avoid predators, too weak to survive the

rigours of their ocean journeys, and too prone to disease. None of which stops them from interbreeding with wild stocks, apparently passing on their bad traits and acquiring few good ones.

Elsewhere, poor logging practices have allowed timber workers, until recently, to cut trees right to the edge of spawning streams, filling rivers with silt and exposing them to sunlight that warms the water to temperatures salmon cannot tolerate. Farms dump animal waste and fertiliser into salmon streams. Developers channel fish-producing streams into concrete culverts. And a gauntlet of hooks and nets awaits any survivors.

Even allowing for all this, no one knows for sure what is wrong. Salmon populations fluctuate wildly at the best of times, perhaps by 50% or more from year to year. On the Columbia river Cecil Andrus, the governor of Idaho, and several environmental groups have suggested increasing water-flow over the river's dams during the months when young fish migrate downstream. The hope is that faster water will carry the fish to the ocean more quickly and in better health.

The idea is bitterly opposed. Aluminium-smelters, for example, need cheap electricity; they would face higher

rates if dam operators allowed water to flow rather than storing it to generate power. Farmers fear that emptier reservoirs would leave irrigation pumps dry, and inland ports worry that fewer ships and barges would be able to operate. In short, the idea of increasing dam flow is so controversial that any tests have been delayed until 1996.

The best answer may, in time, be the least dramatic: repairing hundreds of small streams along the west coast so that fish find them hospitable again. In Oregon, for instance, Knowles Creek on the southern coast once saw 250,000 migrating fish along its 11-mile reach. But old-fashioned timber practices stripped the creek of gravel beds in which salmon lay their eggs, and now only a few hundred fish find their way back home each year.

A green group, the Oregon Rivers Council, aims to restore things by painstakingly adding logs and boulders to the stream and restoring the streamside vegetation that cools and filters the water. In this way, Knowles Creek may become what it once was. But it will take 50 years. How many salmon will be coming that way by then?

The Economist
15 January 1994

8.2 Forest Management: A Key Environmental Issue

Concern about the felling of trees is not just a recent phenomenon. Plato wept when he looked at the bare hills above Athens caused by the expanding city's insatiable demand for timber. However, as demand for forest products has heavily increased in recent decades, there has been a growing fear that the situation has reached a critical stage. The 1992 Earth Summit in Rio de Janeiro brought the earth's forested areas to the forefront of international concern. Forestry is now one of the key battlegrounds between industry and environmentalists.

North America is sometimes referred to as the 'timber-basket' of the world. The continent has huge forest resources, an important source of wealth. It is one of only two areas of the world which have softwood surpluses, the other being Asiatic Russia. The USA and Canada are first and third in the world for the amount of timber cut each year. Canada is the world's largest exporter of forest products with the USA ranking second.

Forestry in Canada

Canada's forests represent about 10 per cent of the world's total forest area. Despite earlier heavy harvesting, the mainly coniferous forest still covers 45 per cent of the country's total land area. In 1991 the inventoried forest area was estimated at 416 million hectares (Figure 8.2.1). A little more than half are considered capable of producing timber. The forest industry is concentrated in British Columbia, Quebec, and Ontario. Many communities in these provinces, and not an insignificant number elsewhere rely on the industry to a considerable extent.

Figure 8.2.1
Canada: area covered by forests

		million hectares
Heritage forests (protected from harvesting by legislation)		22.8
Commercial forests (capable of producing timber and non-timber products)		237.1
• managed forests (currently managed for timber production)	118.9	
• unallocated forests (currently unallocated and unaccessed)	90.7	
• protection forests (unavailable for harvesting by policy)	27.5	
Open forests (small trees, shrubs and muskegs)		156.2
Total forest land		416.2

Trees are being harvested at a rate of around one million ha a year. In recent years the government and the forest products companies have replanted about half of the harvested area with tree seeds or seedlings. However, not all replanted areas regenerate successfully into forest. Replanting and regeneration is important not just for environmental reasons but also to maintain the supply of commercial timber. Thus it is vital to know the amount of timber that can be harvested each year without diminishing the long-term sustainability of the forest. This amount is known as the annual allowable cut, which for Canada is presently 252 million m³. The total harvest in 1992 was 163.8 million m³, mostly taken by clearcutting with an average area of about 60 hectares. Although recent studies show that the national timber supply is sustainable over the next 30 to 60 years, regional supplies vary significantly and some local shortages are apparent.

Model forests

Since Canada's Model Forests Programme was announced in September 1991 a number of new approaches to sustainable forest management have been developed and tested. The 10 model forests (Figure 8.2.2) range in size from 100 000 to 1 500 000 ha, with strikingly different physical, social, environmental and economic conditions. The sites are managed by partnerships involving industry, environmental and conservation groups, aboriginal communities, educational groups, private landowners, outdoor recreation clubs, and all levels of government.

Examples of projects undertaken in the model forests are:

• a study of the northern flying squirrels and pileated woodpeckers in the Fundy Model Forest to determine if they would be useful as indicator species whose numbers could signal the health of old-growth forest ecosystems;

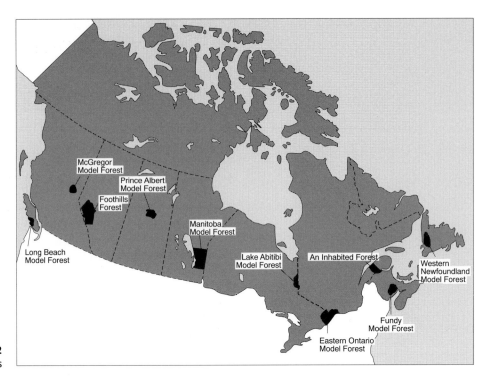

Figure 8.2.2
Canada's model forests

- the First Nations in the Eastern Ontario Model Forest the participation of in a scheme to reestablish and sustainably manage the black ash forests that were the foundation of their traditional basket-making industry;
- research into the seasonal habitat requirements of the elk and caribou in the Foothills Forest, using radio collars attached to the animals. Computerised forest models are being developed to establish the timing and location of logging operations, and to ensure that the animals have enough habitat for their needs throughout the year.

The attitudes of environmental groups

Canadians today are more aware of threats to the environment than at any time in the past and in recent years membership of environmental groups has risen dramatically. The general view of environmentalists is that government action is 'too little, too late'. Considerable concern has been expressed about both large and small scale trends. At the large scale there is growing awareness of the role of northern forests in the carbon cycle (Figure 8.2.3) and with increasing concern about tropical forests the major forestry companies are looking more intensively at the great northern forests. Northern industrial countries have made tropical timber producers agree to sell timber only from sustainably managed forests, in a new International Tropical Timber Agreement. But the northern countries are not keen to sign an agreement which imposes similar restrictions on their own forests. Environmental groups view this stance as gross hypocrisy.

Environmentalists continually stress the importance of biodiversity, arguing that the world's forests are an ecological mosaic, an intricate and dynamic web of plants, animals, insects, fungi and micro-organisms constituting the most complex terrestrial ecosystems on earth. Biological diversity occurs at four different levels: genetic, species, ecosystem and landscape diversity.

Where has all the carbon gone?

FORESTS of conifers cover 1.2 billion hectares of Alaska, Canada, Scandinavia and Russia. This is the great boreal belt of northern woodland, which represents a quarter of the world's forests. Boreal forest takes up more land than any other type of biological community, and for the most part it has been little affected by human activity. Now scientists claim that the boreal forest, and the smaller belt of temperate forest below it, is the key to one of the great mysteries of climate research.

When fossil fuels are burnt they release CO_2 into the atmosphere, where it can influence the world's climate. Most of this CO_2 remains in the air or is absorbed by the sea. But fossil fuels produce more CO_2 than can be accounted for in this way, a discrepancy that has puzzled scientists for decades. CO_2 in the atmosphere has risen from a fairly constant 280 parts per million in the preindustrial period before 1700, to 355 parts per million today. Substantial though this increase is, the amount of CO_2 produced by burning fossil fuels should have resulted in a much more marked increase in atmospheric CO_2, even when increased absorption by the sea is taken into account.

Where has the missing carbon gone? Into the northern forests, says Allan Auclair, a plant ecologist working with the environmental consultancy Science and Policy Associates in Washington DC. Forests absorb carbon by growing and release it when they rot or burn. Auclair's findings predict the presence of a carbon sink in the northern hemisphere to explain the missing carbon.

Before 1890, boreal forests were a source of CO_2, mainly because of forest fires and tree feeling. Deforestation releases CO_2 largely because debris from the forest floor rots more quickly when the trees are cleared. After 1920, a steep increase in tree growth outstripped the losses due to fire and the felling of trees, turning the northern forests from a carbon source into a carbon sink and storing CO_2 from fossil fuel burning for the next fifty years.

But in the late 1970s, the boreal and temperate forests began to lose wood rapidly and once again became a source of CO_2. "If the previous trend had continued, the forest could have absorbed another 15 billion tonnes of carbon between 1976 and the present," says Auclair. "Instead the carbon has stayed in the atmosphere, while the forest is releasing the carbon it absorbed earlier this century." Despite rapid growth in individual trees, massive increases in the amount of timber harvested and the number of forest fires and pests together removed more wood than was added.

In 1990 the Intergovernmental Panel on Climate Change identified positive feedback as the biggest unknown factor in predicting climate change. If Auclair has spotted the first signs of feedback, then changes in the northern forests could have a serious effect on the global climate.

And the northern forests are changing. "World forestry is about to move north," says Bryant. Increasing concern among northern consumers about tropical deforestation and wood depletion in temperate woodlands is already limiting timber production in the south, he says. Only the boreal forests are left.

"Alaska's official policy is to increase the number of trees felled in the interior by twentyfold," says Bryant. Canada, which is facing economic recession, also wants to increase timber production. This year Canadian authorities approved the cutting down of old temperate forests on Vancouver Island despite a storm of protests from environmentalists.

New Scientist
8 January 1994

Figure 8.2.3
Where has all the carbon gone?

British Columbia: the 'Brazil of the North'?

Nowhere has the battle between the forest products industry and environmentalists been more fiercely fought than in British Columbia, where the industry generates 28 per cent of GDP and 200 000 jobs. The forest issue has in fact dominated politics in the province for the last two decades. The industry has been accused of overcutting, damaging the environment and not paying enough for cutting rights. The destruction of the forest has been likened to the situation in Brazil, an accusation that the industry strongly denies. As most forest areas are publicly owned the provincial government has been heavily criticised for allowing such a situation to occur and continue. However, only 20 per cent of the original forest, covering an area of 10.6 million ha, has ever been logged and is being regrown successfully. Another 20 per cent is to be cut and replanted over the next 40 years. The remaining 60 per cent is either set aside in parks or is in inaccessible locations and is unlikely ever to be logged.

The forest industry's practice of replacing a varied natural forest with a managed monoculture has been heavily criticised. Mixed forests of Douglas fir, cedar, spruce and oak are often replaced by Douglas fir alone. The latter is popular because of its relatively fast 50 year maturing time. Opponents argue

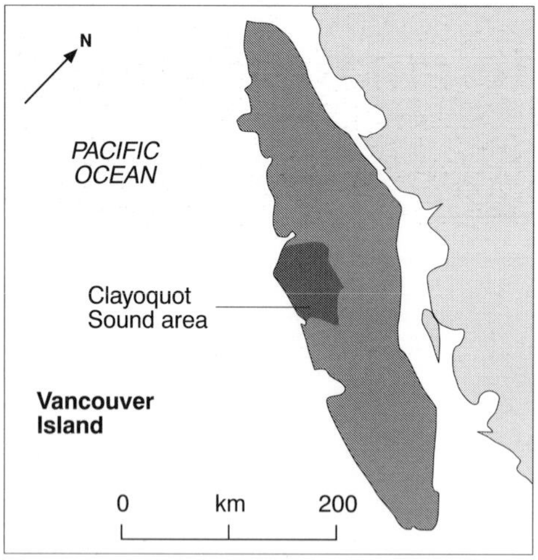

Figure 8.2.4
The position of Clayoquot Sound

that such loss of biodiversity makes forests vulnerable to disease and harms the ecosystem.

The coastal rainforest of British Columbia renews itself very differently from the boreal forest. Its climate is considerably more benign, and fires are relatively infrequent. Most of its tree species, left undisturbed by pests and fire, will reach a great age and size. Unlike the boreal forest, which has large areas with trees of the same age and species, a small patch of coastal rainforest is structurally diverse, comprising trees of many different ages, heights, sizes and species.

The battle for Clayoquot Sound

Clayoquot Sound (Figure 8.2.4), one of the last large untouched forests in British Columbia, encompasses a broad sweep of fir, cedar and spruce trees on the west coast of Vancouver Island. The area, approximately 200 km north west of Victoria, measures 350 000 ha and encompasses Clayoquot Sound, a number of watersheds that drain into the Sound, portions of Strathcona Park, and the Pacific Rim National Park. Since the late 1980s the provincial government has tried to reach public consensus on the controversial issue of land use in the area. The discussions involved the local forest industry and its employees, the local community, environmentalists, the Nuu-chah-nulth Tribal Council (which has a land claim on the area), and representatives of the tourism, fishing and mining sectors.

After various failed attempts to reach agreement, the provincial government imposed a decision in April 1993, permanently protecting 33 per cent of the Sound and dedicating 45 per cent to resource use, including logging. Special management zones were created where some logging is allowed, but the management emphasis is on wildlife, recreation and scenic landscapes. After protests by environmentalists a further reduction in the working forest was proposed, although forest workers argued that this would cause a loss of 3500 jobs.

Central to the latest plan launched in April 1994 was the industry's agreement to an 80 per cent increase in stumpage (cutting) fees paid to the government. The new revenue, C$ 2 billion over the first five years, must by law be used to improve forest management and reforestation, and clean up past damage. Fewer trees will be cut, displaced loggers will be retrained and a new code of logging practice will be enacted. In spite of these changes, environmental groups such as Friends of Clayoquot Sound and Greenpeace are sticking to demands that more old-growth forest be preserved and all clear-cutting banned.

QUESTIONS

1. Outline the differences between the various categories of forest shown in Figure 8.2.1.
2. Why are the northern forests so important in the global carbon cycle?
3. Assess the importance of the forest products industry to the economies of British Columbia.
4. Suggest why Clayoquot Sound has become one of the major battlegrounds in the 'forest debate'.

Manufacturing Industry and Services

9.1 The Manufacturing Belt: Structural and Locational Change

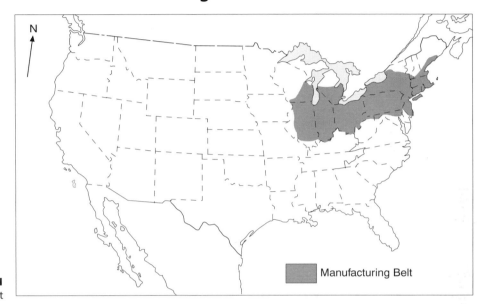

Figure 9.1.1
The Manufacturing Belt: location and extent

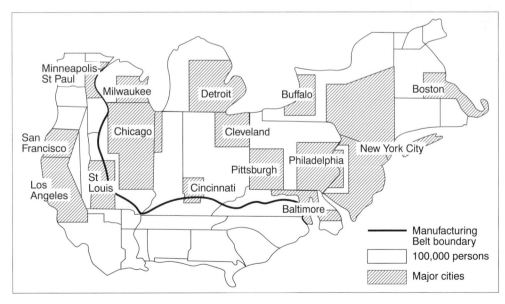

Figure 9.1.2
The Manufacturing Belt, mid 20th century.

In 1890 the United States overtook Britain to become the world's leading manufacturing nation. Starting in southern New England and in the Middle Atlantic coastal area, early American manufacturing consisted of textiles, small metal products and leather goods. After 1860, developments in iron and steel production and related heavy engineering shifted the focus of manufacturing across the Appalachians, first to Pittsburgh and then to the newly-developing interior.

Between 1880 and 1910, the North East and Midwest industrial areas merged to become the nation's dominant industrialised region. In 1927 the term 'Manufacturing Belt' was coined for the whole area extending across the north east of the USA from Boston to Chicago (Figure 9.1.1).

In 1950 the Manufacturing Belt covered only one-eighth of the conterminous USA's area but contained nearly one-half of the American population and a huge 70 per cent of the country's manufacturing employment. The dominance of the Manufacturing Belt at mid-century is dramatically illustrated in Figure 9.1.2, a cartogram in which each state's area is depicted in proportion to its manufacturing employment in 1950.

Figure 9.1.3
Iron and steel production and automotive engineering

Iron and steel production formed the basic industry in the region. It was located at those points where basic raw materials could be assembled most cheaply. Major controls on location were exercised by the enormous Appalachian coalfield stretching from west Pennsylvania to Alabama, with its excellent coking coal at Connellsville near Pittsburgh and at Pocahontas in West Virginia, and the iron ores of the Superior Uplands [Figure 9.1.4]. Major centres of steel production included Pittsburgh, Youngstown and Johnstown on the coalfield, Duluth at the head of Lake Superior, and the break-of-bulk locations of Chicago, Detroit, Cleveland and Buffalo on the Great Lakes. To these major centres must be added the seventy smaller steel-making towns recognised in the region in 1940.

Figure 9.1.4
Iron and steel production

Associated with iron and steel was a wide range of secondary manufacturing activity. By far the most important was motor vehicle production which was concentrated within an 'automotive triangle' with apexes at Buffalo, Cincinnati and Milwaukee, and its centre at Detroit. In 1949, this area contained over nine-tenths of the nation's automotive employees. Michigan alone accounted for 56 per cent of this number, and Detroit and its suburbs for very nearly 40 per cent.

continued page 185

1 With reference to Figure 9.1.3, explain the following terms: basic industry/basic raw materials; secondary manufacturing; break-of-bulk locations; innovative genius; entrepreneurial skill; mass-production methods; economies of scale; industrial and locational inertia.

2 Draw an annotated diagram to represent the changing structure of the US automotive industry.

Access to steel – the basic resource – and to markets in the north-eastern and central USA explains this distribution in general terms; innovative genius and entrepreneurial skill were specific and critical localisation factors. By 1900 there were large numbers of manufacturers in the United States making cars, but there was low output and little product standardisation. By designing a simple and durable internal combustion engine and installing it in a single model fabricated by mass-production methods, Henry Ford gave his company and his home town, Detroit, an unassailable early lead in car manufacture. Foremost among those who followed Ford's approach were R.E. Olds of Lansing, Michigan, who founded the Oldsmobile Company in 1901, and W.C. Durant of Flint, Michigan, who founded Buick.

Geographical concentration was assisted by the existence of very powerful economies of scale in motor vehicle manufacture, which meant that only the very largest companies survived. While total output of vehicles rose rapidly, the number of manufacturers declined sharply. In 1914 there were about 300 motor vehicle firms; in 1923, 103; and in 1927, 44, of which General Motors, Ford and Chrysler produced about 75 per cent of all cars made. The market share of 'The Big Three' rose to 90 per cent in 1939.

This exceptional concentration was both a cause and a consequence of the enormous productive power which was contained in the US Manufacturing Belt, a pre-eminence perpetuated and sustained by powerful forces of industrial and locational inertia. The existence of the Belt in turn testified to the dominant position occupied by manufacturing in mid-century industrial society.

From Manufacturing Belt to Rustbelt

The advantages of the early start enjoyed by the Manufacturing Belt began to evaporate as the twentieth century progressed. Challenges to the Belt's supremacy were first mounted by the development of the South's cotton textile industry in the 1920s, toppling New England from its leading position. After the Second World War, attracted by shifts of population (and thus markets), new power sources, different products, cheaper labour, government contracts and more appealing locations, manufacturing industry spread to many other parts of the USA. Consequently the old massive concentration of industrial activity in the Manufacturing Belt no longer exists.

Figure 9.1.5
US employment trends by economic sector, 1983–2005

Economic Sector	Employees (1000)			Percentage Change		
	1983	1994	2005*	1983–94	1994–2005	1983–2005
Agriculture	3508	3623	3399	3.3	−6.3	−3.7
Mining	952	601	439	−36.9	−27.0	−53.9
Construction	3946	5010	5500	27.0	9.8	39.4
Manufacturing	18 430			−0.7		
Durable goods	10 707	10 431	9290	−2.6	−10.9	−13.2
Non-durable goods	7723	7873	7700	−1.9	−2.2	−0.3
Services	66 407	89 425	107 256	34.7		
Other, self employed, etc.	9161	10 051	11 124	9.7		
US totals	102 404					

*Figures for 2005 are estimates.

Moreover, ever-increasing competition is experienced from foreign countries – especially Japan, which has overtaken the United States as the world's leading manufacturing nation. Such competition has accelerated major structural change in the American economy, with manufacturing industry declining in importance. Such trends seem set to continue into the next century (Figure 9.1.5). Not surprisingly, the Manufacturing Belt itself has been in decline for decades, its obsolescent plant, derelict factories and abandoned industrial landscapes giving it a new title: the Rustbelt.

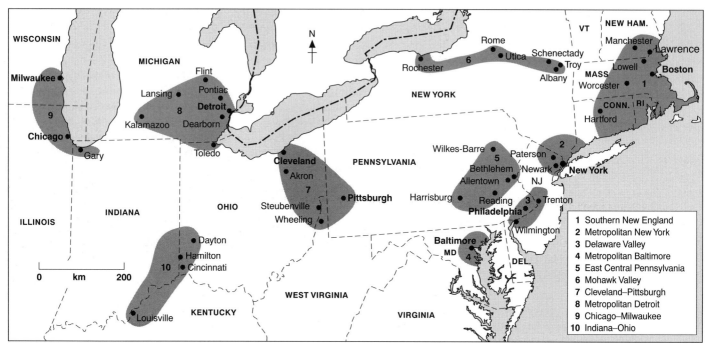

Figure 9.1.6
The Manufacturing Belt: major industrial areas

1	Southern New England
2	Metropolitan New York
3	Delaware Valley
4	Metropolitan Baltimore
5	East Central Pennsylvania
6	Mohawk Valley
7	Cleveland–Pittsburgh
8	Metropolitan Detroit
9	Chicago–Milwaukee
10	Indiana–Ohio

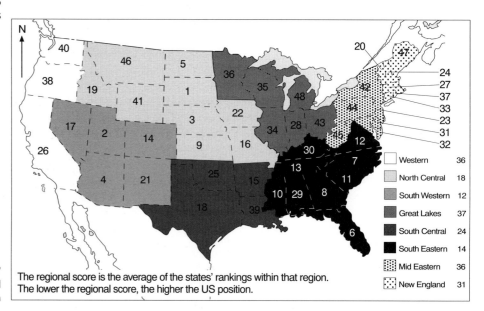

Figure 9.1.7
US: best manufacturing climates by state and region

The regional score is the average of the states' rankings within that region. The lower the regional score, the higher the US position.

☐	Western	36
☐	North Central	18
☐	South Western	12
☐	Great Lakes	37
☐	South Central	24
■	South Eastern	14
▨	Mid Eastern	36
▨	New England	31

All this is not to say that the Manufacturing Belt has disappeared completely. It still employs well over seven million people in manufacturing activities, which are responsible for over $607 billion of value added and the production of goods worth around $1222 billion every year. Within the Belt, manufacturing industry is concentrated in ten areas (Figure 9.1.6). Some areas have responded well to the challenge of changing economic conditions; as old industries have closed, new ones have replaced them. Other areas, however, have found it very difficult to attract new industry; their derelict landscapes and demoralised workforces are scarcely conducive to modern, clean industrial enterprises seeking a bright, fresh image.

Even the basic costs of operating in the Manufacturing Belt tend to be higher than in other parts of the USA, as shown by a recent survey (Figure 9.1.7). The costs which firms considered most important were (in rank order): 1. wages; 2. unionisation; 3. energy. Altogether, 22 factors were evaluated to produce the rankings shown in Figure 9.1.7. Generally speaking, the lower the figure, the more favourable the area for manufacturing location – though some industries (e.g. high-tech) may always be prepared to pay for high-cost locations if they provide the conditions needed for the company to flourish.

QUESTIONS

1 (a) Complete Figure 9.1.5.
   (b) Comment on the changing percentage share of total US employment attributed to (i) manufacturing; (ii) services.

2 (a) Use your answers to Question 1(b) and the data in Figure 9.1.5 to write a concise account of recent and predicted changes in the general structure of the American economy.
   (b) Attempt to explain the principal trends in manufacturing employment.

3 (a) Why were wages, unionisation and energy of greatest concern to manufacturers in deciding where to locate? (Figure 9.1.7).
   (b) Suggest and justify five other factors that would normally be considered when locating a modern factory.

4 (a) Rank the eight regions shown in Figure 9.1.7, placing the region with the best manufacturing climate first.
   (b) Comment on the prospects for the Manufacturing Belt vis-à-vis other regions.

The ten main industrial areas within the Manufacturing Belt: a summary (see Figure 9.1.6)

1. Southern New England

This is the United States' oldest industrial region. Textiles, clothing and leather goods remain important, but the modern emphasis is on specialised, high-quality items rather than mass production. Electrical machinery, jewellery, shipbuilding, paper, rubber and chemicals are among the wide range of manufactured products. The largest producers of machinery are Pratt and Whitney aeroengines at Hartford, and Sikorsky helicopters in neighbouring Stratford. Many light engineering industries, requiring few raw materials and little power, remain – often through inertia – but high-level skills are also important, e.g. in gun making. Modern electronics and high-tech industries have located along Route 128 around Boston and elsewhere in the region. The most characteristic industrial feature is the scatter of manufacturing activities across a large number of small towns.

2. Metropolitan New York

This area contains a wide variety of manufacturing industries (90 per cent of the 500 types of manufacturing listed by the US government) but they are highly concentrated geographically. New York City contains a large number of small-scale, labour-intensive industries such as clothing, printing and publishing, miscellaneous manufacturing, and food and related products. Heavy industry (chemicals, oil refining, steelworking, machinery, shipbuilding and repair, and food processing) is located on the New Jersey shore in old industrial centres like Newark, Elizabeth and Jersey City. Recent additions include electronics and other high-tech industries in suburban locations.

3. Delaware Valley

Extending from Trenton to Wilmington but focusing on metropolitan Philadelphia, the lower Delaware Valley also has a wide variety of industry, but with different emphases. Heavy industry predominates, including shipbuilding and repair, chemicals, and the Fairless iron and steel works at Morrisville, opposite Trenton. This fully integrated works was opened in 1952 and depends on iron ore from Venezuela and coal from Pennsylvania and West Virginia.

4. Metropolitan Baltimore

This area has a similar complex of industries to Philadelphia's. Tidewater access has generated a range of port industries, including sugar and oil refining, copper smelting, fertilisers and Sparrows Point iron and steel works, established in 1887 and now the largest in the United States, using iron ore from Latin America, coal from West Virginia and limestone from Pennsylvania. Baltimore's secondary industries include canning, textiles, clothing, and printing and publishing, but these have declined markedly in recent years.

5. East Central Pennsylvania

Manufacturing in this region originated with the iron ore deposits near Bethlehem, limestone from Allentown and coal from the Scranton/Wilkes-Barre anthracite field. This combination of raw materials led to the early establishment of steelmaking and associated metal-working activities. In addition, textiles, clothing and shoes are products manufactured in the area. However, many of these activities, and the communities dependent upon them, are in long-term decline. Moreover, lacking a central urban core, this region risks losing not only its economic base but its sense of local identity too.

6. Mohawk Valley

This region also lacks a dominant major city, although the chain of industrial centres benefit from their location along the line of easiest access from the Atlantic coast to the Great Lakes and have a range of manufacturing specialisms. The mix of textiles, metals and machinery is similar to that of southern New England. The Albany-Troy-Schenectady industrial complex has long specialised in clothing and machinery, Schenectady being the home of the giant General Electric Corporation. Further west, Utica and Rome have a range of textile, carpet, metals and machinery industries, while Rochester on Lake Ontario is the headquarters of Eastman Kodak and specialises in cameras, optical equipment and electrical machinery.

7. Cleveland-Pittsburgh

This area of eastern Ohio and western Pennsylvania may be regarded as the core region of US heavy industry, with iron and steel being the traditionally dominant activity. Originally based on the juxtaposition of local ores (now depleted) and Appalachian coal, the iron and steel industry later depended on complex movements of raw materials by rail, river and the Great Lakes–St Lawrence Seaway system. Cleveland and neighbouring Lake Erie ports received iron ore from both the Superior Uplands and the Labrador–Ungava field in Canada, the ore then being railed to Youngstown and Pittsburgh – which also received ores from Labrador and Venezuela by rail via Baltimore. Coal from Pennsylvania and West Virginia moved by barge along the Monongahela River to Pittsburgh and by rail to the other steel-making centres. Many other types of manufacturing contributed to this region's dominance in heavy industry, including machinery, rubber, glass and clay products. It should be noted, however, that of all the declining areas within the Manufacturing Belt, this most aptly justifies the 'Rustbelt' tag. Many of the iron and steel centres including

Pittsburgh, have ceased operations entirely, while other industries have suffered a similar fate (e.g. the tyre factories that long made Akron in Ohio the self-styled 'Rubber Capital of the World' have all closed down).

8. Metropolitan Detroit

Detroit, which may be described as a single-industry manufacturing city, is the most important of a number of neighbouring centres in which car, truck and component-part making dominate all other activities. Unlike the vehicle assembly plants elsewhere in the United States, the Detroit area is where engine blocks, bodies, suspension units and trim components are both manufactured *and* assembled. Thus Detroit is a major consumer of steel, much of which is produced within the city. As long as such basic processes as casting, forging and body-pressing remain in the city, Detroit will probably continue to dominate the US automotive industry. Indeed, the dispersal and decentralisation of the industry to other parts of the United States which characterised the 1915–1965 period (up to 200 plants existed outside the Detroit area at one time) has been replaced by contraction back to Detroit in recent years.

9. Chicago–Milwaukee

Metropolitan Chicago is the United States' leading steel-making centre. The almost-continuous line of steel mills, oil refineries and associated processing plants along the Lake Michigan shore between South Chicago and Gary (Indiana) comprises one of the world's largest concentrations of heavy industry. However, as befits the United States' third largest city, regional capital of the Midwest and major transportation centre, Chicago's manufacturing range is extremely diverse and includes electrical machinery, metals, food processing, printing, paper production and furniture. Combined with the industries of Milwaukee and neighbouring towns, which include brewing, farm machinery, leather products and meat packing, Chicago's manufacturing output is second within the USA only to that of metropolitan New York.

10. Indiana–Ohio

This area includes the Miami Valley and central Ohio Valley in south west Ohio, south-east Indiana and adjacent parts of Kentucky. The Miami Valley produces a diversity of manufactures, including iron and steel, machine tools, aircraft, computers, calculators and other products requiring skilled labour. Dayton is the Miami Valley's largest centre, but Hamilton (paper and paper-making machinery) and Middleton (steel and diversified industrial products) are also important. In the central Ohio Valley, Cincinnati is the biggest city; its main products include machine tools, automotive parts, electrical equipment and aeroengines. Downstream, Louisville (Kentucky) specialises in the manufacture of domestic appliances, cigarettes, chemicals and furniture; it is also an important distilling centre.

QUESTIONS

Refer to Figure 9.1.6 and accompanying text.

1 Assess whether industrial inertia has been advantageous or disadvantageous to the recent economic development of southern New England.

2 (a) With reference to the industrial geography of southern New England and metropolitan New York, identify (i) the main similarity; (ii) the main difference.
   (b) Assess the importance of tidewater access to the types of industry which have developed in the lower Delaware Valley and metropolitan Baltimore.

3 Identify those areas which have traditionally depended on coal from the Appalachian and Eastern Interior fields.

4 Account for the Chicago area's status as the leading US steel-making centre, using the following headings: raw materials; transportation links; markets.

5 List the industrial evidence that the Chicago–Milwaukee area is the focus and service centre of a major agricultural region (Midwest and northern Plains).

Figure 9.1.8
US manufacturing: percentage distribution of employment and value added by division; and percentage change in divisional shares, 1950–1991

Division	Employment			Value Added		
	1950	1992	Change	1950	1992	Change
New England*	9.8	6.2	−3.6	8.3	5.7	−2.6
Middle Atlantic*	26.9	14.1	−12.8	26.2	14.2	−12.0
East North Central*	30.0	22.4	−7.6	33.2	22.6	−10.6
West North Central	5.6	7.5	+1.9	5.7	7.6	+1.9
South Atlantic	11.1	16.4	+5.3	9.4	16.0	+6.6
East South Central	4.4	7.7	+3.3	3.8	6.9	+3.1
West South Central	4.0	8.3	+4.3	4.3	9.2	+4.9
Mountain	1.1	3.5	+2.4	1.2	3.6	+2.4
Pacific	7.0	14.0	+7.0	7.9	14.2	+6.3
US	100.0	100.0 (=18.3 m)		100.0	100.0 (=$1429 bn)	

*Constituent divisions of the Manufacturing Belt.

Figure 9.1.9
USA: Manufacturing Location Quotients by state, 1992

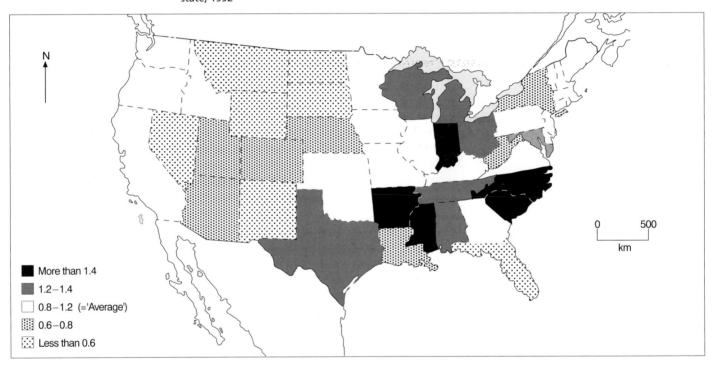

N

- More than 1.4
- 1.2–1.4
- 0.8–1.2 (='Average')
- 0.6–0.8
- Less than 0.6

0   500
km

Division	Employees (1000)	
	(A) Manufacturing	(B) Total
New England*	1091	5967
Middle Atlantic*	2491	16 240
East North Central*	4041	18 851
West North Central	1373	8203
South Atlantic	2983	19 560
East South Central	1432	6378
West South Central	1555	11 069
Mountain	626	6057
Pacific	2481	16 415
US	18 190	108 437

Figure 9.1.10
US manufacturing and total employment by division, 1992

*Constituent divisions of the Manufacturing Belt.

QUESTIONS

1 (a) With reference to Figure 9.1.8, calculate the percentage decline in the Manufacturing Belt's share of (i) manufacturing employment; (ii) value added.

(b) For 1950 and 1991, calculate the Manufacturing Belt's percentage share of total US (i) manufacturing employment; (ii) value added.

(c) Comment on the main trends 1950–91.

2 (a) Referring to Figure 9.1.9, identify those states with an above-average Manufacturing Location Quotient (MLQ).

(b) How many such states lie (i) within the Manufacturing Belt; (ii) elsewhere?

(c) What does your answer to (b) suggest about the changing location of manufacturing industry in the United States?

3 (a) Using the data in Figure 9.1.10, calculate to two decimal places MLQs for all the divisions listed.

(b) Display the results on a choropleth map (like Figure 9.1.9), selecting *three* suitable categories.

(c) With particular reference to the three constituent divisions of the Manufacturing Belt, comment on the divisional patterns shown on your map.

9.2 The South Resurgent: Economic Progress in a Laggard Region

Defining the South

The South is notoriously difficult to define. The Mason-Dixon line was drawn in colonial times to separate 'free' Pennsylvania from the slave-holding areas to the south ('Dixie'). Later the Ohio River became the westward extension of the northern limit of those states where slavery was legal. This remains the boundary of the South region today for census and other official purposes; but this definition includes such states as Maryland and Delaware, as well as Washington DC, which are more obviously part of the north eastern Megalopolis (Section 4.2).

Figure 9.2.1
The South: the old Confederacy

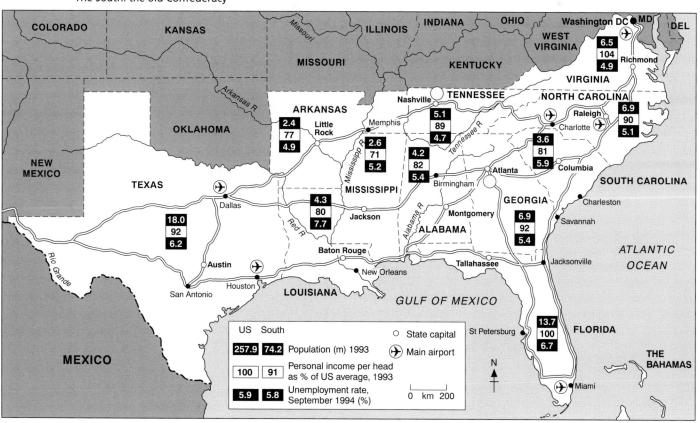

A less contentious definition equates with the area of the old Confederacy (Figure 9.2.1), the 11 states which attempted to secede from the Union during the Civil War (1861–5). These states have a combined area almost as large as that of the European Union, but contain a population rather smaller than Germany's. If this South were a separate country, it would possess the world's fifth largest economy.

Figure 9.2.2
America's new economic powerhouse

It is the land that went to sleep. In 1860, the South was successful and rich. White southerners had the highest standard of living in the world. It was also blisteringly arrogant; arrogant enough, indeed, to try to turn itself into a separate country. When it was defeated five years later, the stuffing was knocked out of it. Most other places flattened by war – Germany and Japan being only the most obvious examples – had picked themselves up within a decade. The American South stayed down for a century; its economy stagnant, its people mired in poverty, its society encased by rigid racial and class barriers.

Now all has changed. Legal segregation is long gone, although racial tensions remain. Thanks in part to the spread of two crucial inventions – the air conditioner, which made summers tolerable, and the mechanical cotton-picker, which removed the need for cheap farm labour – the economy has at last been modernised. Southern cities have expanded and been rebuilt as fast as any in the country; suburbs have mushroomed even faster. The South has, like Sleeping Beauty, awakened from 100 years of slumber.

Over the past two decades economic growth in the South has outpaced most of the rest of America and is now a locomotive powering the American economy, a role previously played by California. In 1993 over half America's new jobs were created in the South, which has only a quarter of the country's population. Eight of the top ten states in terms of growth in manufacturing plants since 1991 are in the South.

In 1940 average income per head in the South was half the American average; and the southern states were at the bottom of the league tables for poverty, illiteracy and education. Income per head is now over 90% of the national average. Indeed, allowing for lower living costs, the average southerner now enjoys a higher standard of living than most other Americans. Two-thirds of the southern states also have lower-than-average unemployment rates.

The South's population is also growing fast. Texas will soon overtake New York to become the country's second most-populous state; Florida is likely to overtake New York early next century. However, the record is not all unalloyed progress. Some southern states have done better than others. Rural areas have lagged. Many of the poor, both white and black, still live in grim conditions.

History shaped the South's disgraceful treatment of black southerners. After slavery was abolished by the 13th amendment, a driving motivational force for the white South became the desire to keep blacks poor and disfranchised. Jim Crow laws and segregation persisted right through to the 1960s, ironically impoverishing whites as much as blacks. It was resistance to racial change more than its long, steamy summers that kept the South sleepy and backward for so long after the civil war.

Now that it is awake, however, the South is shedding the historical baggage that has weighed it down for so long. Far from being trapped in the past, it is marking out a new future. And in doing so it is beating a path for the whole country. In its economy, its politics (President Clinton hails from Arkansas) and even in its race relations, the South is once more full of a long-forgotten confidence.

The Economist,
'A Survey of the American South',
10 December 1994

QUESTIONS

1 (a) Which other states, not high-lighted in Figure 9.2.1, are included in the modern South Region for official census and other statistical purposes?
  (b) With reference to a relief map, what physical characteristics are shared by the states of the Deep South?

2 Referring to Figure 9.2.1, elaborate this statement: 'The record is not all unalloyed progress. Some southern states have done better than others.'

3 What factors have helped to attract manufacturing and other operations to the South in recent decades? (Figure 9.2.3)

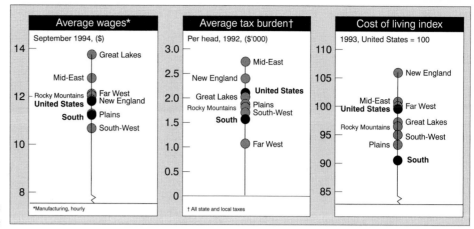

Figure 9.2.3
The South's competitive advantage

Yet the South is more than a geographical expression, and not all the Confederate states are equally 'southern'. South Florida, especially Miami, feels more Latin American than southern, although the rednecks of the Florida panhandle are decidedly southern in their accents and attitudes. Most of Texas – cowboys on the range, ten-gallon hats and crocodile-skin boots – is really in the West; the drawl of east Texans, however, is deeply southern. Northern Virginia is largely a collection of shopping malls and suburban lots attached to Washington, DC and has few links with its southern partners. To many southerners, the true heartland is the six states that stretch from the Carolinas to Louisiana. This 'Deep South' is the warm, humid plainsland area historically associated with cotton growing and the plantation system.

Foreign investment in the South

An especially notable aspect of the modern American economy is the increasing level of foreign investment. This has grown from about 3 per cent of the USA's GDP in 1980 to 7 per cent in 1991 (Figure 9.2.4).

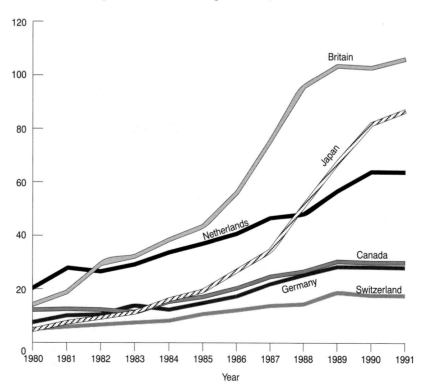

Figure 9.2.4
Foreign countries with highest direct investments in the USA

Two salient features characterise this influx of foreign funding:

- more foreign capital, especially from Japan and Western Europe, has poured into the South than into any other region of the United States;
- with around 40 per cent of all foreign investment, manufacturing industry has proved the most attractive sector for the flow of foreign capital.

Within the South's rapidly expanding manufacturing sector, foreign car makers are leading the drive to revitalise the region's economy. These include Toyota in Kentucky, Nissan in Tennessee, Mercedes-Benz in Alabama and BMW in South Carolina.

South Carolina and the BMW car assembly plant at Spartanburg

South Carolina has proved extremely attractive to foreign investors since the 1970s. Between 1987 and 1993 nearly 5000 companies announced $19.5 billion of capital investment in South Carolina. Nearly a third of this capital was of foreign origin, ranking South Carolina fourth in the United States for such investments. These inputs have fuelled the diversification of the state's economy, boosting not only manufacturing but, increasingly, the headquarters/office facilities, and Research and Development (R & D) sectors (Figures 9.2.5 and 9.2.6).

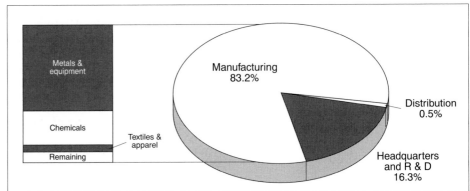

Figure 9.2.5
South Carolina: foreign investment by economic sector, 1992

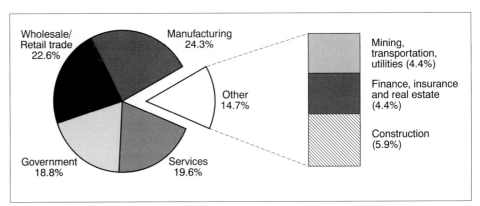

Figure 9.2.6
South Carolina: employment by economic sector, 1992

Historically, Britain, France and Switzerland have each provided about 12 per cent, and Japan 16 per cent of South Carolina's annual foreign investment; but Germany has emerged as the clear leader with 34 per cent overall, rising in 1991 and 1992 to 51 per cent (over $1 billion). In 1992, German investment in South Carolina included what is widely regarded as the most prestigious industrial prize ever gained by any state: the new BMW car plant at Greer near Spartanburg, in the north west corner of the state, the Piedmont or Upstate region (Figure 9.2.7).

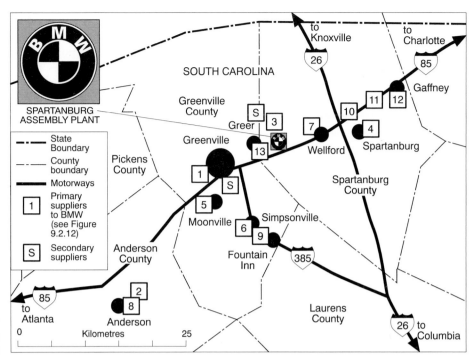

Figure 9.2.7
Location of the BMW assembly plant and its
local suppliers

The Spartanburg factory is BMW's first foreign assembly plant and represents the initial step in the company's plan to expand from a European to a global car maker.

The Spartanburg plant covers an area equivalent to 15 football pitches on a greenfield site of 425 ha and cost $600 million to build (Figure 9.2.8). It incorporates its own power station, waste treatment and recycling plants, and BMW claim that its environmental design is better than any existing car factory in the world. It also has an area reserved for future expansion.

The first car was produced in September 1994. By 1997 the workforce had risen to 2000 (the majority being recruited from within an 80 km radius) and the plant's capacity had expanded to 300 cars a day – an annual output of

Figure 9.2.8
The BMW car assembly plant, Spartanburg

72 000. More than half of this total will be shipped to over 100 countries worldwide – the BMW car plant being the first in the United States designed for such levels of exports. At the same time, however, the Spartanburg plant's Import Car Processing (ICP) section serves as a final quality check-point for all the 100-a-day BMWs imported from Germany. In addition, the ICP installs customised items such as radios and CD-players.

Special features of BMW's operations at Spartanburg

- All employees are known as 'associates' and wear a standard uniform of blue slacks and white jacket (personalised with the individual's name) regardless of their job or salary level. There are no management 'perks' such as reserved parking or separate dining facilities.
- All processes take place under one roof so that visual communication is possible between all workers, emphasising the importance of the team.
- Relatively little use is made of robots and other forms of automation, for two reasons: (i) this strategy allows flexibility to produce multiple models on the same production line, and (ii) it creates an atmosphere of teamwork in the hand-working areas.
- In the paint shop, all the many processes use environment-friendly water-based colours; only the clear top coat is solvent-based.
- The assembly line is where all the myriad parts and components are brought together and integrated into the painted car bodies (Figure 9.2.9). The unusual layout of the assembly line, which is shaped like a lower case 'e', returns the final product to within several metres of all assembly workers. This allows for fast feedback; for example, a problem detected near the end of the assembly line can be quickly referred to the personnel most familiar with that aspect of the car, and any necessary adjustments made.
- All associates are involved in the plant's strategy of Total Quality Control (TQC).

Why South Carolina?

BMW was attracted to South Carolina by the following factors:

- Spartanburg lies in the South's economic 'Boom Belt' – the corridor along the I-85 motorway which leads from the high-tech areas of Virginia in the north (especially around Washington DC's Dulles Airport), via the university centre and 'Research Triangle' of Raleigh–Durham and the financial centre of Charlotte in North Carolina, and through the South Carolina Piedmont region to Atlanta, the international traffic and business centre in the south.
- The state government offered financial inducements, subsidies and 'tax-breaks' amounting to nearly $150 million.
- South Carolina's right-to-work law forbids the 'closed shop' – the compulsory membership of trade unions. Only 2.7 per cent of South Carolina's workers belong to unions, the lowest rate in the United States (the national rate is 16 per cent). South Carolina also has one of the country's lowest work-stoppage rates; between 1981 and 1991 only 0.01 per cent of working time was lost to strikes.
- South Carolina's hourly wage rates have been consistently low. The average manufacturing hourly wage ranks 46th in the nation (92 per cent of the South's average and 82 per cent of the US average). Additional employee benefits, such as superannuation and health insurance, amount to 31 per cent of wages in South Carolina, compared with 39 per cent in both the South as a whole and the United States.
- Other favourable factors are illustrated in Figures 9.2.10 and 9.2.11.

Figure 9.2.9
Inside the Spartanburg factory: BMWs on the assembly line

For BMW these locational advantages translate into low production costs. Each worker at the Spartanburg plant costs the company an average of $40 000 a year – a third less than unionised workers in Detroit and barely half what BMW's own employees cost in Germany. The company reckons that production costs in South Carolina will eventually be 30 per cent below those in Germany. The lesson has not been lost on BMW's compatriot and competitor company, Mercedes-Benz; having been offered a huge $253 million package by Alabama, Mercedes opened a new factory at Tuscaloosa in 1997.

QUESTIONS

1 Using the information in Figures 9.2.5 and 9.2.6, write an appraisal of South Carolina's economy and the role of foreign investment in its diversification.

2 BMW's Spartanburg plant is part of the company's 'globalisation' strategy. What do you understand by this term, and what are the advantages of globalisation to companies like BMW?

3 The Spartanburg facility is only the second European car assembly plant to be built in the United States since the Second World War. The first venture was by Volkswagen, in the north east (Westmoreland County, Pennsylvania), but that plant closed in 1988. Suggest why.

4 (a) Why does the Spartanburg plant need such a large area of land around it? (Figure 9.2.8).
  (b) The crescent-shaped building in the foreground is the BMW Zentrum (Centre). Suggest the Zentrum's main function.

5 What advantages does the BMW factory have over more conventional assembly plants?

Figure 9.2.10
Comparative electricity costs: South Carolina and selected regions, 1992

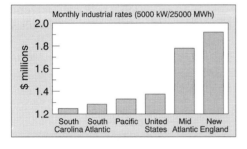

QUESTIONS

1 In what way has Figure 9.2.10 been drawn to exaggerate South Carolina's electricity cost advantage? Redraw the diagram to represent the situation more accurately.

2 When BMW advertised for staff at the Spartanburg plant they received 11 000 applications for 135 office jobs and 100 000 applications for the first 1000 production-line jobs. What does this response indicate about (i) the employment situation in South Carolina; (ii) the quality of the staff actually recruited?

3 Suggest why BMW are reviewing their production operations in Europe. What lessons can be learned from Spartanburg?

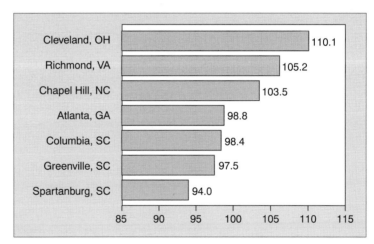

Figure 9.2.11
Cost of living index: Spartanburg and selected major cities

A network of suppliers

The BMW plant has both generated and is dependent upon a network of component and specialist suppliers, not just in South Carolina but extending to other parts of the United States as well as to Canada and Mexico (Figure 9.2.12). Following the investment of over $1 billion and the creation of more than 3000 jobs at the Spartanburg plant and its local suppliers, a survey revealed that consumer spending rose markedly throughout the South Carolina Piedmont, indicating greater confidence in the region's economic future. This kind of positive knock-on effect is known as 'circular and cumulative causation' (or just 'cumulative causation') (Figure 9.2.13).

Figure 9.2.12
North American suppliers to BMW's Spartanburg plant

Supplier	Jobs*	City	State	Part
Aeroquip		Clinton Township	MI	A/C pressure line pipe
Alpine		Greenwood	IN	Radio
ASC		Lansing	MI	Hard top & soft top
Autoliv		Indianapolis	IN	Seat Belts
Autotrim		Chicago	IL	Head liners & package tray
Behr of America		Ft Worth	TX	A/C Condenser & Heater
Bomoro (G)	40	Greenville	SC①	Locks
Bosch (G)	50	Anderson	SC②	Oxygen sensor
Bosch		Toluca	Mex	Starter motor
Brose		Queretaro	Mex	Power windows
Deta Douglas		Winston-Salem	NC	Battery
Electro Wire		El Paso	TX	Cable, harness
Fiamm		Cadillac	MI	Horn
Goodyear		Lawton	OK	Tyres
Hayes Wheels		Gainesville	GA	Aluminium wheels
Hoesch		Hamilton	OH	Springs
HSL		Tecumseh	Ont	Door hinges
Illbruch		Howell	MI	Sound insulation
ITT Automotive		Rochester	NY	Wiper, sound insulation
ITT Automotive		Culpeper	VA	Braking unit
ITT Teves		Morganton	NC	ABS brake system
Karma Huf		Germantown	WI	Door handles
Kenmore		Dallas	TX	Air drier
Kostal Mexicana		Queretaro	Mex	Windshield washer switch
Lear Seating (G)	150	Duncan	SC③	Seats
Lemfoerder (G)	25	Duncan	SC④	Suspension, steering
Lowell Engineering		Alto	MI	Mirrors
Magna Drive (C)	200	Piedmont	SC⑤	Stampings
Michelin (F)		Lexington	SC	Tyres
Miliken Sommer (G)	35	Simpsonville	SC⑥	Trim panels
Modine		Clinton	TN	A/C condenser
Modine		McHenry	IL	Trans fluid air cooler
Morton		Ogden	UT	Airbag
No. Am. Lighting		Flora	IL	Lights
No. Am. Lighting		Salem	IL	Lights
Packard Int (US)	35	Duncan	SC⑦	Wiring harnesses
Petri		Port Huron	MI	Steering wheel
Pioneer Electronics		Springboro	OH	Radio
Plastic Omnium (F)	200	Anderson	SC⑧	Fuel tank, bumpers
Plumley		Paris	TN	Hoses
PPG Industries		Suwanee	GA	Glass
Rieter (G)		Aiken	SC	Heat shield
Saachs Automotive		Florence	KY	Springs, struts
SASA (Hoesch Mexico)		Tlanepantla	Mex	Front & rear stabiliser bars
Schlegel		Maryville	TN	Weatherstrip
Schlegel		Reidsville	NC	Door seals
Siemens Automotive		London	Ont	Pressure shroud
Sommer-INOAC (F/J)	75	Simpsonville	SC⑨	Trim panels
Spartanburg Steel	150	Spartanburg	SC⑩	Stampings
Stabilus		Gastonia	NC	Gas spring bonnet
Stabilus		Colmar	PA	Gas spring lid
Stankiewicz (G)	40	Spartanburg	SC⑪	Insulations
Superior Industries		Fayetteville	AR	Aluminium wheels
Timken Gaffney (G)	35	Gaffney	SC⑫	Wheel bearings
TRW VSSI		Cookeville	TN	Airbag system module
TRW Steering		Marion	VA	Rack/pinion gear
TRW – VSSI		McAllen	TX	Seat belts
United Technologies		Tampa	FL	H/AC control
Valeo		Greensburg	IN	Radiator
Webasto		Rochester Hills	MI	Sunroof frame
Zeuna Staerker (G)	50	Spartanburg	SC⑬	Exhaust system

*South Carolina Piedmont companies only.   Nationality of foreign firms located in South Carolina

SC① – For Piedmont location, see Figure 9.2.7.   C – Canada   F – France   G – Germany   J – Japan

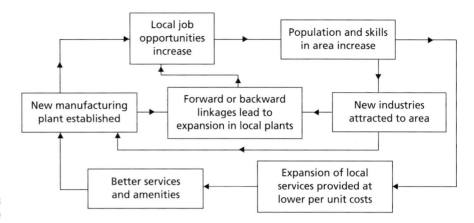

Figure 9.2.13
Circular and cumulative causation (simplified)

QUESTIONS

1 (a) On an outline map of North America (including Mexico) plot as accurately as possible the locations of the Spartanburg plant's main suppliers (Figure 9.2.12).
   (b) Use an appropriate diagrammatic technique to illustrate the number of suppliers in the following regions: South Carolina, Rest of the South, US Manufacturing Belt, Rest of the US, Canada and Mexico.
   (c) Comment on the locational patterns revealed in (a) and (b).

2 List the advantages and disadvantages to BMW of having so many suppliers in so many locations.

3 Considering all aspects from design to retail sales, which elements of BMW cars produced at Spartanburg can be described as German and which American?

4 Explain how the process of cumulative causation in the South Carolina Piedmont may not be wholly beneficial for other regions of the state (Figure 9.2.13).

The South's economic revival: a critical retrospect

Whilst undoubtedly impressive in many ways, the South's revival is associated with three major problems, which can be summarised as follows:

The dependence on financial incentives

- this has been criticised as 'giving the state away', and can be damaging if it leaves too little cash for other important sectors, e.g. local schools;
- companies may play one state off against another to obtain the biggest sums of money;
- companies may think a state has no other advantages to offer;
- companies may expect bigger and bigger incentives in future.

The South's low-cost labour advantage is dubious

- low wages often mean low skills;
- the skills deficit may be hard to make up, since the *overall* standards of school and university education are lower in the South than in the rest of the United States;
- companies tempted to the South by low wages could, under the North American Free Trade Agreement (NAFTA), move on to Mexico where wage rates may be only one-fifth those in the southern states.

The South's economic success has been very patchy

- the cities and suburbs have thrived, especially where served by interstate motorways and airports, e.g. in the eastern 'Boom Belt';
- most rural areas – often with large black populations – have been left behind. In 1990, 9 per cent of black males aged 16 and older were unemployed, compared with 3.3 per cent of white males; and of all children living in poverty in, for example, South Carolina, 72 per cent are black.

9.3 High Technology: Silicon Valley

High technology (high-tech) is North America's fastest-growing manufacturing industry and may be the world's biggest by the end of the century. It involves making or using silicon chips, computers, related peripheral equipment, software, semi-conductor devices, robots, microcomputer-controlled machines, aerospace components and other technically-sophisticated electronic products.

In the early years, the main applications of high technology were in the defence and aerospace fields, and production and supply were dominated by American firms. Increasingly, however, commercial markets have become more important as manufacturing and service industries have adopted electronic equipment more intensively. At the same time, Japan, newly-industrialised countries (NICs) such as South Korea and Taiwan, and West European nations have challenged the supremacy of the United States in both the production and usage of high-tech equipment. Canada also has a thriving high-tech sector.

High-tech origins: Silicon Valley, California

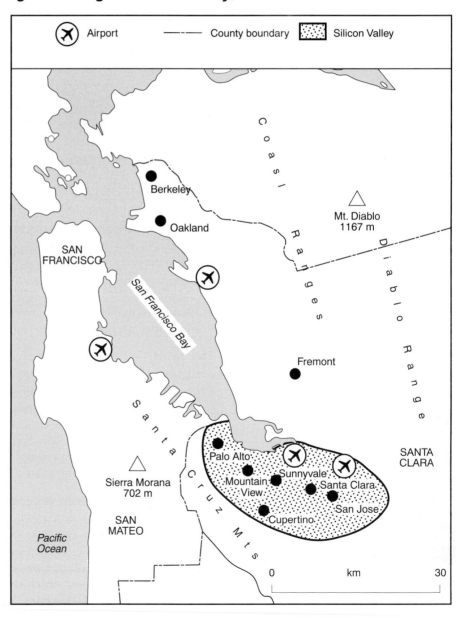

Figure 9.3.1
Silicon Valley, California

The industry originated in 1956, in Palo Alto, California where the very first transistors (electronic switches which superseded valves) were produced in Stanford University's research laboratories. By the early 1960s, companies in Stanford's science park, a joint venture involving the university and the commercial world, had miniaturised transistors so that thousands could be built into a single, tiny silicon chip. The Microelectronics Revolution had begun.

In Santa Clara County, to the south of San Francisco, a 15-km by 25-km ribbon of land between Palo Alto and San Jose has become known almost universally as Silicon Valley (Figure 9.3.1). Here 1845 firms employ over 200 000 people – the most intensive concentration of high-tech activity in the world. Cherry and apricot orchards have been replaced by one- or two-storey laboratories, factories and warehouses belonging to companies such as Amdahl, Apple, Hewlett-Packard, Intel, Sun and Tandem (Figure 9.3.2). These firms have a combined turnover of around $60 billion, or 17 per cent of the worldwide computer industry.

Figure 9.3.2
Microchip factories in Silicon Valley

Since a single speck of dust can ruin a silicon chip, the need for clean, well-lit 'fabs' (fabrication plants) is paramount. Consequently, most fabs are rectangular, lightweight structures with big, clear interior spaces. Good lighting (usually artificial), air-conditioning and filtering devices to purify the working environment are also needed. Such buildings are inhabited by scientists and technicians whose appearance would seem alien to most people. Swathed in surgeon-like garb and speaking of rams, roms, bit and bytes, recent engineering graduates are among the elite work-force who fully understand the ever-advancing technology – and command the high salaries which often go with high-tech employment (Figure 9.3.3)

Although Silicon Valley initially dominated world markets, by the mid-1980s Japanese silicon chip makers had overtaken it not only quantitatively but in the quality and sophistication of their products. Because the American industry was rapidly losing money and jobs, in 1987 the US government and a group of high-tech companies formed Semetech, a research consortium whose mission was to save Silicon Valley by improving the capabilities of silicon chips to previously unattainable levels, thus restoring the USA's technological lead.

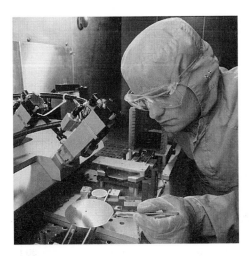

Figure 9.3.3
High-tech worker in a 'fab', Silicon Valley

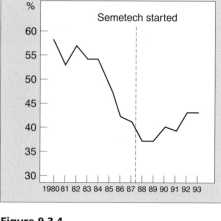

Figure 9.3.4
American manufacturers' share of world silicon chip markets, 1980–1993

Whereas in 1980 American firms still enjoyed 58 per cent of world silicon chip sales, by 1990 this had fallen to only 37 per cent, with Japan's share standing at 49 per cent (Figure 9.3.4). By 1994, however, these proportions had changed to 46 per cent and 41 per cent respectively. Thus Semetech's $1.2 billion investment had paid off, and with Silicon Valley's chip makers world leaders once again, the consortium is scheduled to be wound up in 1997.

Originally all the high-tech companies in Silicon Valley were relatively small-scale specialist concerns, a few of which grew very rapidly. However, as big, acquisitive American and Japanese electronics corporations moved into the valley some of the smaller companies were forced to merge or sell out. The surviving small companies are likely to experience major problems in the future as products become more complex and sophisticated, requiring bigger investment in research and development (R & D). Moreover, new firms will need more start-up funding than in the past, while the skilled-labour shortage will also militate against newcomers.

Despite these problems, the valley's commercial vitality will probably be maintained through the amoeba-like qualities of most high-tech industry: it divides more readily than it combines. That characteristic seems likely to perpetuate the competitive spirit which, with government assistance, has been the key to the continuing success of Silicon Valley.

High-tech location factors

- high-tech industries tend to congregate where a pool of specialised labour is available;
- clustering encourages economies of supply, services and distribution, while ideas, staff and equipment can be swapped – at least in the early days;
- high labour productivity is sought. The existence of a strong work ethic, or readiness to adopt one, is desirable – as is weak unionisation of the local work-force;
- proximity to a major group of universities or research institutions is virtually prerequisite; so too are strong links between the academic and commercial worlds (as exemplified in Stanford University's science park) whereby immediate access to new ideas and research findings can be translated into practical applications and commercial success;
- proximity to major markets is useful but not vital, since high-tech products tend to be high-value/low-bulk goods that can bear the costs of transport – often to overseas markets;
- a favourable tax climate is often sought, together with the encouragement of official agencies in setting up high-tech firms and subsequent support in the form of financial aid and contracts;
- the location should be a pleasant place to live. This is important in attracting and retaining highly-qualified staff who are often able to pick and choose between numerous job opportunities in different parts of the country or abroad.

It should be noted that traditionally important location factors, such as proximity to raw materials and power supplies, are of little significance to high-tech industry. In contrast, access to the latest research in electronic engineering is vital, and the specialist workforce must be prepared to live in the chosen location. Otherwise, high-tech industry is an excellent example of a 'footloose' economic activity: it can locate virtually anywhere.

QUESTIONS

1 (a) Explain why high-tech industry places a high priority on proximity to university research facilities.

   (b) What are the advantages of science parks to high-tech industries and universities?

2 Describe and account for the pattern of high-tech premises shown on Figure 9.3.2.

3 Describe and explain the trends shown in Figure 9.3.4.

4 Explain in detail why high-tech industry can be described as 'footloose' compared with an older industry such as iron and steel production.

5 Why do so many places attempt to attract high-tech industry?

9.4 The Boeing Aerospace Company: America's Biggest Exporter

Origins and growth

William Boeing (1881–1956), the son of a German immigrant, arrived in Seattle in 1908 to purchase fresh stock for his father's timber business in Minnesota. However, he stayed on in Washington State, learned to fly, and in 1916, helped by a friend, built two wood-framed seaplanes.

Thus began the Seattle-based Boeing Company, which by 1928 had become the United Aircraft and Transport Corporation, a huge conglomerate of aircraft manufacturers, airlines and kindred enterprises (Figure 9.4.1). Broken up by the federal government's anti-trust laws in 1934, the conglomerate spawned several large air-transport-related firms, of which Boeing, Pratt & Whitney (the engine makers) and United Airlines have survived to the present day.

Figure 9.4.1
Boeing facilities in the Seattle area

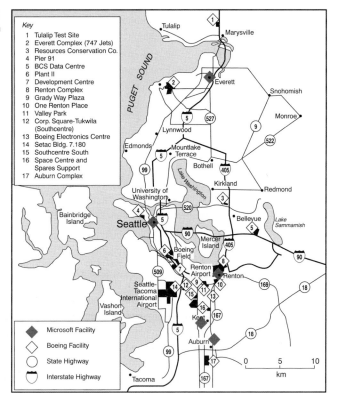

Before and during the Second World War, Boeing produced several famous bomber aircraft, including the B-17 Flying Fortress (first manufactured in 1935), of which a staggering 12 726 were produced. Ironically, many of these planes were used during the Second World War to blast the homeland of William Boeing's own ancestors. The B-29 Superfortress (completed in 1942) delivered fire-storms to Tokyo and the atom bombs to Hiroshima and Nagasaki that forced Japan's surrender in 1945.

After the war, the company continued its military output (e.g. the B52 bomber) but also developed important commercial aircraft. These included the Boeing 707, America's first jet airliner, which went into transatlantic service in 1958. By the 1960s Boeing had become a major aerospace company, designing and producing conventional aircraft, helicopters, moon rockets (the first stages of Apollo and Saturn), lunar orbiters, missiles and ground support systems. In addition, it has diversified into such areas as hydrofoils, railway carriages, computer services, and civil engineering for power stations and for water treatment and supply.

The Boeing 747 jumbo jet: wonder of the modern world

Commercial aircraft still account for 90 per cent of Boeing's business. Its most famous product is the 747 jumbo jet airliner, which has been acclaimed as one of the seven wonders of the modern world (Figures 9.4.2 and 9.4.3).

Due mainly to the success of the 747, Boeing has long held around 60 per cent of the international civil jet market and is the United States' biggest single exporter. In 1994 the company received a further boost when, with its smaller compatriot McDonnell Douglas, it was chosen by the government of Saudi Arabia to re-equip the national airline in a deal worth $6 billion.

The traveller

SEEN from the gallery where most tourists are herded, a jumbo jet on the production line at Boeing's Everett plant in Washington state is an oddly disappointing sight: it looks big, but not overwhelming. This *trompe l'oeil* is achieved only because its surroundings—a wonder in their own right—are even bigger. The 747 factory is the largest building, by volume, in the world, encompassing 13.4m cubic metres of space—enough room to accommodate about 550 Westminster Abbeys.

However, unlike its offspring, this factory cannot fly. The enormous objects it produces can—and have therefore shrunk the world. Boeing 747s have carried 1.4 billion passengers, the equivalent of a quarter of the world's population. They have flown 29 billion kilometres, almost 200 times the distance from the earth to the sun. They have made faraway places more accessible, not just for the rich but for Everyman. They have begun to fulfil the lofty vision of the first man to buy one, Juan Trippe, chairman of Pan Am: he called the plane "a weapon for peace", because there can be "no atom bomb potentially more powerful than the air-tourist."

Range and size are the reasons why the 747 made the world smaller. Both depend on power. A single modern jumbo-engine can generate more thrust than the four-engine array on an old 707, the jumbo's predecessor. That is why a modern 747 can carry up to 570 passengers or 122 tonnes of cargo.

Assembled from 6m parts, not counting rivets, the jumbo jet is not just the biggest machine produced on an assembly line, but the most complex too. So far there have been 15 different models, but no two of the 1,000 or so planes are identical. Each airline has its own preferences for the configuration of seats, and Boeing's engineers make other changes all the time. Thanks to extra fuel tanks (including one stuffed into the horizontal tail fin), a lighter fuselage and bent-tip "winglets", the 1,000th 747, delivered to Singapore Airlines earlier this year, has a range of 13,340 kilometres. The first 747, introduced in 1968 (and still flying), has a reach of only 8,510 kilometres.

The 747 which nowadays cost around $125m each, has been Boeing's most successful aircraft. Yet it was built as a stop-gap. Boeing and its first jumbo customer, Pan Am, both reckoned that the 747 would soon be pushed aside by supersonic travel. It may yet be. In the meantime, it continues to shrink the globe.

The Economist
December 1993–January 1994

Figure 9.4.2
The traveller

Figure 9.4.3
Boeing 747 jumbo jets under construction

Flying through the storm

Despite this background of success, several dark clouds appeared to be gathering on Boeing's commercial horizon, threatening a much stormier future:

- the early 1990s were a period of deep recession in air travel, with airlines losing around $10 billion worldwide from 1990–2;
- airlines scrapped fewer than 100 planes a year, compared with the expected 300 so new orders failed to materialise. Moreover, about 1000 redundant aircraft (one-tenth of the world's total commercial fleet) were placed in storage – over 200 were parked on desert airfields like Tucson, Arizona and Mojave in southern California (Figure 9.4.4).
- major new competitors for 747 markets appeared in the form of Europe's Airbus Industrie and several Asian NICs which were lining up to enter the aerospace business. Airbus Industrie (which includes France's Aérospatiale, Germany's Daimler-Benz Aerospace, British Aerospace and CASA of Spain) claimed nearly 40 per cent of the world passenger aircraft market by 1996;
- in 1993, Boeing's sales were down 16 per cent (to $25 billion) and its net profits down 20 per cent (to $1.2 billion). In addition one of the company's two 747 production lines was closed down, as monthly output fell from three jumbo jets to two. Overall, Boeing's production declined to only 20 aircraft a month, little more than half the 1992 output;
- in terms of employment, the all-too-familiar Boeing Boom–Boeing Bust syndrome came into effect. In 1989, Boeing's work-force numbered over 100 000 but this fell dramatically with 22 000 employees being laid off in 1993–94. Even so, Boeing remained Washington's largest single employer and the mainstay of the state's economy.

However, the fortunes of Boeing could, once again, be on the threshold of change. At the time of writing, Boeing were negotiating a takeover bid of McDonnell Douglas (a military aircraft manufacturer) worth $12.5 billion. If successful the advantages of the merger would include:

Figure 9.4.4
Surplus airliners parked in the Mojave Desert, southern California

- the saving of thousands of jobs at McDonnell Douglas, though some head-quarters staff may be lost. The new company would have a work-force of nearly 200 000 in 26 states.
- Boeing could utilise spare design, research and production capacity at McDonnell Douglas plants now that the depressed civil market is rapidly picking up. (In 1996, Boeing hired 20 000 workers to help it meet record demand for the 777 and other civil jets [1287 plance worth about $80 billion].)
- Boeing could finalise plans to build the new super-jumbo, a version of the 747 that will carry 550 passengers, 100 more than the original aricraft.
- Boeing could afford to exclude any immediate collaboration with Europe's Airbus Industrie, leaving the latter to face much more formidable US competition in the world's civil-jet market.

QUESTIONS

1 (a) What location factors originally influenced the establishment and early development of the Boeing Company in the Seattle area?

(b) What new factors have become important as the company has expanded and diversified the range of its products?

2 The high-tech company Microsoft, America's biggest software firm, is also based in the Seattle area (Figure 9.4.1). What mutual advantages for Boeing and Microsoft stem from this shared location?

3 Industry in Seattle (and the Pacific North West region as a whole) has generally been disadvantaged by its geographical remoteness from major markets. Assess the extent to which such isolation affects the Boeing Company.

4 What evidence is there in Figure 9.4.2 that justifies the description of the Boeing 747 as both a wonder and a shrinker of the modern world?

5 The number of Boeing employees has fluctuated considerably over the years as the following data show:

1939–4000	1945–15 000	1989–105 000
1941–30 000	1967–148 000	1995–72 000
1944–50 000	1971–38 000	1996–92 000

(a) Explain the 'boom-and-bust' pattern of Boeing's employment record since 1939.

(b) Which categories of workers are most likely to be 'hired and fired' most frequently?

6 (a) Why are desert locations favoured by airlines for parking their redundant planes?

(b) What disadvantages are there?

Boeing's future – the 777

Unveiled in 1994, the Boeing 777 will be the world's biggest twin-engined jet,

Figure 9.4.5
The Boeing 777

capable of carrying 300 passengers more than 6400 km (Figure 9.4.6). Later versions will carry 370 and fly almost twice as far. It cost Boeing $4 billion to develop – the company's biggest project since the 747.

Several processes that were innovatory for Boeing were incorporated into the development of the 777:

- major customers (including British Airways) were invited to participate in the plane's design and layout, whereas largely 'off-the-peg' aircraft have been offered in the past;
- computer-aided design (CAD) allowed potential problems to be identified in the pre-assembly stages, thus avoiding expensive re-working. There was no 777 prototype: the first plane produced was ready to operate commercially;
- practices already common in other industries were adopted; for example, 'total quality control' (TQC) – delegating responsibility for quality to all shopfloor workers, and 'just in time' (JIT) – arranging for suppliers' components to arrive precisely when needed rather than expensively storing them on Boeing's premises;
- setting an ultimate cost-cutting target of 25 per cent, thus improving profit margins.

Together, these management changes should guarantee the 777's success. Even so, although the 777 is capable of being 'stretched' to meet future demands, Boeing will soon have to decide whether to build a completely new 'super-jumbo', able to carry more than 500 passengers. But the $10 billion development costs and limited production run of about 500 aircraft (half that of the 747) will force Boeing to collaborate with rival aerospace companies – including Airbus (with whom feasibility studies have already been undertaken) and, perhaps, Asian and Russian manufacturers.

Thus, if Boeing is to maintain its position as industrial champion of the United States, even greater flexibility will be demanded as it faces the challenge of building the next generation of airliners.

QUESTIONS

1 Of all the new practices introduced by Boeing in making the 777, TQC and JIT probably had the biggest impact on the assembly process. Explain how.

2 For operating reasons, Boeing's customers are divided over the '777 or super-jumbo' issue. For example, American Airlines favour aircraft capable of long-range, direct flights which save passengers having to change planes at crowded hub-airports. On the other hand, British Airways and Japan Air favour aircraft capable of making fewer flights with more passengers, especially in Asia where airports have a shortage of landing slots. Which airline(s) are likely to order Boeing 777s and which super-jumbos, and why?

3 'The possible Boeing–McDonnell Douglas merger is mainly about horizontal integration.' Discuss and exemplify this statement.

4 Within Airbus Industrie, British Aerospace specialises in wing construction while Aérospatiale undertakes assembly work. Explain why the British partners are more eager than the French to collaborate with Boeing over the super-jumbo issue.

9.5 Post-Industrial America: The Service Sector

Traditionally, economic activity has been divided into three distinct sectors:

Primary: activities concerned with the direct exploitation of natural resources, e.g. farming, fishing, forestry, mining and quarrying.

Secondary: activities that convert the products of primary activity into more finely processed goods with more value added at each stage, e.g. manufacturing and construction.

Tertiary: activities concerned with the exchange and consumption of goods and services, e.g. trade, transport, distribution, business, personal and professional services. Tertiary activities are also known collectively as the 'service' sector.

The appropriateness of this simple tripartite division is now rather questionable. In highly developed countries like Canada and the United States a steadily diminishing percentage of the work-force is employed in primary and secondary (or goods-producing) activities. At the same time, the service sector has grown to dominate the 'post-industrial' employment scene and itself has become variously divided into different functional sectors.

One such sub-division is into *consumer services* and *producer services*. The former serve the consumer directly, e.g. shops, hospitals, cinemas, restaurants and public transport, while the latter assist other firms to operate more efficiently or by marketing and distributing their products.

Another sub-division of the services recognises increasingly sophisticated third, fourth and fifth levels of operation:

Tertiary: includes non-technical services, e.g. food, cleaning, personal; transport and trade.

Quaternary: includes office work, data-processing and other medium-grade activities which may be said to 'service the services'.

Quinary: includes higher levels of professional expertise, e.g. legal, financial and management consultancy; decision-making, and intellectual inputs from research and government agencies which play important roles in managing and controlling service and other activities.

A special category of *advanced producer services* (APS) is also recognised, combining elements of the sub-divisions already mentioned. APS may be defined as complex, knowledge-intensive business services designed as direct inputs into clients' operations. Typical examples include high-technology engineering, communications consulting, human resource management, strategic marketing, financial portfolio advice and specialised legal services. In short, APS assist other companies through the generation and transfer of knowledge, thus providing client firms with a competitive edge in their business operations.

Although the North American service sector contains many well-known companies of national or even global stature (e.g. airlines and banks) often employing thousands of people internationally, rather surprisingly, small firms run by one owner, functioning in one location and employing fewer than 20 people dominate the employment scene in both the United States and Canada.

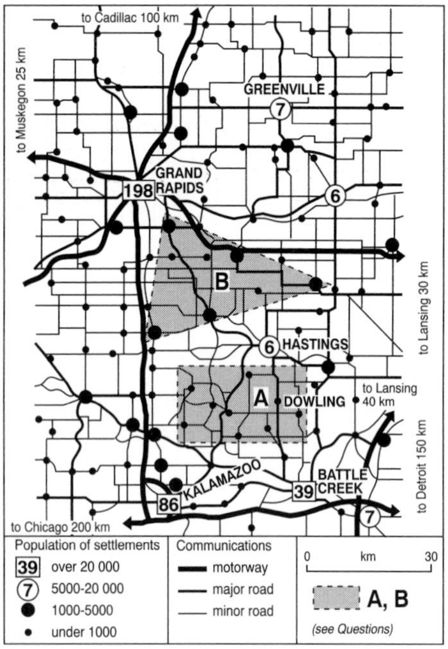

Spatial aspects

Important location factors affecting services include:

- the distribution and density of population;
- demographic factors such as age and gender;
- people's purchasing power;
- availability of labour with appropriate skills;
- proximity to other service activities.

Service centres in the Middle West

Generally speaking, services are grouped together in 'central places' which can be hierarchically arranged in terms of the services offered: places with larger populations offer a wider variety of services than smaller places.

In the Midwest of the United States most settlements exist essentially to meet the needs of the region's farming communities. Medium-sized centres usually have factories which specialise in manufacturing fertilisers, machinery and other agricultural equipment. Others buy up the produce of local farms and process it into food ready for the table. For example, Battle Creek (Figure 9.5.1) is the original home of Kellogg breakfast cereals. A drive to such centres also gives farm families an opportunity to visit bigger stores and brighter places of entertainment than are normally available nearer home.

Figure 9.5.1
Service centres in part of southern Michigan

For most farm families, however, day-to-day requirements are met by the much smaller country towns where the population usually numbers hundreds rather than thousands. In these centres the main buildings are the church, town hall, fire station, bank and school. Other services usually include a garage, a farm machinery repair shop and perhaps a co-operative creamery or grain elevator, together with a limited range of shops and offices.

Connected to the surrounding farms by a grid of minor roads, these small service centres are as characteristic of the Middle West as the farms themselves. For many of these once-prosperous service centres, however, the future looks very bleak. As increasing numbers of farm-dwellers have left the countryside for the city, and as the network of motorways has extended across the region, so the small towns have experienced a steady decline in their local importance.

QUESTIONS

1 (a) In the rectangle marked A on Figure 9.5.1, what is the average distance between each service centre with less than 1000 population and its 'nearest neighbour' of similar size?

(b) Make similar calculations for (i) the service centres with populations of 1000–5000 in the triangle marked B; (ii) the centres with populations of more than 20 000.

(c) What general relationship exists between size and distance apart?

2 (a) If you were living on a farm at A on Figure 9.5.1 which service centre would you probably visit for each of the following purposes:

- to buy weekly groceries
- to have dental treatment
- to buy a new combine harvester
- to see a top-class baseball game
- to have a tractor serviced
- to see a film
- to select an engagement ring

(b) What general relationship exists between a town's size and (i) the range of services it offers; (ii) the extent of its 'sphere of influence' (service area) over the surrounding countryside?

(c) Why does the area have so many small service centres but only a few large ones?

Services in cities

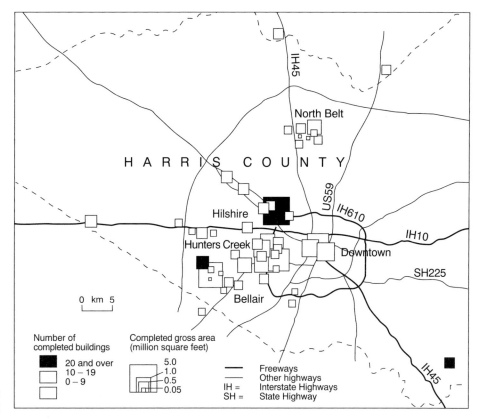

Figure 9.5.2
Office parks in Houston, Texas

QUESTIONS

1 Explain and exemplify why, in cities, consumer services usually occupy prime sites in the CBD core while producer services are usually found in the CBD frame (Figure 4.1.11).

2 Attempt to explain the locational pattern of office parks (Figures 9.5.2 and 4.1.11).

The Central Business District (CBD), despite its relative decline in recent decades, remains the favoured location for around two-thirds of all office jobs in North American cities – the skyscrapers reflecting high land values generated by competition for limited space. Consumer services often occupy prime sites in the CBD core while producer services are usually found in less expensive locations in the CBD frame.

Increasingly, however, office jobs are being decentralised into purpose-designed complexes known as *office parks* or *business parks*. These are located in the suburbs or in open countryside where offices are grouped together in carefully landscaped areas of lawns, shrubbery and woodland to provide an attractive working environment. By clustering together, companies can economise by sharing the cost of essential services. Planning laws normally exclude other land uses, but large numbers of office workers often attract consumer services such as cafes, restaurants, bars and shops which locate nearby.

APS providers, originally heavily concentrated in the CBD, have followed their customers out to suburban office parks and further afield to smaller cities and locations around and between major urban centres.

Houston (1994 population 1 702 000; metropolitan area 3 653 000), now the USA's fourth largest city, has 54 office parks (Figure 9.5.2). Most are between six and 16 km from the CBD. They range in size from four ha to 530 ha, the largest park having 61 buildings. Although the office park concept originated in the United States, many other countries have adopted the idea, resulting in a significant de-urbanisation of office employment.

10

Transport Developments

10.1 Continental overview

Transport is the movement of goods and people from one place to another. It can be viewed as the prerequisite that makes all other economic activity possible. Efficient transportation systems are vital to the high standard of living enjoyed in North America. The continent's highly developed networks are the result of considerable cumulative investment over a long period of time. Most places in both the USA and Canada can be reached more easily, more quickly, and more cheaply in real terms from other places than at any time in the past. However, as the scale of movement and the size of transport facilities have increased, a host of almost inevitable problems have come to the fore – among them pollution, congestion and saturation.

Recent trends

Transportation is a major sector of the economy employing 3.5 million in the USA and 450 000 in Canada in 1991. The industry has a huge turnover, with great rivalry for custom both between the different modes of transportation and companies offering similar services. Post-war trends in both countries have been generally similar although some important variations are apparent with regard to both passenger and freight movements.

Figure 10.1.1
USA: Volume of domestic intercity freight traffic, (% share)

Mode	1950	1970	1980	1993
Railroads	57.4	39.8	37.5	38.1
Truck	15.8	21.3	22.3	28.1
Water	14.9	16.5	16.4	15.1
Oil pipelines	11.8	22.3	23.6	18.4
Domestic airways	0.03	0.2	0.2	0.4
Total traffic volume [billion tonne km]	1571	3117	4004	4762

Figure 10.1.1 shows changes in US domestic intercity freight traffic between 1950 and 1993, during which time the relative contribution of the railroads declined from 57.4 per cent to 38.1 per cent. However, this decline, due to increasingly intense competition from road haulage and pipelines, began much earlier. In 1916 railways moved 77 per cent of all US intercity freight. At this time the dependence of the national economy on the railroads was overwhelming although it must be kept in mind that the relative decrease in the railways' contribution actually masked a doubling of freight carried between 1916 and 1980. Despite the continuing problems of the railways they are clearly still the major freight carrier in the USA and likely to remain so for some time.

Road haulage has relentlessly increased in importance, its market share almost doubling between 1950 and 1993 because of its cost-competitiveness and high degree of flexibility. Road transport has benefited considerably from a

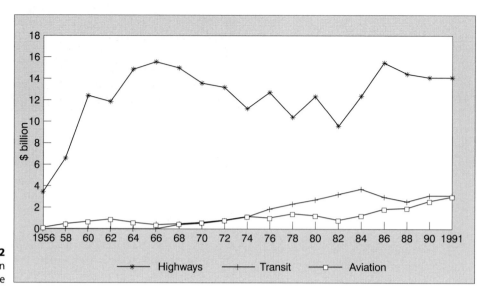

Figure 10.1.2
USA: federal capital investment in transportation infrastructure

huge federal capital investment in infrastructure (Figure 10.1.2). In contrast, the era of rapid pipeline construction ended in the 1970s and so it is not surprising that this mode has passed its peak in relative terms.

The network of domestic water transport routes was already well established by 1950, handling mainly high-bulk, low-value products. The value to volume ratio of such products, many of which are unprocessed raw materials, frequently makes transport by other modes prohibitive. Over 40 000 tonnes can be moved in a single tow on inland waterways. In terms of cost, only pipelines are cheaper but they are suitable only for the movement of a limited number of products. The cost-effectiveness of water transport for bulk movement is illustrated by the long-standing attraction of coastal, lake, and major riverside locations to heavy industry. Shallow-draught units can easily be dropped off at various locations along a waterway with the main tow being sent on its way in a short time. Facilities for berthing barges need not be extensive; the Mississippi River system alone contains 60 inland ports and about 1000 marine terminals. The Great Lakes–St Lawrence Seaway practically provides a fourth coastline for both nations, although traffic has not lived up to early expectations (see section 10.6). However, the USA does possess a number of other important waterways, especially the Mississippi–Ohio system, which boast impressive annual traffic flows, but in Canada movement by this mode of transport is very limited outside the main Great Lakes–St Lawrence Seaway.

Figure 10.1.3
USA: volume of domestic intercity passenger traffic (% share)

Mode	1950	1970	1980	1993
Private automobiles	86.2	86.9	82.5	80.8
Domestic airways	2.0	10.1	14.9	17.4
Buses	5.2	2.1	1.8	1.1
Railroads	6.4	0.9	0.8	0.7
Total traffic volume [billions of passenger km]	818	1901	2362	3423

In terms of US passenger transportation, post-war changes, while being significant, have been of a lesser magnitude than for freight (Figure 10.1.3). This is mainly because a major transition had already taken place in the preceding inter-war period with the great surge in motor car ownership. In recent decades the dominance of the motor car has been maintained while the railways have continued their relative decline. The main development during this latter period has been the tremendous growth of the airlines as intercity passenger carriers, increasing their market share from 2 per cent in 1950 to over 17 per cent in 1993.

Data for Canadian freight transport is not directly comparable with that for the USA, but some differences are clearly apparent. In Canada the railways have maintained a stronger grip on freight traffic compared with their counterparts in the USA, while pipelines also account for a greater share of movement in Canada the difference is not great. As in the USA, intercity passenger movement is dominated by the private car. However, the relative decline of bus and rail passenger transport has been less severe. While the use of air transport has risen rapidly, per capita movement remains below that of the USA.

Increasingly intermodal

Overall there has been a trend towards complementarity as termini for different modes have been linked for both passenger and freight transport, with each mode functioning where it is most cost-effective. Intermodality has undoubtedly been the most important trend in North American transportation in recent years and will continue to be so for some time to come. For freight, intermodal transportation is primarily based on the fast and efficient switching of containers from one form of transport to another. Containers are boxes made of steel or aluminium into which goods are packed. They are 2.4 metres high, 2.4 metres wide and can be 3, 6, 9 or 12 metres long. They can be carried on railway flat cars, lorries, aircraft, barges and ships and moved quickly from one to another. The general use of standard-sized containers began in the early 1960s and now all major transport terminals in the USA and Canada have container facilities.

Deregulation

From the late 1970s there has been a significant reduction in government control of transport services in North America. This process, known as 'deregulation' began in 1978 in the USA with the airlines, to be followed by road haulage and segments of the rail network in 1980. Canadian airlines were deregulated from 1984 and in 1988 the government passed a new National Transportation Act. The Act was designed to lessen economic regulation, to improve operating efficiency, to increase the range of services available by promoting competition within modes as well as between them, and to strike a new balance between the needs of users and carriers.

Deregulation has thus brought market forces to the fore in the transport sector. It is now much easier for companies to enter or leave the market. It has allowed carriers to design more flexible and demand-responsive services and clearly shown that some companies had been uneconomically protected by the previous regulations. In 1980 there were 298 Class 1 intercity motor carriers of general freight in the USA but by 1990 the number was down to 191. During the same period the number of Class 1 motor intercity passenger carriers fell from 48 to 21. An obvious benefit to consumers is that the real cost of transport in general has declined but some experts are concerned about the implications for safety in an increasingly competitive market.

Government funding

In the United States transport funding comes from federal, state and local sources. In 1992 the federal government spent $34 billion on transport. 62 per cent of the total was destined for land transportation with 27, 11, and 0.8 per cent for air, water and 'other' transportation respectively. In Canada funding comes from federal, provincial and municipal governments. In 1992 an estimated C$7078 million was spent on highway, road, street and bridge construction and repair.

Environmental impact

There is almost universal recognition that the environment is extremely vulnerable to the development of transport infrastructure and the use of the networks provided. The difficult job of government at all scales is to attempt to balance the economic and environmental perspectives. Environmental groups feel strongly that the latter has not been given a high enough priority.

The advent of the Clinton administration brought the issue of energy taxes centre-stage. Initially the plan was for new taxes on all forms of energy except renewable sources. In June 1993 the proposals were modified to include only those fuels used in transportation. The transport industry has vehemently opposed such a move. Figure 10.1.4 shows the pollution impact of transportation in the USA in 1992 with this sector of the economy responsible for over 80 per cent of carbon monoxide emissions.

Environmental groups advocate greater investment in public transport as one strategy for tackling the considerable pollution problem (Figure 10.1.5). However, in recent years there has been a relatively limited investment in urban public transport with a significant disinvestment in rural services as competition, higher fuel and labour costs, and rising car ownership have forced many small companies out of the market.

Figure 10.1.4
US air pollutant emissions caused by transportation, 1992

	Transportation % of total
Carbon monoxide	80.3
Sulphur oxides	4.9
Volatile organic compounds	36.4
Particulates	3.6
Nitrogen oxides	44.8
Lead	29.8

CARS, COACHES AND CARBON MONOXIDE

What produces more air pollution — 43 cars and their drivers, or a bus carrying 43 people? That's what Environment Canada asked when it compared the exhaust emissions of a 43-seater diesel bus with those of 43 new cars. Their findings? The cars together produced about 8 times more pollution than the bus — 54 times the carbon monoxide, 8.4 times the hydrocarbons and 2.4 times the nitrogen oxides — all known to contribute to the greenhouse effect. The bus, however, because it used diesel, produced more 'particulate matter' (tiny particles of pollution) than the cars.

Canada Year Book, 1994

Figure 10.1.5
Cars, coaches and carbon monoxide

Regional variations

The transport system in North America exhibits distinct spatial variations. Such differences can be examined by using a basic model (Figure 10.1.6a) which is concerned with two important dimensions of transportation. The first dimension relates to the maturity of the transport service. A system can be 'mature' in that the basic facilities are in place, offering a high level of service as represented by quality and frequency; or it can be undeveloped with relatively low quality and frequency of service. The second dimension relates to the degree of competition, both inter-modal and intra-modal. The level of competition may vary from high inter-modal with many modes and several carriers for each mode, or high intra-modal with basically one mode and several carriers, to the most basic service of one mode and one carrier. Figure 10.1.6b applies the model to the movement of bulk commodities in Canada. Eastern forest products can be moved by a variety of modes, with rail, truck and water available. In contrast, Arctic resources are generally located in remote areas where the transport infrastructure is relatively undeveloped with little competition. The model can be applied to passenger transport with equal validity.

The fundamental question is why such variations occur. The causal factors are multivariate but the major influences on the development of transport networks have been the distribution and density of population and the physical environment. The effect of the physical environment on transport routing and operation has weakened with advances in technology but, nevertheless, it remains strong. This inter-relationship differs markedly between the various modes of movement according to their technological limitations and the degree of inertia existing in each system.

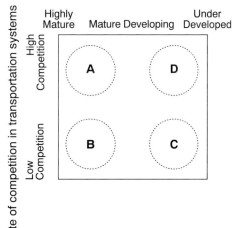

(a) State of maturity of transportation systems

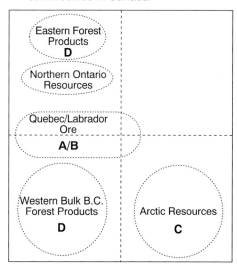

(b) Model applied to the Movement of bulk commodities in Canada

Figure 10.1.6
The maturity of transport systems: a model

The northlands: isolation and transportation

The great distances separating the generally small communities in the northlands and the severe environmental conditions prevalent in this vast region have created substantial economic, engineering and maintenance difficulties for transportation development. Immense areas of the northlands are lacking in surface communications. Not one of the eleven railway lines extending into the northlands crosses the Arctic Circle. The most northerly is the Alaska railroad from Seward, on the south coast, terminating at Fairbanks (c. 64.5°N). Construction began in 1915 and it is the oldest of all the Northland railways. Only five of the other lines were constructed prior to the 1950s. The most recent railway, that to Dease Lake was completed in the late 1970s.

The road system is also very sparse, the most important elements of which have been the construction of the Alaska, Mackenzie and Demster highways. The Alaska Highway, an all-weather gravelled road 2450.5 km long, extends north west from Dawson Creek to Fairbanks. The project was completed during a ten month period in 1942 as an extension of an existing Canadian road between Dawson Creek and Edmonton. The northern limit of the Mackenzie Highway is Yellowknife on the northern shore of the Great Slave Lake. The town, which has a population of over 6000, was founded in 1935 after the discovery of rich deposits of gold.

The Demster Highway which was opened in 1979 begins near Dawson and cuts through tundra and mountains to Inuvik. In winter an ice road continues the highway to Tuktoyaktuk on the coast of the Beaufort Sea. Vehicles cross the Peel and Mackenzie rivers by ferry in summer. Ice bridges make it possible to drive across in winter. The highway is closed during the spring break-up (April–May) and the autumn freeze-up (October–November). The times when the highway is closed roughly coincide with the yearly migration of over 100 000 caribou between the Yukon and Alaska. A metre-thick layer of gravel and shale insulation under the highway keeps the permafrost from melting and breaking up the surface. In winter, drifting snow is a major problem. When drifts get too high (up to six metres) the snow-ploughs bypass them and make an auxiliary winter road.

The use of water transport in the region is dictated by location and season but the growing importance of air transport in the post-war period has greatly reduced the significance the former had to native communities in earlier times.

1 Account for the changes in US intercity passenger and freight movement between 1950 and 1993 (Figures 10.1.1 and 10.1.3).

2 Describe the trends in US federal capital investment in transport infrastructure shown in Figure 10.1.2.

3 What are the reasons for the increasing intermodality of transport?

4 (a) What is deregulation?

(b) Why has deregulation been an important aspect of government policy in both the US and Canada in recent years?

5 Comment on the pollution data provided in Figures 10.1.4 and 10.1.5.

6 Examine the reasons for regional variations in transport provision.

7 Account for the provision of transportation in the Northlands.

10.2 Railways

The role of the railways in both passenger and freight transportation has been steadily undermined in the twentieth century, although a combination of historical, geographical and economic factors has enabled Canadian railways to prove more resistant to competition from other modes of movement than their counterparts in the USA. Transportation developments have invariable diffused more slowly through Canada. In 1850 only 106 km of track existed in the whole country compared with 14 520 km in the USA. Conversely the relative decline of the railways in Canada has been a much more recent phenomenon. Canadian railway passenger traffic peaked in 1967 while in the USA the zenith was reached 51 years earlier.

The railway is most competitive over long journeys and average haul distances are significantly greater in Canada. In addition, while severe winter conditions are a hazard to all modes of transport, railways are less vulnerable to disruption than is road haulage. Thus the greater severity of Canadian winters has given the railways a certain competitive advantage in such conditions. An associated factor is that in the USA alternative transportation systems have received proportionately higher federal funding, resulting in the development of more sophisticated road and air networks. The economics of railway rates is often complex but in general Canadian railways have operated in a more favourable regulatory climate with greater government sympathy. As a result Canadian railways have given higher returns on capital, acting as an important stimulus to further investment.

America's shrinking passenger rail network

In 1916, the peak year for US railways in passenger traffic, a total of 56 billion passenger km were logged. This accounted for 98 per cent of all intercity passenger business. However, after the First World War a sharp decline occurred in the demand for passenger and freight rail transport. By 1966 only 27 billion passenger km were logged with passenger trains running on only about one-third of the network. Passenger use fell even further to 16 billion passenger km in 1978.

The motor vehicle dominated the first phase of competition, signalled by the establishment of the first federal highway programme in 1916. In the 1920s intercity motor transport soon surpassed movement by rail, and by 1930 was six times as great with 23 million cars registered in the USA. The success of the motor vehicle was virtually guaranteed as it could provide the flexibility and convenience that was impossible for the railway system. The great dispersal of population created by the automobile further alienated the railway from an increasing proportion of the population and strengthened reliance on motor transport.

The second major phase of competition emerged strongly after the great technical advances of the US aircraft industry in the Second World War. The scale of the US, in terms of physical size and economic activity, has given the airlines an obvious and increasing advantage over the railways for passenger movement.

Amtrak: passenger rail's last chance

It is only since the early 1970s that substantial government involvement with the railways has been apparent and only then to prevent the virtual disappearance of the network from large parts of the country. By 1970 intercity rail passenger traffic had dwindled to only 7 per cent of the total for public carriers with only 450 daily intercity passenger trains remaining in service. Some companies had totally discontinued services while others performed acute surgery on their unprofitable networks.

In a dramatic bid to halt this downward trend the Rail Passenger Service Act was passed. It came into operation on 1 May 1971, creating the first nationwide passenger rail system in the US. A National Railroad Passenger Corporation known as Amtrak was established to take over passenger operations on the systems of 22 railway companies. Without this action services would have disappeared from many parts of the country. Amtrak scrapped many poorly used services but routes linking all the major cities were kept. Much of the rolling stock has been modernised but a lot of government money has been needed to keep Amtrak going. However, in recent years Amtrak has been able to reduce its level of subsidy. Travel on Amtrak in 1992 was running at 9.6 billion passenger-km a year, up from 8 billion in 1970 (Figures 10.2.1 and 10.2.2).

TRAIN spotters in the Chicago area like to bumble down to Dolton Junction. Nowadays there are fewer trains than in Dolton's heyday, but at least the ones passing through are running at a reasonable profit. Add to this a string of big railway mergers and some impressive increases in productivity and America's iron horses are looking more lively than they have for years.

Graeme Lidgerwood-Dayton, an analyst at cs First Boston, reckons that operating income at the 12 biggest railway firms, which account for over 90% of freight traffic, will increase 16% to around $5 billion this year on revenues of $31 billion. Rail freight has been winning business from the roads. In the first eight months of this year the 12 biggest firms carried 11.9m car loads of goods—5.5% above last year's figure. "Intermodal" traffic (ie, containers and trailers) increased from 4.7m to 5.4m units.

The railways firms claim great leaps in customer service—particularly "schedule integrity", which is railspeak for being punctual. But the revival owes as much to vigor-ous pruning of track networks and personnel. In 1993 the firms employed 193,000 workers, in 1980 460,000. The companies have at least ploughed money into improving the track they have retained and, more recently, updating engines and rolling stock. Already, $4.6 billion in capital spending has been announced this year, up 9.5% over 1993. Some 50,000 new freight cars may be delivered this year.

According to the Bureau of Labour Statistics, productivity in the railway industry grew at an annual lick of 7% between 1987 and 1992, placing it fifth among the 172 industries measured. However, this achievement stemmed almost entirely from the privately owned freight carriers. State-owned Amtrak, which accounts for all intercity passenger traffic, has seen ridership remain flat at around 22m passengers a year. Its greatest achievement has been to reduce its level of subsidy—only 20% of its 1993 budget against 28% in 1990. But Amtrak is now accepting bids to supply over 20 high-speed trains to cut travel time in the north-east corridor.

Now the private rail industry is consolidating. Two couplings announced this summer and now waiting for approval from federal regulators would tie Burlington Northern to Santa Fe (which would create the country's biggest railroad with $7 billion in combined revenues and 33,000 miles of track), and Illinois Central to Kansas City Southern. Federal regulators are also considering a request from Union Pacific to exercise voting rights in stock held in the Chicago and North Western railway. Many speculate that Norfolk Southern is trying to revive a ten year-old interest in Conrail.

The railway firms' biggest problem is that they have won business from the roads only by cutting prices savagely. Hence the move towards partnerships with trucking and other transport firms. The talk is of offering clients "seamless service" between road and rail through one-stop shopping. Maybe the best way to make money as a railway company is to stop being one.

The Economist
17 September 1994

Figure 10.2.1
A railway revived

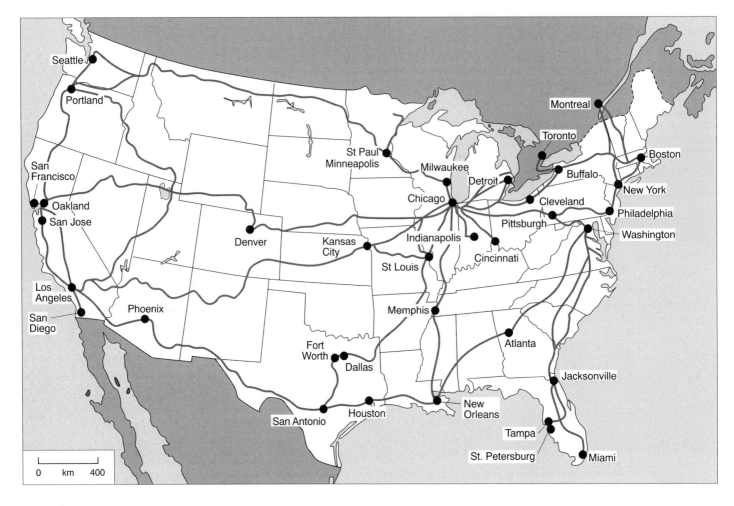

Figure 10.2.2
The Amtrak passenger rail system in the USA

On the east coast Amtrak has captured 42 per cent of the travel market between New York and Washington.

Rail freight in the USA

The railways have been most vulnerable to competition on short and medium haul routes and thus the most profitable companies have tended to be those in the south and in the west. The shorter hauls of the north east Manufacturing Belt have suffered from much fiercer competition from the truck, resulting in bankruptcies among railway companies and cuts in the overall network.

Five years after the establishment of Amtrak the federal government began its second major excursion into railway operations under the Railroad Revitalisation and Regulatory Reform Act of 1976. The Act established the Consolidated Rail Corporation, known as Conrail, by acquiring six bankrupt railway companies. Its operational area covered over 27 000 route km in 16 states with a few routes extending into Canada. As with Amtrak the federal government was forced to intervene under threat of the virtual disintegration of the freight network in the north east. However, although freight provision in the region has been rationalised, Conrail continues to incur substantial losses, largely because it is faced with the same difficulties as the six independent companies it replaced.

In recent years rail freight has made something of a comeback (Figure 10.2.1), with substantial gains in productivity since deregulation in 1980. The increasingly popular technique of 'double-stacking' (i.e. putting containers on top of each other on flat cars) has reduced costs and enabled carriers to offer more competitive prices. In the mid-1980s west-coast shipping companies pioneered double-stacking inland from ports such as Los Angeles, Oakland and Seattle. The industry has also improved its telecommunication tracking of individual freight cars, a big concern for customers.

The railways have played an increasing role in intermodal traffic involving either old-fashioned containers that have to be hoisted by crane on to and off cars, or the new box-trailers that come with interchangeable sets of road and rail wheels. Intermodal traffic now accounts for about a quarter of the railways' total revenues of around $28 billion (Figure 10.2.3). Such big money has resulted in a number of partnerships between rail and trucking companies who talk of offering clients a 'seamless service' between road and rail.

As companies and regions take a fresh look at rail transport a number of important improvements to infrastructure are in the pipeline, the largest of which is the Alameda Corridor project (Figure 10.2.4).

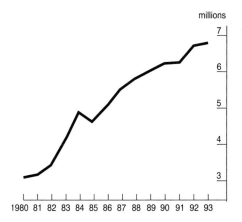

millions

Figure 10.2.3
Trailers and containers transported by rail

Figure 10.2.4
The Alameda Corridor project

NEVER underestimate an incumbent. The office of Governor Pete Wilson of California, who is running for re-election in November, has just announced that it is seeking approval from the California Transportation Commission to start work on a massive rail project linking Los Angeles to its twin ports 18 miles [29 km] to the south. Voters approved Proposition 116, which earmarked $80m to start work on the so-called Alameda Corridor project, back in 1990. The $1.8 billion Alameda Corridor—the biggest construction project west of the Mississippi—is expected to provide thousands of building jobs over the rest of the decade.

The rail link has even greater significance for Los Angeles in terms of international trade. Business has picked up impressively at the twin ports of Long Beach and Los Angeles (at San Pedro) since the NAFTA agreement went into

effect in January and since China had its most-favoured-nation status renewed in May. Both ports expect their cargo volumes to increase by 6–7% this year and even more next year. That means lots of new jobs; and they are good jobs, too, paying 17% above average.

The biggest brake on further expansion of the ports—and the number of new jobs—is the link between the ports at Long Beach and San Pedro and the warehouses, factories and rail

way yards in downtown Los Angeles. No fewer than 34 level crossings delay the trains as they creep through the districts of Watts, Lynwood, Compton and Carson. Trains can take anything up to eight hours to make what is a 30-minute journey by road. The Alameda Corridor would put the trains in a trench for half the route and provide overpasses for road traffic for the rest, cutting the journey time dramatically.

The $80m of state money that Mr Wilson is seeking to release will allow work to start immediately on five big overpasses without having to wait for final approval of the whole project. So far, the Alameda Corridor Transportation Authority has managed to secure $1.1 billion of the money needed. The federal government is expected to stump up the remaining $700m.

The Economist,
24 September 1994

Canada's Via Rail

Via Rail is Canada's national passenger rail company. In 1993 it operated 435 trains weekly on 14 000 km of track, serving more than 400 communities. Carrying more than 3.6 million passengers, the company earned more than C$164 million in total operating revenue. Economic impact studies estimate that more than 23 000 full-time jobs depend on Via Rail's operations and related off-train passenger spending and that Via Rail generated activity contributed approximately C$1.8 billion to Canada's GDP. However, in 1989 it was announced that the federal subsidy of Via Rail would be reduced by more than 40 per cent over five years. The Via cutbacks, including the cancellation of the historic daily Transcontinental, sparked an emotional national debate. In April 1993, further reductions to the Via Rail subsidy were announced as part of the deficit reduction measures contained in the federal budget. Via services in the Quebec City–Windsor corridor attract the highest ridership and generate the greatest revenue. Via is following closely the work of the joint federal–Quebec–Ontario task force on the possible development of a high-speed rail system.

Canadian rail freight

The two transcontinental giants, now known as CN North America and CP Rail System, dominate railway freight traffic. Together they operate 89 per cent of the country's main and secondary lines and handle a similar proportion of rail freight traffic. The basic networks of these two large carriers are supplemented by a number of smaller regional companies. CN (Canadian National) is a Crown corporation formed over the period 1919 to 1923 from the collapse of four railways. CP (Canadian Pacific) was the first transcontinental railway. In 1990 they operated networks of 31 400 and 20 300 km of line respectively. There are 23 other railways in Canada; these include six US companies serving Canada and the provincially owned British Columbia Railway and the Ontario Northland Railway. CP and CN are currently examining how their networks east of Winnipeg might be merged so that both companies can reduce running costs and offer more competitive services.

Figure 10.2.5
The St Clair tunnel under construction

Major improvements to the infrastructure of rail freight have been completed in recent years in an effort to improve the functioning of the system. Such works include:

- CP's Vaughan Terminal in Toronto, which became operational in 1991. The terminal, rated by CP as the most modern inland intermodal terminal in North America, boasts 2.4 km of working track, over 3 km of storage and make-up tracks, and three computer controlled gantry cranes which can unload a 1.5 km train in ninety minutes. The terminal, strategically located near major highways, covers an area equal to 133 football fields.
- The new St Clair Tunnel (Figure 10.2.5). This giant 9.5 metre diameter tunnel under the St Clair river links Canada with the double-stacked rail network of the USA. The 2 km-long tunnel was completed in 1995 at a cost of $155 million, and links Sarnia, Ontario with Port Huron, Michigan. It is a major advance in the creation of a 'seamless' North American freight transportation network as it accommodates oversized railcars, which previously had to be taken across the river by barge.
- Rogers Pass. The high passes of the Western Cordillera were a great challenge to the first railway engineers. That challenge returned recently as freight traffic to the Pacific ports increased to such an extent that major improvements became necessary. Rogers Pass crosses the Selkirk Mountains in British Columbia, curving through a steep-sided valley between peaks towering 2800 metres high (Figure 10.2.6). The first track through Rogers Pass was laid in the 1880s. Avalanches from the snow-capped peaks were a frequent and dangerous problem. To avoid this hazard, the 8 km Connaught Tunnel, opened in 1916, was built 150 metres below the summit of the pass. Now a new and longer tunnel exists to cope with the increasing number of trains between Calgary and Vancouver.

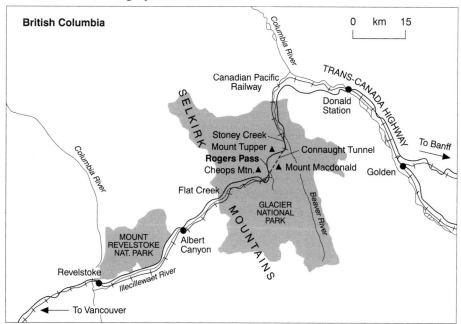

Figure 10.2.6
Rogers Pass in the Selkirk Mountains

The pass is used mainly by unit trains carrying bulk materials such as coal, potash and grain. The coal trains often consist of over a hundred wagons hauling up to 10 000 tonnes. Such loads are pulled by six diesel engines over gentler gradients. But up to six more engines were needed to push the load over 12 km up the steepest part of the track to the Connaught Tunnel. Once the train was on a gentler gradient the pusher locomotives returned downhill for the next push. This happened up to ten times a day, causing considerable delays to traffic.

QUESTIONS

1 Analyse the Amtrak passenger rail network shown in Figure 10.2.2.

2 To what extent has rail freight halted its 'decline'?

3 Why was Conrail established?

4 Discuss the rationale of the proposed Alameda Corridor route (Figure 10.2.4).

5 Why was the upgrading of the Rogers Pass route so important to Canadian rail freight?

6 Assess the environmental impact of rail transport.

Rogers Pass had a capacity of 15 westbound trains a day. By the early 1980s this limit had almost been reached. The solution was to build 34 km of new track for westbound traffic, including a new 14.5 km tunnel, the longest in North America (Figure 10.2.7). This has reduced the gradient and speeded the flow of traffic. Rogers Pass was the last phase of the Canadian Pacific Rail project to eliminate the steep inclines facing westbound trains by reducing the gradient to a maximum of 1 per cent. It was completed in 1989 at a cost of C$600 million.

Figure 10.2.7
The Rogers Pass Project: work on the Macdonald Tunnel

NAFTA

The recent signing of the North American Free Trade Agreement (NAFTA) between Canada, the USA and Mexico brought the continentalisation of the economy to the fore. The increase in trade between the three nations and projected future expansion has resulted in new attention being given to the development of transportation services. Within Canada and the US it has heightened questions about the implications of further growth in north–south trade for a railway system that has depended primarily on east–west traffic.

Environmental considerations

Railways are generally regarded as having a relatively low impact on the environment. The advantages of this mode of transport include efficient fuel consumption, low emission levels, and minimal land use. Of course most people avoid living close to railways if they have the choice, and environmental groups carefully monitor proposals for improvements and extensions. However, if traffic has to move by either road or rail, the latter is seen as a much lesser evil.

10.3 Atlanta's MARTA Mass Transit System

Atlanta (population 396 000; metropolitan area population 3 331 000) has always been an important transport centre. Founded in 1837 and originally named 'Terminus' because several railways converged there, Atlanta grew rapidly and then was virtually destroyed for its strategic importance during the Civil War. As the modern regional focus of road, rail and air networks, Atlanta has become a major administrative, distribution and convention centre (Figure 10.3.1).

Figure 10.3.1
Atlanta's CBD

Atlanta's varied industries include motor vehicles (General Motors at Doraville), aerospace (Lockheed at Marietta), textiles, chemicals and wood products. The city gave the world Scarlett O'Hara and Martin Luther King, and is the home of the world's largest soft-drinks company (Coca-Cola) and its main 24-hour news TV network (CNN). Symbolic of the 'New South', Atlanta is the capital of Georgia and the unofficial capital of the south east region.

MARTA: an integrated bus and rail system

Continuing the city's transport tradition, the Metropolitan Atlanta Rapid Transit Authority (MARTA) was created in 1965 and became an operating agency in 1972 when it purchased a privately-owned bus fleet and constructed a 100-km rapid transit rail system, or metro, funded by a 1 per cent local sales tax.

Figure 10.3.2
MARTA train traversing Atlanta's CBD

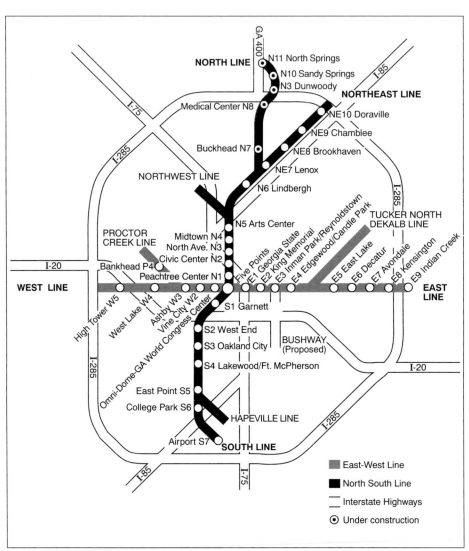

Figure 10.3.3
The MARTA metro system, 1994

Since 1979 about 65 km of the metro network and 33 stations have been opened, at a cost of $2.3 billion, of which 43 per cent has been raised locally and 57 per cent is federal funding. By 1994 the metro system remained confined within Atlanta's ring road, Interstate Highway 285 (I-285) (Figure 10.3.3). There are, however, plans to break through that boundary to reach dynamic, high-growth areas in the north and east.

As new metro lines open, MARTA's policy is to integrate its bus and rail systems by diverting bus routes to feed rail lines and by charging a standard low fare ($1.25 in 1996) with free transfer between buses and trains throughout MARTA's service area. To encourage motorists to use the metro, MARTA has provided over 25 000 free parking spaces, mostly at outlying park-and-ride stations. Ultimately, MARTA hopes to operate a series of commuter rail corridors linking communities throughout northern Georgia (Figure 10.3.4).

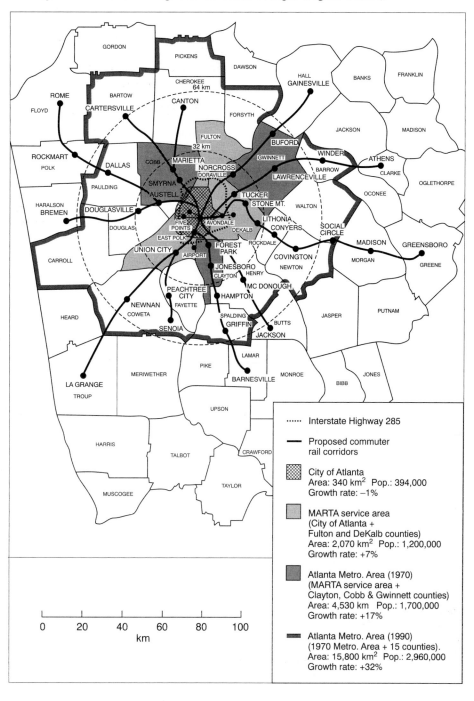

Figure 10.3.4
Proposed commuter rail corridors in northern Georgia

Figure 10.3.5
MARTA: operating data for 1994 ($ million)

Operating costs	191.1	Fare revenues[1]	72.2
		Local sales tax[2]	92.6
		Other local income[3]	10.1
		Federal subsidy	8.9
Total:	191.1	Total:	183.8

[1]Must, by law, cover at least 35% of operating costs.
[2]By law, not more than 50% of the total $185.2 million tax may be used to cover operating costs.
[3]Includes interest on capital, advertising revenues, etc.

The 1996 Olympics

Atlanta has more than its fair share of the problems that beset every large American city. The core city has lost over 100 000 people since 1970 as the middle classes, both black and white, have fled to the prosperous suburbs. Two-thirds of central Atlanta's residents are black, and close to one-third fall below the poverty line. Thus, central Atlanta is steadily losing its tax base and cannot afford to repair its decaying urban infrastructure. Most worryingly, with 404 violent crimes for every 10 000 people, Atlanta is the most violent city for its size in the United States. Even Miami, Washington and New York are safer.

Despite these problems, Atlanta was chosen to host the 1996 Olympic Games, not least because of the excellent transport facilities provided by MARTA. The metro's South Line reached Hartsfield International Airport in 1988. Trains leave the Airport Station every eight minutes on weekdays and give access to the most distant MARTA rail terminal in less than 40 minutes. Since some two million visitors were expected during the 17 days of Olympic competition, MARTA's ability to provide such good links with Atlanta's busy airport and aviation hub was crucial in the city's successful bid for the '96 Games.

Figure 10.3.6
The plan to avoid gridlock

QUESTIONS

1 (a) Discuss (i) the actions that MARTA might take to remedy the deficit shown in Figure 10.3.5; (ii) the difficulties attached to each suggested action.
  (b) Why should MARTA expect to receive little financial help from Atlanta's city government?

2 (a) Referring to the proposed rail link between Atlanta and Athens (Figure 10.3.4), for each successive section of the metropolitan area, describe and explain (i) its demographic and socio-economic characteristics; (ii) the actual and prospective mass transit situation.
  (b) Suggest the most likely source of funding needed to extend the MARTA system to the area shown on Figure 10.3.4.

The plan to avoid gridlock:

'Take the bus' — or just try walking

It's an incredibly intricate plan, involving as many as 10,000 workers, 2,000 buses, 2 million spectators and 13 venues, but A.D. Frazier Jr, sums up the 1996 Olympics transportation system for Atlanta in five words:

"Park and take the bus."

Of course, it's not quite that simple, the Atlanta Games chief operating officer is quick to add. The transportation plan unveiled Thursday is a logistical mind-blower that must coordinate everything that moves inside I-285 during the 17 days of the Atlanta Olympics.

If the plan works — and it must, to ensure a successful Games — Atlanta will be converted from a region of freeway junkies to a city reliant on mass transit and shoe leather.

Meanwhile, Olympics officials say that it will make sure that everyone who needs to get to work will still be able to, although they may have to change their routes and timing.

The transportation plan contains a number of firsts. No other Olympic city has attempted to orchestrate transportation to this extent, in part because no other Games has been so concentrated in a city centre.

For the first time, the cost of transportation will be included in the price of a ticket to an Olympic event, in the form of a $2.50 built-in charge. Each ticket will be good for rides on the Olympic bus system and MARTA for the entire day of the event.

Around-the-clock schedules

The Atlanta Games will also mark the first time that dozens of cities lend their newest buses for a major event elsewhere; in all, 2,000 are coming from around the country. Fetching, maintaining and returning the buses will cost $10 million, paid for by the federal government.

Olympic park-and-ride lots will ring the city from as far downtown as Alpharetta, Marietta and Stockbridge. From there, visitors will take shuttle buses to MARTA stations. When the trains overflow, buses will take patrons directly to venues.

MARTA will run trains and buses around the clock. The rail system will operate at peak capacity, using every rail car available for as long as 20 hours a day. As the centrepoint of the entire system, MARTA rail will carry as much as three times its usual daily load of 184,000.

Inside the Olympic ring downtown, private cars will be more a liability than an asset. Several main streets will be closed to cars and turned into pedestrian malls. Others will be exclusively bus routes. People who work in the area will be issued special passes that will allow them to drive.

"There will be no way for healthy people to get around but to walk," said Joel Stone, chief transportation planner for the Atlanta Regional Commission. Stone oversaw the writing of the plan.

On the highways, high-occupancy-vehicle lanes will be dominated by Olympic express buses. A command centre operated by the Georgia Department of Transportation will monitor traffic and coordinate signal lights across the region.

In Los Angeles, a larger region in which the venues were less concentrated, freeway traffic dropped by 5 percent during the Olympics.

"I'm shooting for 10 per cent," Stone said.

The Atlanta Journal
5 August 1994

Another positive factor was the ability to stage nearly all the various sports activities – at venues like the purpose-built Olympic Stadium – within the 'Olympic Ring' in Atlanta's downtown area. Two existing venues, the huge Georgia World Congress Centre and the Georgia Dome, were built adjacent to the Omni Station, expressly to avoid the need for massive parking facilities. In response, MARTA renamed its station there and constructed an additional entrance and walkway to provide easier access to these Super Bowl and Olympic-proportioned facilities. Even so, cramming 350 000 commuters into Atlanta's CBD (e.g. 5000 at Coca-Cola's offices alone) while simultaneously moving crowds of Olympic spectators into and out of the same area of the city presented a major problem for Games organisers (Figure 10.3.6).

In retrospect the transport arrangements for competitors and spectators alike were judged one of the weakest aspects of the '96 games. Many of the specially 'imported' drivers simply could not find their way around Atlanta.

QUESTIONS

1 The 1996 Olympics cost Atlanta an estimated $3.6 billion, with operating profits from the Games going to the International Olympic Committee (IOC), not to the city.
(a) What did Atlanta hope to gain from hosting the Games?
(b) Discuss the 'pros and cons' of Atlanta as an Olympic venue from the IOC viewpoint.

2 'Mass transit and shoe leather' (Figure 10.3.6). Elaborate this summary of Atlanta's strategy for dealing with the transport problems generated by the Olympics.

3 (a) Suggest how Atlanta's employers and employees might have helped to ease commuting problems during the 17 days of the Games.
(b) Why were local commuters asked not to switch from their cars to MARTA buses and trains during the Games?

10.4 Road Transport: a continent on wheels

Figure 10.4.1
The intersection of a busy interstate highway

Considering the whole spectrum of movement in North America, road transport is indisputably the most essential facility, being of vital importance to the economic, political and social unity of the continent. The motor vehicle is an indispensable part of North American life, with a resultant pattern of land use in both urban and rural environments markedly different from less motorised nations. With motor vehicle registrations exceeding 205 million in 1991, the continent is the most vehicle-oriented society in the world.

Highway systems

The sophisticated highway network which links the various regions of North America is a relatively recent phenomenon, the emergence of which was dependent upon the construction of the Interstate Highway System and the Trans-Canada Highway (Figure 10.4.2). The importance of fast and efficient inter-regional highways became more and more apparent in the war years. In 1944 the Federal Highway Act was passed in the USA authorising 'a national system of Interstate Highways not exceeding 40 000 miles in total extent, so located as to connect by routes as direct as practicable, the principal Metropolitan areas, cities and industrial centres to serve the national defense and to connect at suitable border points with routes of continental importance in the Dominion of Canada and Republic of Mexico'.

The 1944 Act was in fact a declaration of intent and it was the 1956 Federal Aid Highway Act which gave the effective go-ahead for the system to be built, with the federal government paying approximately 90 per cent of the total cost. The 1956 Act also added 1600 km to the proposed system. A further 2400 km was designated in 1968. Initially the entire system was expected to be completed by 1971. However, by 1980 over 4000 km of essential links were still not open to traffic, and it was not until the early 1990s that the system was finished.

With the completion of the interstate system, a new National Highway System (NHS) that would link the interstates with primary arterials for pur-

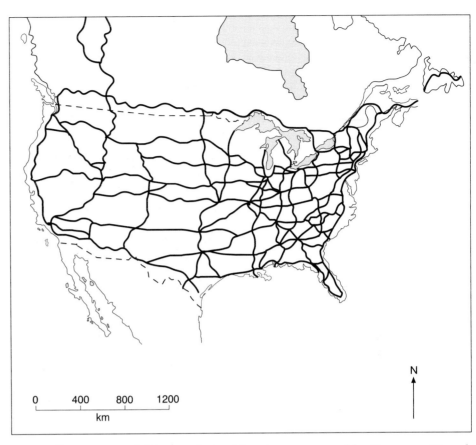

Figure 10.4.2
The US Interstate Highway system and the
Trans-Canada Highway

poses of commerce, defence and travel has been proposed by Congress. Signed into law on 18 December 1991, the $151 billion Intermodal Surface Transportation Efficiency Act (ISTEA) is the world's largest authorisation for surface transportation programmes, highways and transit. According to proponents, ISTEA has given state and local officials unprecedented flexibility to spend their allocations of federal money. In addition to the NHS, ISTEA provides authorisation for the six year, $23.9 billion Surface Transportation Project, which includes all roads except rural minor arterials and local roads; $16 billion for the Bridge Replacement and Rehabilitation Programme; and $17 billion for the Interstate Maintenance Programme.

The Intermodal Dimension

Another key purpose of ISTEA and NHS is intermodal expansion. According to ISTEA, intermodality involves 'using a coordinated flexible network of diverse but complementary forms of transportation to move people and goods in the most energy-efficient manner'. It also emphasised that intermodality should be a key element of future transportation policy, stating, 'If we seriously expect to be competitors in the global economy, we can no longer afford to keep our highways separate from mass transit, keep railways isolated from ports and have inadequate transportation connections to airports'.

Canadian Highways

In Canada, as in the USA, highway provision relates strongly to population density although in recent decades much has been done to improve road transport in isolated regions. The most important route in Canada is the Trans-Canada Highway, 7821 km in length, which was completed in 1970. It is the longest national highway in the world.

In 1990, there were 12.6 million passenger cars in Canada, compared with 10.3 million in 1980. Total motor vehicle registrations reached nearly 17 million in 1990 with 86 per cent of those old enough to drive holding a driver's licence. In 1992 expenditures on highways in Canada totalled over $7 billion. To limit the burden on taxpayers, provincial governments are turning to private developers who will recoup their investments by charging tolls.

Mounting opposition to new road schemes

New road proposals have met with increasing opposition in recent years. One of the most controversial issues was the plan to build a road bridge to link Prince Edward Island and New Brunswick. The C$850 million, 13 km bridge was to replace the ferry service which operated between the two provinces. A determined group of Prince Edward Island residents, calling themselves the Friends of the Island, opposed the scheme on the grounds that:

- greater accessibility would attract more visitors and general activity which would seriously disrupt their largely rural way of life;
- 500 ferry jobs would be lost;
- the island's fishing industry would be adversely affected.

Supporters of the scheme argue that:

- more jobs will be created than lost;
- commuters will save an average of 85 minutes per trip;
- vehicles crossing the bridge would use five times less fuel than it would take for the ferry to carry them;
- there would be no significant environmental impact.

The Friends of the Island took their case to the federal court early in 1993 and were successful in halting the project. In the same ruling the federal court also ruled that eliminating the ferry service between the two provinces would be unconstitutional. In 1873, when Prince Edward Island joined the Confederation, the federal government of the time promised, and entrenched in the Constitution, an efficient steam service to and from the island. The federal government is now negotiating with the province for a change in constitutional wording that would guarantee general, not just steam, transportation service in future. In mid-1993, Public Works Canada released an evaluation summarising all the environmental issues that had been raised regarding the bridge. The primary concern had been the bridge's design, which has now been refined to discourage ice jams. Initial studies had sparked concern about the effects of ice scour on marine biota. Although the Friends of the Island won the first round, the bridge was eventually completed in November 1996.

Elsewhere new highway developments have been opposed by an increasing number of people concerned that:

- they encourage greater vehicle use resulting in even more air pollution, noise and eventual congestion;
- they take up a lot of space and greatly influence the way that land around them is used;
- highways are extremely expensive to build and maintain;
- road transport is a large user of the continent's resources and is not as energy-efficient as alternatives.

Road transport has brought many benefits but the population in general and decision-makers in particular are now much more aware of the costs.

The air pollution hazard.

Figures 10.1.4 and 10.1.5 have already shown the pollution impact of transportation which is caused overwhelmingly by motor vehicles. Although the problem remains considerable, passenger car emissions of hydrocarbons and carbon monoxide have been reduced by 96 per cent and those of nitrogen oxides by 76 per cent since the introduction of emission controls in Canada. The basis of the latter was the introduction of the catalytic converter in 1971 and the move to more fuel-efficient cars. As part of Canada's 'Green Plan', the Canadian Council of Ministers of the Environment has developed a 10-year national plan to manage emissions of nitrogen oxides and volatile organic compounds. As in Canada, fuel consumption for all types of vehicles in the USA has declined steadily and further progress in this direction will undoubtedly be made in the future.

By the end of the century many cars in the USA will be radically different. Most will still use petrol, but they will share the roads with vehicles that run on alternative fuels such as methanol, ethanol and natural gas. Other cars will run on electricity supplied by rechargeable batteries. Vehicle emission regulations adopted by California and several north eastern states require that thousands of non-polluting electric vehicles will be available in late 1997, with more to follow in subsequent years. In September 1990 the California Air Resources Board (CARB) adopted strict vehicle emission rules to compel production of a new generation of ultra-clean cars and fuels. Under CARB regulations, 2 per cent of all new car sales in model year 1998 must be zero-emission vehicles – that is, electrics. By the year 2000, 10 per cent or 200 000 of all new car sales will have to be electric (unless another equally non-polluting fuel is developed). Shortly afterwards the federal Clean Air Act Amendments of 1990 were signed into law.

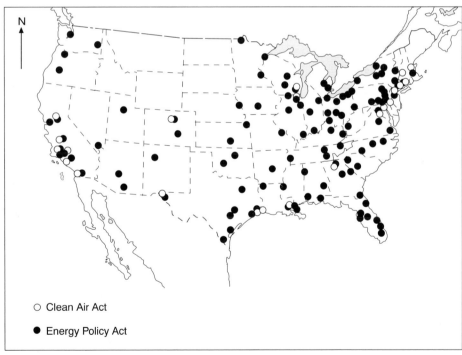

○ Clean Air Act

● Energy Policy Act

The Clean Air Act Amendments of 1990 required operators of private and government fleets in 22 major metropolitan regions (white circles) to buy 'clean fuel' vehicles starting in October 1997. Under the 1992 Energy Policy Act, federal fleet operators in major metropolitan areas (dark circles) were required to purchase 5,000 alternative-fuel vehicles (AFVs) in 1993, 7,500 in 1994 and 10,000 in 1995. For state fleets, the AFV programme began in 1996, when 10 percent of all new vehicles were required to run on alternative fuels.

Figure 10.4.3
How zero-emission laws affect the USA

They included numerous provisions affecting motor vehicles, the leading source of urban air pollution. Much attention has centred on vehicle emission rules aimed at the 22 major metropolitan regions rated by the Environmental Protection Agency (EPA) as serious, severe or extreme 'non-attainment areas' for federal ozone and carbon monoxide standards (Figure 10.4.3). Private and government fleets are expected to purchase most of the early electric vehicles. The key question is whether enough individual motorists will purchase them given the cars' limited range and performance and the inconvenience of recharging.

Transportation has become an increasingly important factor in people's perception of the quality of life. In 1990, Bremerton, Washington, a picturesque city on Puget Sound was named by 'Money' magazine as the country's best place to live. Bremerton received a perfect 100 score for its transportation system, one of the criteria used in the rankings. Then people started to move in and, by 1992, transit and city planners were watching 40 000 car commuters jockey daily for the city's limited parking spots. Bremerton is trying to persuade more people to use public transport by making the latter quicker and more attractive.

Intercity travel by bus in the USA fell below 2 per cent of the total in the early 1970s and is now just above the 1 per cent level. The world renowned Greyhound' (Figure 10.4.4) remains popular with tourists and students over long distances but is used by most Americans mainly on shorter intercity runs.

Road freight in North America has developed from what used to be almost totally a short haul service to provide a wide range of services over short, medium and long haul routes. As late as 1948 the distance beyond which Canadian railways possessed a cost advantage over road freight was estimated at 56 km. Although certain delivery advantages allowed road freight to compete beyond this range, the role of this mode of transportation was essentially a very limited one. Since the 1960s hauls of over 1500 km have become common while operations over 3000 or 4000 km are not unusual. Although the construction of the Interstate and Trans-Canada Highway Systems were not solely responsible for the huge increase in the volume of road transportation, it was the fundamental catalyst. As the railways strive to maintain their share of the freight market it remains to be seen if roads can be kept from the number one spot.

Figure 10.4.4
Greyhound bus on the US Interstate Highway system

Figure 10.4.5
US Domestic motor fuel consumption: average km per litre

	Cars	Buses	Lorries
1970	4.79	1.96	2.79
1980	5.47	2.11	3.38
1985	6.44	2.07	3.47
1992	7.65	2.23	3.84

10.5 Air Transport

North America has the most highly developed air transport system in the world. The general affluence of the two nations allows an extremely high per capita usage of air travel to cover the great distances which often separate major cities and make this mode of transport so popular on long haul passenger routes. Almost 65 per cent of North Americans have experienced air travel compared with about 1 per cent of the population in Asia. Air freight has increased significantly in recent decades but is still very limited in volume when compared with the other transport modes. Many isolated communities in the Northlands rely totally on air services for at least part of the year. In fact flying provides the only transport to and from almost half of Canada. In Alaska, one in every 40 people has a pilot's licence and there are more private planes per person than anywhere else in the world, one for every hundred people.

In 1992, 473 million revenue passengers were enplaned in the USA, up from 382 million in 1985. During the same period total revenue passenger km increased from 541 billion to 753 billion while mean passenger trip length rose from 1418 km to 1625 km. The industry was also encouraged by a sharp rise in freight movement from 9706 to 17 685 tonne km between 1985 and 1992. In the latter year the air transport industry as a whole employed 540 000 people.

QUESTIONS

1 Describe and explain the connectivity of the Interstate and Trans-Canada Highway systems.

2 Why is the proposed National Highway System thought to be necessary in the USA?

3 What are the reasons for mounting opposition to more road building?

4 Assess the air pollution problem caused by motor transport.

Since the Airline Deregulation Act of 1978 US air carriers have gained considerable freedom on market entry and exit compared with the time the industry was under the control of the Civil Aeronautics Board. The Act established a 'zone of reasonableness' for airline rates which gave carriers far more freedom in rate setting. Route addition regulations were also relaxed. The inevitable result of deregulation has been greater competition. In order to remain competitive, most airlines embarked on route restructuring to improve the efficiency and cost-effectiveness of their flight operations.

Hub-and-spoke networks

The hub and spoke network model has been adopted by many air carriers as the main strategy for organising their route structures. In a hub-and-spoke network, hubs serve as central locations which collect and distribute passengers between a set of nodes connected to hubs. One major advantage of this system is the savings in the number of direct flights necessary to connect all nodes on a network. The large passenger volumes channelled through hubs allow air carriers to achieve economies of scale and fly larger and more economical aircraft. Such economies enable carriers to offer more frequent flights. However, hub-and-spoke systems can cause inconvenient transfers and lengthy routes for passengers as well as creating congestion at hubs. Thus, decisions concerning the number and location of hubs must be made with great care. The largest US carrier, American Airlines utilises Dallas/Fort Worth and Chicago as national hubs, followed by Nashville and Raleigh/Durham as regional links (Figure 10.5.1). American Airlines' major US competitors in rank order are United, Delta, Northwest, USAir and Continental. The twelve largest carriers control 97 per cent of the US market.

Figure 10.5.1
American Airlines network showing hub and spoke system

Airport capacity problems

Many major airports have struggled to cope with increasing traffic volume which impacts in three ways:

- the rate of aircraft operations is affected by the layout of runways, the paths through the airspace leading to and from the airport, the air traffic control rules, weather conditions and the mix of aircraft using the airport;
- the number of passengers who can be accommodated at any one time depends on the terminal facilities and their design;
- the volume of motor traffic that can be taken on access roads to the airport.

A number of alternatives have been suggested to ease the capacity problem:

- build new airports;
- expand existing airport facilities;
- make better use of existing airport capacity. For example, improved technology could reduce the spacing between aircraft on final approach;
- manage demand so that activity is more evenly distributed by time of day and among airports.

Probably the greatest problem facing air transportation is public opposition to airport expansion and new airport building. Most people want to avail themselves of the benefits of air travel but do not want to be affected by the negative externalities. Los Angeles International (LAX) airport underwent major redevelopment between 1981 and 1984 at a cost of $700 million. However, the redevelopment was held up for ten years by public opposition until the courts finally gave the go-ahead for work to commence.

Air transport in Canada

In Canada, 32 million passengers were enplaned in 1992 compared with just under 3 million in 1955. Deregulation began in 1984 and, as in the USA, it has had a major impact on the industry. To compete effectively in the domestic market the two major airlines, Air Canada and Canadian Airlines International, have dropped a number of shorter routes that were difficult to service profitably with large jet aircraft. Instead, both airlines have entered into agreements with affiliated 'feeder' airlines, which schedule flights between smaller airports and the main airports. In 1992 there were more than 800 smaller carriers providing scheduled services. Feeder agreements enable Air Canada and Canadian Airlines to compete more effectively on longer-haul domestic and international routes.

Canadian carriers have struggled to survive in an increasingly competitive international market with conditions made much worse by the significant recession of the late 1980s and early 1990s. Worldwide the airline industry lost $10 billion between 1990 and 1992, and during this same period Canada's two major airlines lost almost $1.5 billion.

An increasing number of Americans and Canadians go to work by air partly because people tend to live much further from work than in Europe. Even small communities have airstrips for commuter aircraft. Cities have centrally located airports to handle such traffic, for example Toronto Island which is adjacent to the CBD. This leaves large airports free to link up with the rest of North America and abroad.

Environmental impact

Airports and aircraft affect the environment in a substantial way: airports occupy huge areas of land, cause high volumes of traffic on routes to city centres, and cause noise pollution. However, recent concern has centred on engine emissions which may contribute to the greenhouse effect and the depletion of the ozone layer. The emission of NOx in particular, could cause ozone depletion in the upper atmosphere and has been linked to acid rain and smog in the lower atmosphere.

Pittsburgh and Denver: superhubs

Pittsburgh in 1992 and Denver in 1994, the first new airports in the USA to be built since the 1970s, are being hailed as superhubs – the airports of the future. These are the first airports to be tailor-made to the hub and spoke system which became so popular during the 1980s. Previously airports have had to make do with existing airport facilities which have been adapted, often at great inconvenience to travellers, from the original basic terminal. Now all travellers departing from Pittsburgh check in and undergo security scanning in the main concourse. Train-type people-movers then transport departing passengers to the X-shaped boarding terminal holding the four gate concourses marked A–D (Figure 10.5.2). Each arm has shops and restaurants. A computerised baggage ticketing system has been installed which is much faster than conventional baggage retrieval systems.

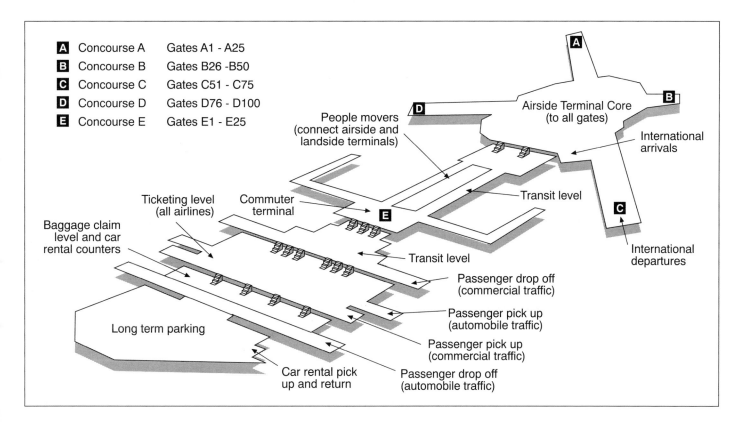

A Concourse A   Gates A1 - A25
B Concourse B   Gates B26 -B50
C Concourse C   Gates C51 - C75
D Concourse D   Gates D76 - D100
E Concourse E   Gates E1 - E25

People movers (connect airside and landside terminals)

Airside Terminal Core (to all gates)

International arrivals

Ticketing level (all airlines)

Commuter terminal

Transit level

Baggage claim level and car rental counters

Transit level

International departures

Passenger drop off (commercial traffic)

Passenger pick up (automobile traffic)

Long term parking

Passenger pick up (commercial traffic)

Car rental pick up and return

Passenger drop off (automobile traffic)

Figure 10.5.2
Pittsburgh Airport

The new airport which took more than 20 years to plan and 5 years to build is four times bigger than its predecessor. At present the airport deals with 18 million passengers a year but it has the ability to cater for around 24 million annually. It has become the main hub for USAir, offering connections to 70 cities in North America. 63 per cent of the airport's passengers are in transit. A new four-lane motorway connects the airport with downtown Pittsburgh, about a 40-minute journey. However, once the Maglev (magnetic levitation) transport system opens the journey time will be cut to ten minutes. The new airport has been financed by bonds which will entail paying off nearly $1 billion over the next 30 years.

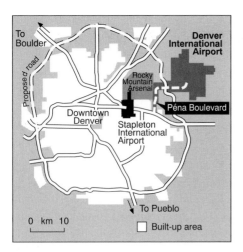

Figure 10.5.3
Position of Denver International Airport

Figure 10.5.4
Denver International Airport

Even larger is the new Denver International Airport (DIA) covering 135 km² on a site north east of Denver (Figures 10.5.3 and 10.5.4). Like Pittsburgh, DIA's design is based on the separated landside/airside concept. Its five huge runways, arranged in pinwheel fashion around the terminal, can land three aircraft at once. When all stages of the project are completed, some time next century, it will have 200 gates and 12 runways. The delay-plagued airport took four years to build at a cost of $4.9 billion, considerably more than the original estimate of $1.7 billion. The main delay factor involved problems with the automated baggage handling system, regarded by DIA management as 'an enormous leap forward'. The total cost is meant to be recouped over 30 years from landing fees, user fees, building rentals, car rentals, parking and concessions. By 2010, DIA is expected to employ around 34 000 people.

As one of the biggest public-works projects ever completed in North America, DIA has its critics. When the old airport, Stapleton, was built in 1929, five miles east of the city centre, passengers complained about the distance they had to travel. The new airport is 37 km away. There is also concern about the time lag before commerce, and particularly corporate headquarters, are attracted to the vicinity of the airport. Dulles International sat isolated outside Washington DC for 20 years before the motorway between them was flanked by office buildings hotels and high-tech industries. Some detractors argue that there is nothing much wrong with Stapleton; despite the problems of the airline industry, 1993 was the old airport's third busiest year.

The optimists – officials of the city of Denver and the state of Colorado, chambers of commerce, bankers and economists – believe that the decision to build the new airport was 'the single most important decision made in Colorado this century'. The Metro Denver Chamber of Commerce is looking at 500 companies, mainly in telecommunications, computers and financial services to locate in the area. Some local economists estimate that DIA will help create 50 000–75 000 new jobs over the next 20 years.

QUESTIONS

1 Suggest why (i) air travel is cheaper in North America than elsewhere in the world; (ii) so little freight is carried by air.

2 Why do many airlines operate hub and spoke systems?

3 Examine the arguments for and against the construction of Denver International Airport.

10.6 The Great Lakes–St Lawrence Seaway System

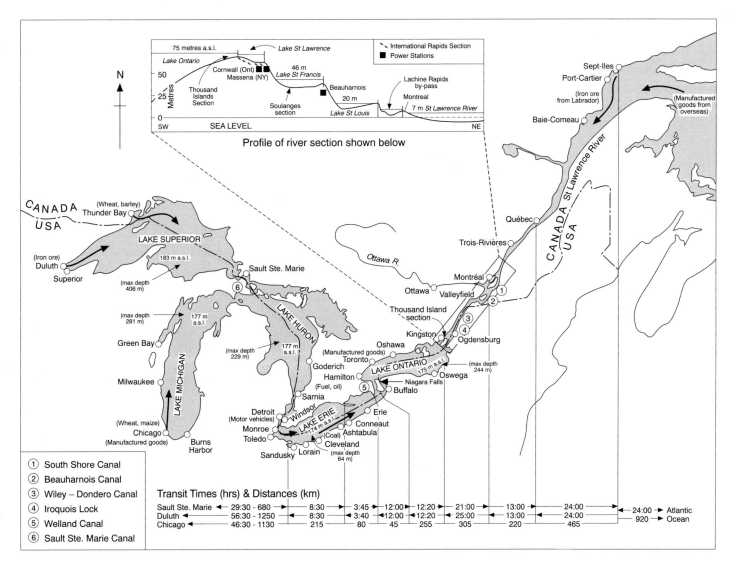

Figure 10.6.1
The Great Lakes–St Lawrence Seaway system

Background to the St Lawrence Seaway

The St Lawrence River–Great Lakes system had long been an Indian thoroughfare before it was used by European explorers and fur-trappers to gain access to the interior of North America (Figure 10.6.1). But as a commercial waterway between the Atlantic and the continental heartland it had severe limitations:

- upstream from Quebec City (the original head of navigation), the St Lawrence River was characterised by narrow, sand-choked channels;
- between Montreal and Lake Ontario, the Lachine, Soulanges and International rapids occurred where the river crossed outcrops of ancient resistant rocks. This succession of obstacles to navigation presented a combined rise of 67 metres;
- between Lakes Ontario and Erie rose the formidable Niagara Escarpment with its spectacular 50-metre Falls;
- between Lakes Erie and Huron the Detroit and St Clair river channels required constant dredging;
- between Lakes Huron and Superior lay the Sault Ste Marie Rapids and a rise of six metres.

Clearly, before the commercial potential of this waterway could be realised many improvements were necessary. Work on the upper St Lawrence commenced as early as 1779 and continued periodically until 1904, by which time the Lachine, Soulanges and Cornwall canals had been cut and deepened to 4.3 metres. The first Welland Canal was constructed in 1829, and successive improvements took place until the fourth canal (7.6 metres deep, with eight locks) was completed in 1932 (Figures 10.6.2 and 10.6.3). At Sault Ste Marie, the first Soo Canal (1789) was improved in 1855 and again in 1904 when new deep canals and locks were built by the US and Canadian authorities.

By the late 1940s the St Lawrence–Great Lakes route, although busy with traffic, could be likened to a dumb-bell in terms of its capacities: vessels of 25 000 tonnes could navigate upstream to Montreal, while between Kingston and Duluth a channel averaging seven metres in depth took vessels of 20 000 tonnes; but between Kingston and Montreal the narrow canal system along the upper St Lawrence River could barely accommodate vessels of 4000 tonnes.

Figure 10.6.2
The Welland Canal: twinned flight locks

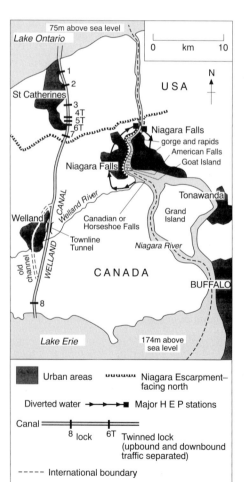

Figure 10.6.3
Niagara River and Welland Canal

Canadian–US cooperation

Although the St Lawrence is essentially a Canadian river, it forms the boundary with the United States along part of its course, and early treaties made it available to American shipping. However, the exact boundary line and rules of navigation were not defined until 1871, which delayed international agreement upon improving navigation between Montreal and the Great Lakes. In 1897 a joint Canadian–US Deep Waterways Commission reported in favour of improving the upper St Lawrence, but the US Senate failed to ratify both this and a series of similar agreements drawn up during the first half of the twentieth century.

With continuing economic development of the continental interior, the growth of foreign trade, the inability of rail transport to meet all requirements and, above all, the increasing urgency to harness the river's turbulent waters to generate HEP, the Canadian government signalled that it was prepared to take unilateral action. In 1951 it set up the St Lawrence Seaway Authority to construct an all-Canadian 8.2-metre-deep waterway as far west as Lake Erie, and passed the International Rapids Power Development Act to exploit the St Lawrence's HEP potential. Only then did the United States begin serious negotiations to resolve the political problems associated with the scheme – an additional stimulus being the increasing need to transport Labrador iron ore

QUESTIONS

1 (a) How are the modern Great Lakes divided politically between Canada and the United States? (Figure 10.6.1)
(b) Which lake is, politically, anomalous?

2 (a) Look at Figures 10.6.2 and 10.6.3. In which compass direction was the photograph taken?
(b) Explain the number, concentration and complexity of the locks constructed in the St Catherines area.

3 Explain why Canada agreed to bear 75 per cent of the total costs of Seaway construction and most of its operating expenses.

upstream to US steel centres around the Great Lakes.

In 1954 the US Congress established a separate St Lawrence Seaway Development Corporation to construct on American territory all the 8.2-metre navigation facilities needed to clear the bottleneck on the International Rapids Section. At the same time Canadian–US co-operation began on the St Lawrence Power Project, and the first electricity was generated in 1958. After a number of compromises and accommodations, the Canadian authorities accepted responsibility for five locks and the Americans for two locks on the Montreal–Lake Ontario section of the waterway, while sole responsibility for the Welland Canal around Niagara Falls was accepted by Canada. The completed Seaway was officially opened by Queen Elizabeth II and President Eisenhower on 26 June 1959.

The St Lawrence Seaway: benefits and problems

Strictly speaking, the international Seaway extends only from Montreal to Lake Ontario (approximately 305 km), but even this relatively short section is the world's longest artificial waterway. Canadian Seaway legislation also includes the Welland Canal, which underwent a major realignment in 1973 (Figure 10.6.3). In its broadest sense, however, the term 'Seaway' includes the St Lawrence River below Montreal and the Great Lakes System west of the Welland Canal. It is thus applied to the whole of the navigable waterway (minimum depth of 8.2 metres throughout) extending from the Atlantic Ocean to Duluth at the head of Lake Superior. This Great Lakes–St Lawrence Seaway system can accommodate 25 000 tonne ships measuring up to 225 metres in length and 23 metres in width.

The industrial and agricultural heartland of the continent is now accessible to ocean-going ships, in effect giving Canada and the United States another coastline and greatly reducing freight costs from Rotterdam, for example, to Chicago. The area served by the Seaway is larger than all of western Europe and contains nearly a third of the combined Canadian and US populations. The Seaway's contribution to the Canadian economy alone is incalculable; it has created thousands of jobs in countless related industries. Its future ought to be assured by the fact that it is essentially a bulk-cargo waterway strategically located along the border of two countries which are world leaders in both agricultural and mineral resources (Figure 10.6.4).

Figure 10.6.4
St Lawrence Seaway (proper): traffic by commodity groups, 1993

Commodity Group	Percent of Total Commodities in Transit (41 million tonnes)	Percent of Commodity Tonnage in Transit	
		Upbound	Downbound
Wheat	16.9	0.0	100
Other Grains	4.9	1.5	98.5
Other Agricultural Products	6.0	2.3	97.7
Total Agricultural Products	27.8	0.6	99.4
Coal and Coke	10.8	1.3	98.7
Iron Ore	26.6	91.3	8.7
Other Minerals	11.4	33.3	66.6
Total Minerals	48.8	57.9	42.1
Iron and Steel Products	10.9	92.9	7.1
Petroleum Products	2.7	51.6	48.4
Other Manufactured Products	9.7	44.4	55.6
Total Manufactured Products	23.3	84.0	16.0

	GRAND TOTAL	100.0	48.0	52.0

1 Suggest how the Seaway authorities make good use of the annual three-month shut-down.

2 (a) Study Figure 10.6.1. For a ship passing through the entire waterway system from the Atlantic Ocean to Duluth, calculate (i) the total distance travelled; (ii) the total time taken; (iii) the total height the ship would have ascended.

 (b) For each of your three answers, what percentage is represented by the Seaway proper (Montreal–Lake Erie)?

 (c) Comment on the importance of the Seaway proper.

 (d) Explain the difference in transit times for upbound and downbound vessels.

3 With reference to Figure 10.6.1, describe and attempt to explain the principal commodity movements shown in Figure 10.6.4. (N.B. The same ship may carry completely different commodities on its upbound and downbound journeys).

Despite these benefits and advantages, the Seaway's operations and economic viability are hampered by considerable problems, three of which are briefly examined below:

Winter closure

'General Winter' is the Seaway's oldest adversary. While the St Lawrence River has been open to navigation up to Montreal since the 1960s, ice in the canals and along the lake shores closes the rest of the system for three months every year. However, the navigation season is gradually being extended through structural improvements to the 'lakers' (specially constructed bulk carriers of ore, coal and grain – the staples of Seaway traffic) and by applying a number of ice-management techniques to the system's channels and locks. Methods currently in use include:

- air-bubbler and heating devices designed to prevent ice build-up on and around the lock gates;
- compressed air 'curtains' to reduce the quantity of ice pushed into lock chambers by downbound vessels;
- diversion channels that steer floating ice around, rather than through, locks;
- ice booms and artificial islands to control the formation and break-up of ice in the navigation channels;
- ice-breaking by the Coast Guard, particularly to hasten the opening of the navigation season.

Since the 1960s the navigation season has been extended by some five weeks; in 1993 the season lasted 272 days – from 30 March to 26 December. Future technological developments may extend the season still further.

Increasing size of lakers

Whilst the Seaway remains a remarkable engineering achievement, with hindsight it is perhaps regrettable that its construction was not on an even larger scale. The lakers have to be very large to operate most economically. To the west of the Seaway proper, ships carrying cargoes of 70 000–80 000 tonnes can move through the Soo Locks between Lakes Superior, Huron and Michigan. But these 'superlakers', over 300 metres long, cannot pass through the Welland Canal and so are 'imprisoned' on the western Great Lakes, forever shuttling between Superior and Buffalo.

A similar situation exists at the other end of the Seaway, with lakers of 180–200 metres length (Figure 10.6.5) having to transfer their cargoes to ocean-going vessels at the mouth of the St Lawrence River. Thus technology and economics – which have jointly created very large ships since the Seaway was opened – appear to have restored the 'dumb-bell' problem that the 1959 improvements were intended to resolve.

Adverse economic conditions

To the Seaway's climatic and technological problems must be added an economic one. Changing world market conditions have, since the mid-1970s, adversely affected tonnages carried and, in turn, toll revenues collected by the Seaway authorities. Despite special tariffs to encourage new trade (e.g., upbound steel slabs from Europe and downbound coal from the Powder River Basin), on-going reductions in staff, and the sale of capital assets such as bridges, the Seaway has operated at a loss for years and seems likely to do so well into the foreseeable future.

Figure 10.6.5
Canadian lakers in the Thousand Island sector of the Seaway

11

Recreation and Tourism

11.1 Continental Overview

Recreation

Recreational activities may be classified in the following ways:

- active/passive – depending upon the amount of physical exercise involved;
- formal/informal – depending upon the level to which participation is organised;
- resource-based/user-oriented – depending upon the degree of reliance on the natural environment, purpose-built facilities and special equipment.

While some passive recreational activities, such as reading or watching TV, normally involve minimal effort and are undertaken mainly in the home, more active pursuits usually take place away from the participant's home. Strictly speaking, however, if the recreational activity involves a stay away from home of at least one night, then recreation becomes tourism. In other words, recreation only includes activities of less than 24 hours away from home.

In North America recreation has become an increasingly important aspect of people's lives, for several reasons:

- increasing personal wealth for most people in two countries which themselves rank among the wealthiest in the world leaves higher disposable incomes;
- increasing leisure time as Americans and Canadians gain longer annual holiday entitlements and, for some, a shorter, four-day working week;
- increasing personal mobility (usually related to high levels of car ownership and improved road networks) allows more and more people to visit recreational areas and facilities that were previously remote and inaccessible;
- increasing levels of health education have made more people aware of the dangers of stress and physical inactivity and the benefits of mental relaxation and regular physical exercise.

Recreation in the United States

Recreation has become a major North American business. In 1994, in the United States alone, recreational spending amounted to nearly $370 billion, up 294 per cent from 1970 in constant dollar terms, while recreation's percentage share of all personal consumption expenditures rose from 4.3 to 8.3 per cent (Figure 11.1.1). With around one billion visits a year, cinema-going is by far the most popular spectator activity in the United States.

Whatever the season, 'going to the ball game' is a very important feature of American life, and attendances at major spectator sports have increased almost across the board since 1980 (Figure 11.1.2).

While spectator sports retain their mass appeal, active pastimes, involving more physical exercise, are becoming increasingly popular. At the same time, widespread affluence is reflected in the increasing use of machines and machinery in US recreational activities (Figure 11.1.3).

Figure 11.1.1
USA: recreational expenditures, 1970 and 1991

Category	1970	1994
	($ billions)	($ billions)
Books and maps	12.8	19.1
Magazines, newspapers & sheet music	16.7	22.5
Toys, sports & photographic supplies and equipment	22.5	77.5
Video and audio products, computer equipment and musical instruments	6.2	89.0
Spectator and commercial amusements	18.6	51.2
Other	17.0	110.6
Total	93.8	369.9
Percentage of total personal consumption expenditure	4.3	8.3

Figure 11.1.2
USA: attendance at major spectator sports, 1980 and 1991 (1000s)

Major Sport	1980	1993
Baseball	43 746	71 237
Basketball	10 697	19 117
American football		
professional	14 092	14 772
college	35 541	34 871
Ice hockey	10 534	15 714
Horse-racing	74 690	45 688
Greyhound-racing	20 874	28 003*

*1992

Figure 11.1.3
USA: selected recreational activities, 1980 and 1994

	1980	1994
	Participants (1000s)	
Softball	30 000	42 000
Golf	11 245	24 300
Tennis	10 655	18 700
Tenpin bowling	72 000	79 000*
	Items in use (1000s)	
Boats	11 800	16 600
Outboard motors	8241	12 502

*1993

In addition, in 1991, 35.6 million anglers (19 per cent of the total US population) on 511 million days hung up 'Gone fishing' signs, spending $24 billion on their pastime. Equivalent figures for American hunters (mostly for big game) were 14.1 million (7 per cent of the population), 236 million days, and $12 billion.

A recent national survey in Canada indicated the percentage of people aged 10 and over who had participated at least once in sports activities during the preceding 12 months. A similar survey in the United States questioned people aged 7 and above who had participated in sports more than once during the previous year. The results can be compared in Figure 11.1.4.

Levels of participation in these sporting activities vary considerably, the most active proportionately being the young, the better-off, single people, and those living in western provinces and states.

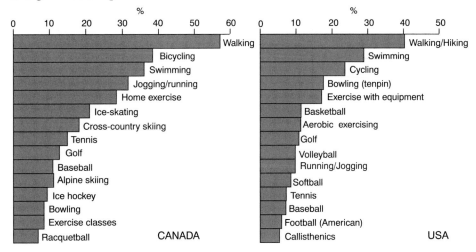

Figure 11.1.4
Participation in sports activities, Canada and USA, 1991

Recreational venues – national parks

To cater for the millions of North Americans who wish to pursue recreational activities in the continent's 'great outdoors', the governments of the United States and Canada have created a system of national parks and other areas that offer opportunities as yet unmatched elsewhere in the world.

The world's first national park was Yellowstone, located mainly in the American state of Wyoming. Yellowstone was designated in 1872 when members of a scientific expedition brought back photographs and persuaded the federal government to preserve this part of the great western wilderness in its unspoilt natural condition. The park's 10 000 spectacular thermal features, including the famous Old Faithful geyser, its mountains, rivers, waterfalls, forests and wildlife were thus protected from all commercial exploitation such as logging, mining and settlement. At a time when the pervading ethos was to tame, conquer and profit from such wilderness areas, the idea of preserving Yellowstone for the benefit and enjoyment of the general public and future generations was very unusual indeed. Even so, the idea spread quickly and most countries now have several national parks within their boundaries.

National Parks in Canada

Canada designated Banff, Alberta, as its first national park in 1885; like Yellowstone, this area also centred around thermal springs and magnificent Rocky Mountain scenery. Canada's early parks (Banff, Yoho, Glacier and

Jasper) were all served by the Canadian Pacific Railway (CPR), which was given major commercial concessions such as long leases on hotels in the park areas. The suspicion thus arose that the parks were created to benefit the CPR rather than to preserve the environment for all Canadians. However, competition from the motor car and the designation of more parks in the twentieth century have all but severed the links between passenger rail services and Canada's national parks.

Subsequently, Canadian national parks were created to encourage economic development in laggard regions. For example, Cape Breton Highlands, Nova Scotia, opened in 1936, was the first of seven parks which have stimulated tourism and brought money into the economically depressed Atlantic provinces. Political tensions within Canada have ensured that every province and territory has acquired at least one national park (Figure 11.1.5). However, since the early 1970s the main criterion for the creation of new parks has been the preservation of unique environments. Today, Canada has 36 national parks which cover more than 200 000 km² of territory – almost 2 per cent of the country's vast area and 1.5 times the size of England. Visits increased sevenfold between 1953 and 1981 but have stabilised in the last decade or so.

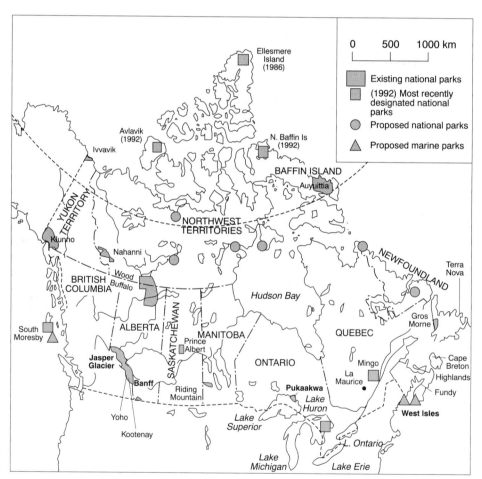

Figure 11.1.5
Canada's national parks

In addition to the terrestrial parks, two national marine parks protect 285 km² of Canada's coastal environment, and more are planned; more than 100 national historic parks and historic sites have been created; and, at regional level, over 165 000 km² of provincial parks play an important role in preserving Canada's natural and cultural heritage. All told, about 7 per cent of the country currently has some form of protection, but the federal government's Green Plan calls for 12 per cent of Canada's total area to be set aside as protected space.

The US National Park Service

In the United States, the National Park Service (NPS) was set up in 1916 and has become the model for similar organisations around the world. Some idea of the great variety of areas for which the NPS has taken responsibility can be gained from Figure 11.1.6. While national parks themselves account for a total area 1.4 times larger than England (Figure 11.1.7), when all NPS areas are included the figure rises to nearly 2.4 times larger than England. Adding the 4.6 million ha of state parks augments the federal area by some 15 per cent.

Figure 11.1.6

US National Park Service areas: recreation visits, millions, 1960 and 1990

	1960	1994
National Parks	26.6	60.1
National Monuments	10.7	26.5
National Historic, Archaeological and Commemorative areas	21.8	59.5
National Parkways	9.0	29.3
National Recreation Areas	3.7	52.3
National Seashores and Lakeshores	0.5	24.0
National Capital Parks	6.9	5.4
Other areas	n.a.	10.5
Totals	79.2	267.6

Figure 11.1.7

US national parks

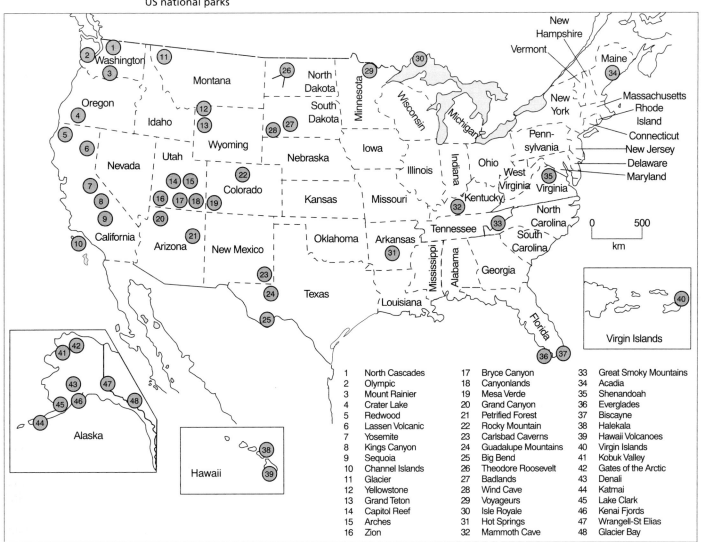

1	North Cascades	17	Bryce Canyon	33	Great Smoky Mountains
2	Olympic	18	Canyonlands	34	Acadia
3	Mount Rainier	19	Mesa Verde	35	Shenandoah
4	Crater Lake	20	Grand Canyon	36	Everglades
5	Redwood	21	Petrified Forest	37	Biscayne
6	Lassen Volcanic	22	Rocky Mountain	38	Halekala
7	Yosemite	23	Carlsbad Caverns	39	Hawaii Volcanoes
8	Kings Canyon	24	Guadalupe Mountains	40	Virgin Islands
9	Sequoia	25	Big Bend	41	Kobuk Valley
10	Channel Islands	26	Theodore Roosevelt	42	Gates of the Arctic
11	Glacier	27	Badlands	43	Denali
12	Yellowstone	28	Wind Cave	44	Katmai
13	Grand Teton	29	Voyageurs	45	Lake Clark
14	Capitol Reef	30	Isle Royale	46	Kenai Fjords
15	Arches	31	Hot Springs	47	Wrangell-St Elias
16	Zion	32	Mammoth Cave	48	Glacier Bay

Forest recreation

The US Forest Service, a branch of the US Department of Agriculture, was created in 1891 when the Shoshone National Forest (originally the Yellowstone Park Timber Reserve) was established in Wyoming. The system now incorporates 155 national forests (Figure 11.1.8).

Figure 11.1.8
US national forests

Figure 11.1.9
US national forest recreation use, 1981 and 1993

	1981	1993
	(% visitor-days)	
Mechanised travel and viewing scenery	27.0	33.7
Camping, picnicking and swimming	31.7	26.8
Hiking, riding and water travel	8.5	9.0
Winter sports	4.8	6.5
Hunting	7.0	5.8
Resorts, cabins and organisation camps	6.3	5.8
Fishing	7.2	5.5
Nature studies	0.9	0.9
Other (team sports, attending talks and programmes, gathering forest products, etc.)	6.6	6.0
Total	100.0%	100.0%
Recreation visitor-days (1000)	235 709	295 473

While the remit of the Forest Service is primarily related to the production of timber, in law it also runs to the provision of livestock grazing, watershed control, wildlife preservation and, not least, outdoor recreation.

Although national forests are usually less scenically spectacular than national parks, they are increasingly popular as recreational venues, partly because there are fewer restrictions than in national parks and partly because campgrounds and other facilities in many parks have reached full capacity. In 1993 recreation use in national forests totalled 295 million visitor-days, a 26 per cent increase over 1981 levels (Figure 11.1.9).

Wilderness areas

In response to the concerns of environmentalists, and under the terms of the 1964 Wilderness Act, the Forest Service has designated large areas of its land as wilderness areas, where strict regulations control access and usage. The 1964 Act prohibits some actions to preserve 'outstanding opportunities for solitude and unconfined types of recreation'. The Act specifically prohibits permanent or temporary roads, motor boats or vehicles, and mechanised equipment or transport, including bicycles, trail bikes, all-terrain vehicles (ATVs), snowmobiles, hang gliders, parasails and parachutes.

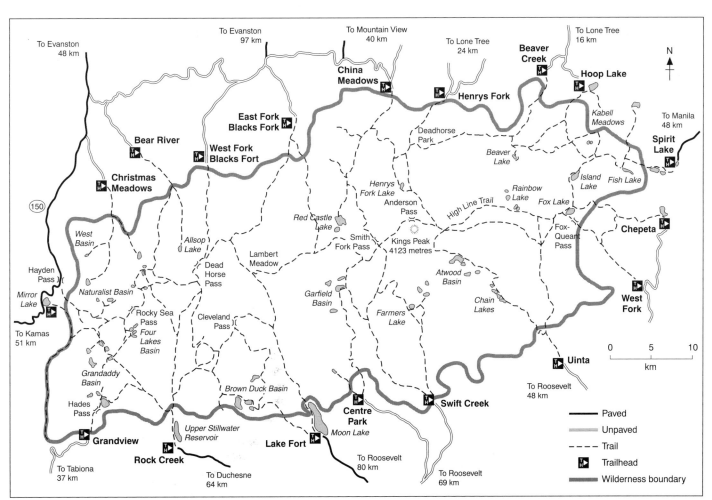

Figure 11.1.10
The High Uintas Wilderness, Utah

In the High Uintas Wilderness, Utah (Figure 11.1.10), in order to protect natural resources and to enhance the visitor's wilderness experience, the following actions are banned:

- entering or being on the High Uintas Wilderness with a group size exceeding 14 persons and 15 head of stock;

- terrain permitting, camping within 60 metres of any occupied campsite, trail, lake, pond, stream, spring, or any other water source;
- camping for a period of more than 14 days at an individual site;
- shortcutting a trail switchback;
- disposing of debris, garbage or other waste. Non-burnable materials including tin foil, bottles and cans must be packed out. Human waste must be buried in soil at least 15 cm deep;
- bedding, tethering, or hitching a horse or other saddle or pack animal for longer than two hours within 60 metres of lakes or springs;
- hitching or tethering a horse or other saddle or pack animal directly to a tree for more than two hours. They must be moved sooner if damage to the tree, soil or vegetation at the base of the tree is occurring.

Recreation areas

Another aspect of the Forest Service's work is the administration of some of the federal recreation areas, whose management is shared among several government agencies. Most national recreation areas are associated with rivers, lakes and reservoirs – the areas run by the Forest Service being the most popular.

QUESTIONS

1 With reference to Figures 11.1.5 and 11.1.7, account for the main similarities and differences in the distribution of national parks in the United States and Canada.

2 (a) Which type of NPS area (Figure 11.1.6) has recorded the highest rate of increase in visits between 1960 and 1990?
  (b) Suggest why the number of visits to national parks themselves has tended to decrease slightly in recent years.

3 Describe and account for the distribution of national forests in the United States (Figure 11.1.8).

4 Summarise the possible impact of each statutory function of the Forest Service on other land uses in the national forests.

5 What measures can the Forest Service take to reduce the environmental damage caused by recreational use?

6 With reference to wilderness areas, present the main arguments in the 'access v. preservation' debate.

Tourism in Canada

By the end of the twentieth century tourism may well have become the world's most important industry. In Canada in 1992 tourism:

- contributed $25 billion to the national economy, accounting for 4 per cent of GNP;
- was the third biggest earner of foreign exchange, preceded only by motor vehicles and automotive parts;
- provided the government with $11 billion in tax revenues;
- was a major employer – 554 000 jobs in 60 000 Canadian businesses depending directly or indirectly on tourism.

Large sums of money are spent each year by Tourism Canada, a federal agency, trying to persuade Canadians to spend their holidays exploring their own country, as well as foreigners to visit Canada. Advertising is carefully aimed at certain groups of people and at particular countries, the United States being the main target.

Tourism is also significant for all provincial and territorial economies. In 1990, for example, tourism accounted for between 3 per cent of gross provincial product in Quebec and 6.8 per cent in Prince Edward Island. The provinces often target their own advertising on neighbouring American states, promoting their part of Canada as distinctively different from the USA and as an all-year-round holiday destination.

While Americans remain the most frequent foreign tourists, over four million visitors now arrive from overseas. Western Europe is Canada's second largest source of tourists, but in recent years increasing numbers have come from countries such as Japan and Korea. Indeed, Asia is seen as the best potential source of new tourists to Canada in the future.

For Canada as a whole, however, all this is only a partial success story. Not only do more than two-thirds of Canada's tourist revenues come from Canadians themselves but, since 1968, Canadians have spent more on international travel services than they have received; the deficit in 1992 was nearly $8.3 billion.

For the large numbers of Canadian tourists who visit the United States every year (160 000 in 1993), the lure of Florida is most powerful. In 1991, Montreal–Fort Lauderdale, with 236 000 passengers, was the most popular 'city-pair' for charter flights – Florida appearing in seven of the top ten Canada–US city-pairs. Another favourite rendezvous with the sun is Cancun (Mexico) which retained its top position as Canadians' preferred Third World destination, with 95 000 passengers flying from Toronto and 87 000 from Montreal. In Canada–Europe traffic, Montreal–Paris retains its top city-pair ranking with 198 000 passengers; Toronto–London is second with 66 000 passengers.

Tourism in the United States

Tourism is an even bigger industry in the US than in Canada. In 1993, the American travel and tourism industry:

- attracted around 16 million foreign tourists (another 30 million visits to the US are made for business purposes, though many of these include elements of holiday, recreation and pleasure);
- was the nation's largest business services export, contributing $74 billion of visitor spending to the US economy and a net trade surplus of $22 billion;
- generated $56 billion in tax revenues;
- accounted for 6 per cent of GNP (13.4 per cent if indirect expenditures are included);
- was the second-largest employer (after health services), providing 6 million jobs directly and another 5 million indirectly.

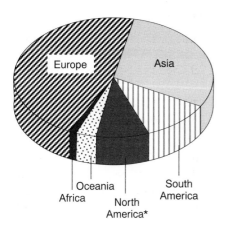

Figure 11.1.11
USA: foreign tourists, 1985–1994

Figure 11.1.12
USA: overseas tourists by continent of residence, 1993

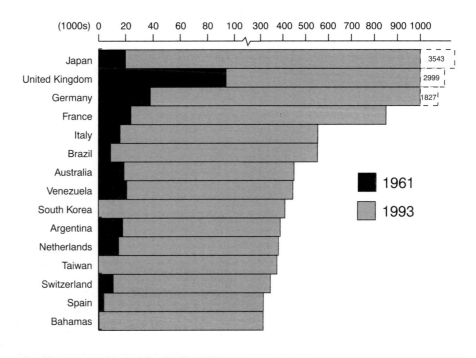

Figure 11.1.13
USA: overseas tourists by country of residence (top 15), 1961 and 1993

Figure 11.1.14

US: top ten destinations for tourists from Asia and Europe

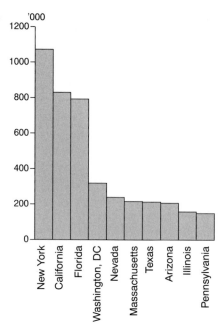

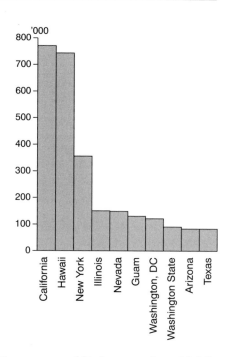

Figure 11.1.15

US: domestic travel expenditure – top ten states, 1993

	$ million	% of US total
California	42 480	13.1
Florida	28 629	8.9
Texas	19 950	6.2
New York	20 215	6.3
Illinois	13 804	4.3
Nevada	12 539	3.9
New Jersey	11 134	3.4
Pennsylvania	10 060	3.2
Virginia	9076	2.8
Georgia	9186	2.8
Total, Top 10	177 073	54.8
US Total	323 272	100.0

Figure 11.1.16

US: domestic tourism data, 1994

Total trips (millions)	434.3
Average number of family members	1.8
Average nights on trip	3.7
Average km per trip	1322
Primary mode of travel (%)	
by road	78
by air	17
Rental car used during trip (%)	9
Stayed in hotel during trip (%)	40

As recently as 1961, however, the United States earned little more than $1 billion from international tourism, almost 90 per cent of all foreign visitors coming from its immediate neighbours – Canada and Mexico. The immense changes that have occurred since then are due to several factors:

- the advent of the Boeing 747 jumbo jet in 1969 reduced both the time and cost of transoceanic travel, ushering in the age of mass tourism;
- the overall expansion of the tourist market and increasing competition within it have combined to lower costs generally, e.g. through package tours;
- after 1971 the use of the microcomputer created a global reservations network which greatly simplified foreign travel arrangements;
- higher incomes and longer holidays, especially in Europe and Japan, have widened the pool from which the USA attracts visitors;
- the cultural dominance of the United States, exerted through Hollywood films and American TV programmes, provides global publicity and projects a generally attractive image of the USA.

A picture of modern tourism in the United States can be constructed from Figures 11.1.11 to 11.1.14.

Important as foreign tourism to the United States undoubtedly is, it should be remembered that, as in Canada, domestic tourism (i.e. Americans holidaying in their own country) comprises an even bigger economic sector (Figures 11.1.15 and 11.1.16).

QUESTIONS

1 Explain the pattern of Canadian tourism as represented by the 'city-pair' data (in the text).

2 Under each of the following headings, provide at least two examples of tourist-related activities that provide jobs in the United States: Travel, Transport, Accommodation, Food and drink, Entertainment, Shops, Finance, Public facilities.

3 With reference to Figure 11.1.13, for each country calculate the percentage increase in visitors to the United States. Using the new data, rank the 15 countries and compare this rank order with that shown in Figure 11.1.13.

4 (a) Which destinations appear in both top ten lists (Figure 11.1.14)? Explain the differences between the two lists.
  (b) For each destination suggest what might be the major tourist attraction(s).

5 (a) Using the data in Figures 11.1.15 and 11.1.16, write a short statement about the holiday patterns of Americans exploring their own country.
  (b) Account for any significant differences in the most popular destinations chosen by Americans and by foreign visitors (Figures 11.1.14 and 11.1.15).

11.2 Zion National Park, Utah

The Physical Background

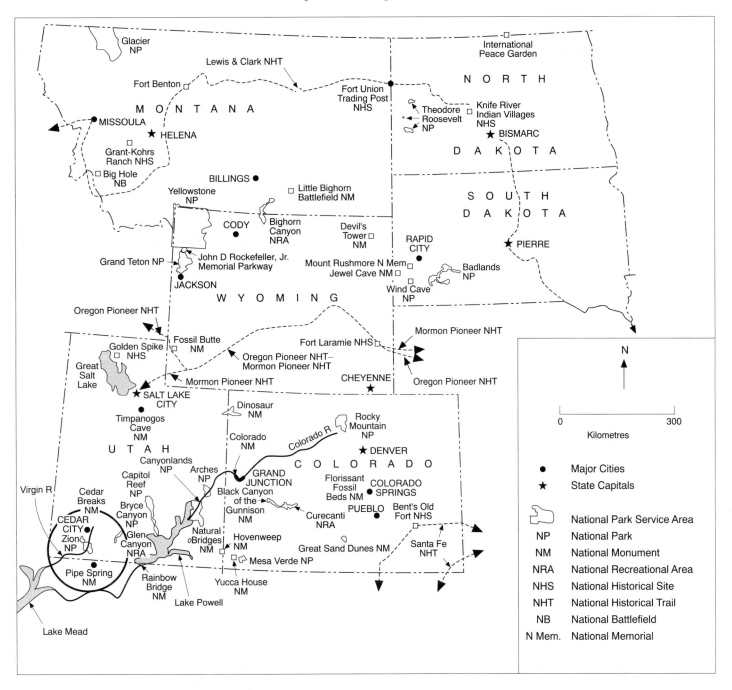

Figure 11.2.1
National parks of the Rocky Mountain region

Administratively, Zion National Park forms part of the Rocky Mountain Region of the US National Park Service (NPS) (Figure 11.2.1). This region contains a range of units which are indicative of the wide variety of historic and scenic attractions that draw large numbers of visitors to the 'Recreational West', especially during the summer months. Zion's natural orientation, however, is south towards the Colorado River and its tributaries which dissect the Colorado Plateau. An arid climate, spasmodic rainfall and, at Zion, the Virgin River, have created a canyon landscape of extraordinary scientific interest and scenic beauty.

Figure 11.2.2
Composite stratigraphical column, Zion National Park

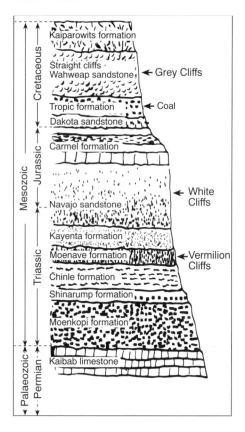

The oldest geological formation in the area is the Permian Kaibab limestone, which is the youngest formation along the rim of the Grand Canyon; thus where the Grand Canyon story ends, the geological history exposed at Zion begins (Figure 11.2.2). The story continues through younger and younger strata northwards to the spectacular erosional features displayed at Cedar Breaks and Bryce Canyon.

The most famous of the geological formations at Zion is the Navajo Sandstone, which forms the spectacular cliffs within the national park. This near-ubiquitous formation, which reaches its maximum thickness of 670 metres in the park, consists almost entirely of desert dunes. Ancient dunes truncated one another, resulting in cross-bedding and the fantastic patterns on the sandstone surfaces now exposed (Figure 11.2.3).

Figure 11.2.3
Navajo Sandstone Cliffs

Initial uplift of the Colorado Plateau began perhaps 25 million years ago. About 13 million years ago the uplift resumed, causing an increase in river gradients and load-carrying capacities, which in turn resulted in greater downcutting and widening. Gradually the younger deposits were eroded away, leaving the spectacular series of step-like terraces and plateaus which are encountered northward into the Utah from the Grand Canyon.

Zion Canyon, the centre-piece of the national park (Figure 11.2.5), has been cut into the White and Vermilion Cliffs. The principal agent of erosion has been the Virgin River (a tributary of the Colorado River), which rises at an elevation of 2700 metres and flows southwestward for a distance of 320 km to where it enters the artificial Lake Mead, at an elevation of about 100 metres. In places within the canyon the gradient is twice as steep. After heavy summer thunderstorms or winter rains the Virgin River becomes a raging torrent, cutting down through the soft Navajo Sandstone with harder rocks and boulders picked up from above.

Uplift, which triggered the down-cutting, was interrupted by still-stands when the Virgin River ceased to cut downwards and instead cut sideways, allowing tributaries to enter the main stream at the same level. Resumption of uplift caused the Virgin River to renew its downcutting, but the tributaries could not keep pace and their valleys have now been left hanging at levels 330–390 metres above the main stream.

Widening of Zion Canyon has been influenced principally by the soft shale beds immediately below the Navajo Sandstone. Groundwater percolates through the sandstone until it reaches the impervious shale, whence it emerges as a line of springs. It also erodes the shale, undermining the sandstone and causing the rock-falls and slides which are quite common within the park area. Surface run-off after heavy rain also washes away the shale and undermines the overlying rock. At the Temple of Sinawava the junction of the Navajo Sandstone and Kayenta shales is at river level. Upstream the river has not yet exposed the shales and the canyon is narrow; downstream the shales are exposed, are washed away and the canyon is widened.

QUESTIONS

1 With reference to Figure 11.2.1, attempt an alternative classi-fication of the physical and human phenomena in the care of the NPS, based on the areas' names and your own general knowledge.

2 (a) What does the profile of the stratigraphical column (Figure 11.2.2) indicate about the characteristics of the different rocks?
 (b) Account for the unusual names applied to many of the geological formations in this region of the United States.

3 (a) Calculate the average gradient of the Virgin River from its source to Lake Mead (see the account).
 (b) What other factors contribute to the river's erosive power?

4 Referring to the account, draw annotated diagrams to illus-trate (i) the formation of hanging valleys along Zion Canyon; (ii) the process of canyon widening.

The national park's objectives and problems

Zion's erosional features first received a measure of official protection in 1909, and the national park was designated ten years later. The park managers' objec-tives can be summarised thus:

- to preserve the natural processes that have created an outstanding example of canyon erosion;
- to preserve and protect the scenic beauty and unique geological features, including the labyrinth of remarkable canyons, the dramatically coloured strata and rare sedimentation;
- to preserve the archaeological features relating to the area's pre-Columbian inhabitants;
- to preserve the entire area – including its rare and endangered species – for the purpose of scientific research;
- to provide opportunities for visitors to learn about and enjoy the park's scenic and other resources – including clean air, brilliant night skies and nat-ural tranquillity – without destroying or degrading them.

This last objective forms the nub of Zion National Park's problems. Consistently increasing numbers of visitors – nearly all of whom arrive by car (about 750 000 vehicles in 1993) – certainly reflect the attractiveness of the park but have also resulted in overcrowding, over-use and degradation of facilities, damage to nat-ural and cultural resources, and diminished visitor experience and enjoyment. Such problems are now common throughout the NPS system (Figure 11.2.4).

Figure 11.2.4
Visits to Zion and other national parks

Visits (millions)

	1960	1970	1980	1990
All US national parks	26.6	45.9	60.2	57.7
Zion National Park	0.6	0.9	1.2	2.3
Zion visits as % of all national parks	2.3	1.9		

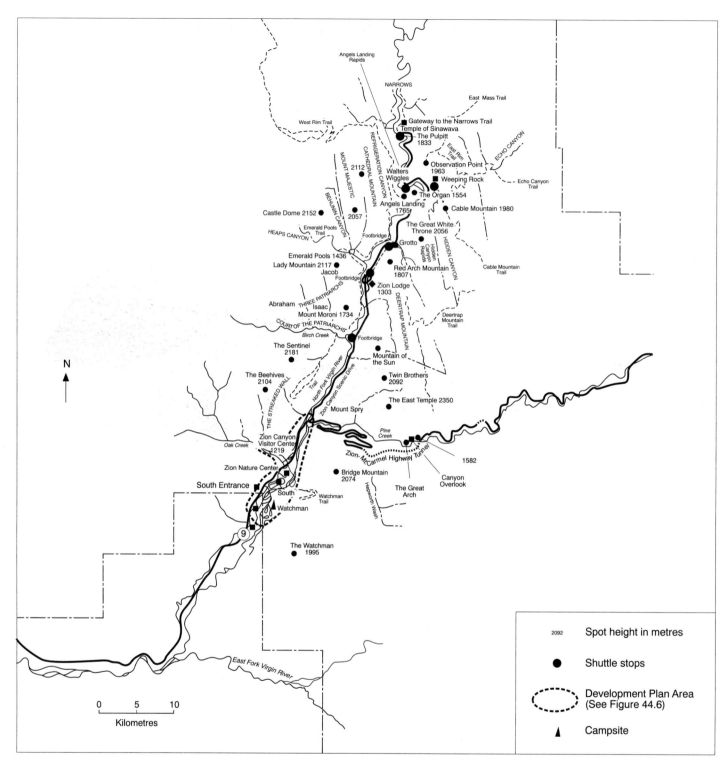

Figure 11.2.5
Zion National Park

Owing to its spectacular scenic attractions, hiking opportunities and lodging facilities, the ten km drive through Zion Canyon itself has always been the main visitor focus of the park (Figure 11.2.5). However, many visitors cannot find parking spaces at the trailheads or points of interest because of the sheer volume of traffic. Consequently, visitors either park illegally at the side of the road (built without curbing to facilitate run-off of rain or snowmelt), damaging the vegetation and creating potential traffic hazards, or leave the canyon without ever enjoying the features they came to see. This is obviously not the visitor experience that the park managers wish to promote.

The Development Plan for Zion National Park, 1994

Figure 11.2.6
Zion National Park: Development Plan area

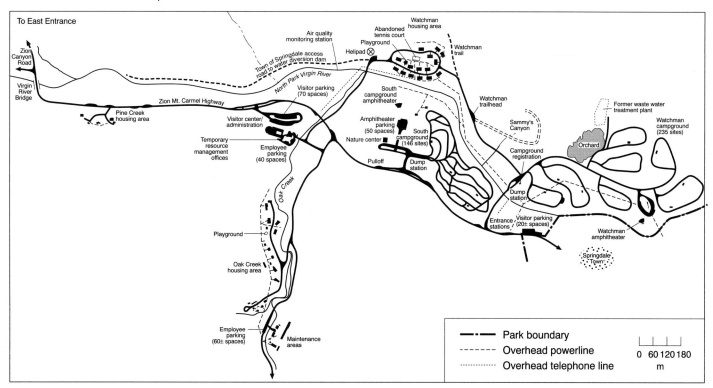

JAN	FEB	MAR	APR	MAY	JUN	JUL	AUG	SEP	OCT	NOV	DEC
54 096	52 279	108 002	208 349	222 529	337 745	349 966	378 242	202 693	260 201	94 457	55 921
54 096	106 375	214 377	422 726	645 255	983 000	1 332 966	1 711 208	2 003 901	2 264 102	2 358 559	2 414 480

Figure 11.2.7
Zion National Park: monthly and cumulative visitor numbers, 1993

To address the problem of traffic congestion and other issues relating to serious overcrowding, in 1994 the NPS approved a Development Plan for Zion National Park at an overall cost of $30 million (Figure 11.2.6). During the eight month peak visitor season (Figure 11.2.7) the following measures are now operative:

- a mandatory shuttle system ('park and ride') in Zion Canyon;
- a limit to the capacity of the car park for shuttle passengers;
- a ban on tour buses and oversize recreational vehicles;
- a campground reservation system, with prior booking of places.

Although similar systems have been imposed at other US national parks since the 1970s, the mandatory shuttle system is still the most controversial of the measures for Zion. It operates as follows:

- upon entering the park, visitors are directed to a new Transit/Information Centre which will provide basic orientation, safety and park interpretive information;
- the centre has 575 parking spaces, a shuttle loading/waiting plaza (with shade structure and bicycle racks), a picnic area and even a designated pet-exercise area;
- the shuttle route within the park (there is a feeder service from the nearby town of Springdale) is approximately 13 km one way – from the Transit Centre to the Temple of Sinawava – and takes about 44 minutes;

- passenger numbers average about 4300 a day;
- the shuttle operates 14 hours a day (from 7.00 am to 9.00 pm) but these times can be extended depending upon demand;
- thirteen vehicles (plus two reserves) costing $1 740 000 are available to cope with peak demand during the four busiest months;
- the shuttle vehicles are open-air 'trams' (rail-less), each capable of carrying 60 passengers;
- the trams use propane, which is preferred to petrol, diesel, compressed natural gas or electric power;
- the shuttle is operated by a private company on contract to Zion National Park;
- the shuttle system costs approximately $250 000 a year to operate and maintain;
- the only alternatives to the shuttle allowed in Zion Canyon itself are pedestrians and cyclists.

Figure 11.2.8
Hiking and bridle trails in Zion National Park

Trail	Starting point	Distance (round trip)	Climb	Average Time (round trip)	Remarks
Gateway to The Narrows	Temple of Sinawava	3.2 km	17 m	2 hours	Easy, no steep grades. All-weather trail. Fine view of river flood plain. Trailside exhibit near Temple of Sinawava.
Weeping Rock	Weeping Rock parking area	0.8 km	30 m	$\frac{1}{2}$ hour	An easy, surfaced, self-guiding trail. Water drips from overhanging cliff; springs issue from it. Hanging gardens; travertine deposits.
Emerald Pool	Zion Lodge or Grotto picnic area	3.2 km	21 m	2 hours	Cross river on footbridges. Small pool formed by two falls. Loop or one-way trail.
Canyon Overlook	Parking area, upper end of large tunnel	1.6 km	49 m	1 hour	A self-guiding trail. Mostly easy walking to top of Great Arch. Excellent view of Pine Creek Canyon and west side of Zion Canyon.
East Rim	Weeping Rock parking area	11.3 km	652 m	5 hours	Fairly strenuous foot and horse trail. Carry water, lunch. Cross footbridge and climb to East Rim. Trail sign.
Hidden Canyon	Weeping Rock parking area	3.2 km	258 m	2 hours	Fairly strenuous. Hidden Canyon represents Zion's 'Shangri-la,' an almost inaccessible canyon of quietness and solitude.
West Rim	Grotto picnic area	20.1 km	931 m	8 hours	Strenuous foot and horse trail. Carry water, lunch. Cross river on footbridge.
Angels Landing	Grotto picnic area	8.0 km	451 m	4 hours	Strenuous foot trail; experienced hikers only; steep climb. Half of trail hard-surfaced. Cross footbridge over river. Excellent view of Zion Canyon.
Watchman Viewpoint	River bridge above South Campground	3.2 km	112 m	2 hours	View of The Watchman, Springdale, and Zion and Oak Creek Canyons. Changing vegetation.

QUESTIONS

1 (a) Complete the 1980 and 1990 percentage data in Figure 11.2.4.
  (b) What do these statistics indicate about Zion's relative attractiveness?

2 (a) Judging by the 'Remarks' in Figure 11.2.8, rank all the trails *qualitatively* in the order that you would like to visit them.
  (b) Produce a *quantitative* ranking of all the trails using the Distance/Climb/Average Time data. For example, a simple addition of the figures for East Rim would be: 11.3 + 652 + 5 = 668.3; for Canyon Overlook the sum would be 51.6.

(c) Compare the two rankings – one qualitative, the other quantitative – by applying an appropriate test, such as the Spearman Rank Correlation Coefficient.
(d) Comment on the results obtained.

3 (a) Argue the case for and against the new shuttle system from the viewpoint of (i) the average car-driving visitor; (ii) the park manager responsible for introducing the scheme to Zion.
  (b) List the advantages and disadvantages of using (i) open-air vehicles for the shuttle system; (ii) contracting a private company to operate it.

11.3 Quebec: Tourism in an Historic City

Quebec City (1991 population: city 167 500; metropolitan area 645 500) is the capital of predominantly French-speaking Quebec Province and Canada's eighth-largest city. With the European rivalry for control of North America, Quebec became a crossroads for French and British culture, endowing the modern province with a unique heritage – a North American lifestyle with a French accent. This special legacy makes Quebec City, which is 96 per cent francophone, a particularly attractive tourist destination – as the province's Ministry of Tourism emphasise in the following publicity literature.

Figure 11.3.1
Quebec City, tourist map

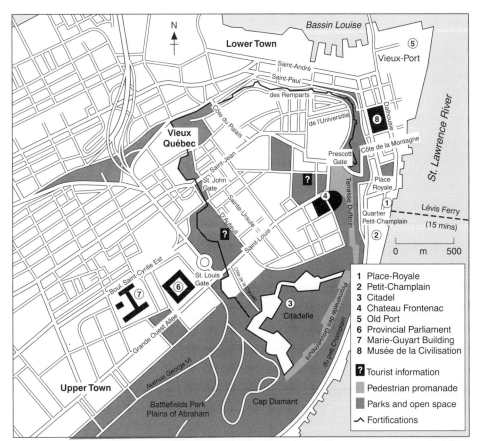

1 Place-Royale
2 Petit-Champlain
3 Citadel
4 Chateau Frontenac
5 Old Port
6 Provincial Parliament
7 Marie-Guyart Building
8 Musée de la Civilisation

? Tourist information
Pedestrian promanade
Parks and open space
Fortifications

Figure 11.3.2
Quebec City, panoramic view

Quebec City – A Joy Forever
(Numbers in text refer to numbers on Figure 11.3.1)

Backed by the ancient Laurentide Mountains, Quebec, the only fortified city in North America, is perched atop Cap Diamant, a massive headland dominating the majestic waters of the St Lawrence River and its tributary, the Saint-Charles. The romantic charm of the city invariably leaves travellers spellbound.

In recognition of its invaluable historic wealth, UNESCO has proclaimed Old Quebec (Vieux-Québec) a 'world heritage treasure'. Quebec is the only North American city to be so honoured. Filled with vast green parks, the cradle of New France proudly displays its age-old buildings, testimonials to its French and English origins. This splendid setting opening out on breathtaking panoramas promises great discoveries.

Refusing to withdraw nostalgically behind its ramparts, Quebec City is also a dynamic government and financial centre focused on the future. Busy and animated, it has created major cultural events and fun-filled festivals for the enjoyment of residents and visitors alike.

The capital embodies the legendary *joie de vivre* of its people, for whom hospitality and gastronomy are a vital tradition. Quebec City is much more than a wonderful place to visit. It's a meeting with history to be enjoyed in the midst of modern living.

Roots

Four centuries of history can be read in the monuments, stone buildings, coiled stairways, and narrow winding streets hugging the steep rocky contours of the oldest city on the continent. To spend time here is to understand the birth and evolution of a unique and original nation.

The setting takes you back to 1608 when Samuel de Champlain, finding the site appealing, pictured it as an impregnable fortress. He built a first dwelling, which grew into a trading post, village and city.

Nothing tells the story of Québec City better than its fortifications. Growing threats transformed the first palisade erected by Champlain into an impressive wall with four archways. Meanwhile, the colony flourished. At the foot of the cape, in Lower Town, Place-Royale[1] and the Petit-Champlain quarter[2] grew into a prosperous village. Simultaneously, on the promontory, Upper Town began to take shape and major structures were erected: the Ursuline Convent (1639), Hôtel-Dieu Hospital (1644), and the Séminaire (1678). Grande Allée, a magnificent avenue lined with trees, monuments and Victorian homes was, and still remains, one of the busiest in Upper Town.

New Masters, New Style

In the summer of 1759 Québec City was besieged by British General James Wolfe. Québec City was defended by an army under the Marquis de Montcalm.

The defeat of General Montcalm on the Plains of Abraham sounded the knell of New France, as the Treaty of Paris (1763) ceded the territory to the British Crown. Soon threatened by the Americans, the British rebuilt the fortifications and constructed the Citadel[3]. The last battle fought at Québec City occurred in 1776, when an American army under Benedict Arnold was repulsed by the British defenders. The city then began to take on its present-day appearance. Tightly huddled Normandy-style houses were built alongside religious buildings influenced by English architecture. By the mid-19th century, neoclassical stood next to Victorian structures, creating a highly original cityscape: neither quite French nor quite English. And then came the famous Château Frontenac[4] (1893), complete with turrets, crenels and parapets, lending a medieval appearance to the overall picture.

Figure 11.3.3
Chateau Frontenac, Quebec City

Parks have always been vitally important in this city, undeniably one of the most beautiful in North America. As they erected monumental structures, the religious orders and civil institutions of New France cultivated large plots of land. And rich Upper Town merchants surrounded their opulent villas with English gardens. In the early 20th century, the Plains of Abraham were converted into a public park.

Between North America and Europe
In this 'European city in North America,' you can safely go about on foot or in a horse-drawn *calèche* and contemplate past and present attractions. Street corners and winding alleys offer pleasant surprises at every turn. Museums, churches and chapels contain countless artistic and historical treasures. Rue du Trésor, where settlers came to pay their dues to the Royal Treasury, has become an outdoor art gallery. Contemporary culture blooms in intimate cafés and bistros, lively jazz clubs, art galleries, handicraft shops and theatres big and small.

There's never enough time to soak up Québec City's irresistible ambience and explore its manifold attractions: the newly restored Old Port[5] and its summer activities; the Citadel and its daily Changing of the Guard ritual; guided tours of the Parliament buildings[6], seat of the National Assembly; the exhibits and breathtaking panorama on the 31st floor of the government's Marie-Guyart Building[7] (Rue de la Chevrotière); the stately Victorian townhouses of Grande Allée, converted into small hotels, restaurants and outdoor cafés.

A stroll on Terrasse Dufferin opens out on the river's wide expanse; an outing on the Lévis Ferry to the south shore of the St. Lawrence reveals an equally spectacular panorama, the escarpment of Cap Diamant and the majestic Château Frontenac.

From Terrasse Dufferin, the funicular or the *Casse-cou* stairway takes you right down to the oldest part of Québec City, Place-Royale and the Petit-Champlain quarter, with their restored 17th- and 18th-century buildings converted into residential and commercial locations. Historical homes, a church, restaurants, boutiques and workshops proudly welcome travellers.

Major Events in Quebec City
An exciting summer schedule of major cultural events involves parks, streets, bars, nightclubs and theatres. Performances and street entertainments are everywhere: a summer festival of performing arts in July, the largest international French event in North America, drawing record audiences; and Expo-Quebec in late August, a trade and country fair with exciting rides and shows.

In winter the city is covered in deep fresh snow. This is the time of the world's greatest winter carnival, the *Carnaval de Québec* – 10 days of February festivities which have been a highlight of winters in Quebec since the 1950s. In the parks and public places of Quebec City, magnificent snow-structures appear – fairy-tale palaces, gigantic snowmen, awe-inspiring abstract shapes and animals of all kinds – their pristine whiteness broken only by the blazing red of the Canadian flag and the brilliant blue of the flag of Quebec. Skaters cut intricate patterns on the surface of frozen rivers and ponds, and the gaily-decorated streets are thronged with people enjoying the spectacle of a parade, as the city celebrates the *Carnaval de Québec*.

Tourisme Québec

Figure 11.3.4
Quebec City in winter

Q U E S T I O N S

1 Identify, in the first section of the tourist literature, at least three examples of the 'overblown' phrases that might be regarded as typical of such publicity. Rewrite the phrases more prosaically, emphasising geographical accuracy.

2 Identify the facts and characteristics that support Quebec's claims to be both a unique North American city and a 'European city in North America'.

3 Which of Quebec City's attractions stem from (i) its physical setting; (ii) its special history?

4 With reference to Figures 11.3.1 and 11.3.2, from which vantage point was the photograph taken, and in which compass direction was the camera facing?

5 Referring to Figures 11.3.1 and 11.3.2, name the following features: A – mountain range; B and C – rivers; D – suburb; E and F – city districts; G – pedestrian thoroughfare; H and I – buildings.

6 Suggest what is the principal function of the Chateau Frontenac building (Figure 11.3.3). Why is it widely regarded as an asset, visually, to Quebec's cityscape?

7 Given that most visitors come from the United States and other parts of Canada, discuss whether Quebec City's predominantly francophone culture is likely to help or hinder its tourist industry.

8 How has the city turned its severe winters to its commercial advantage (Figure 11.3.4)? What limitations are there to winter tourism in terms of potential visitors?

11.4 Florida Tourism: Sunshine and Shadows

Florida, the most southerly state of the continental USA, is located just north of the Tropic of Cancer between 24.5° and 31°N. Key West is only 145 km from Havana in Cuba. Florida thus benefits from warm winters, hot summers and long hours of sunshine. It also possesses excellent beaches, lagoons, sand bars and warm blue seas, providing an attractive coastal environment. Small wonder that the Sunshine State has developed a highly profitable tourist industry (Figure 11.4.1).

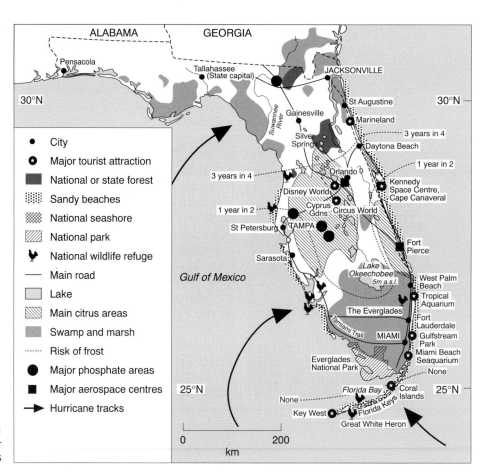

Figure 11.4.1
Florida: selected tourist attractions and other economic features

Florida attracts 41 million visitors a year, almost seven million of whom come from abroad, including 2.4 million from Canada and 1.1 million from the UK. Tourism is annually worth around $20 billion. It is the largest sector of the state's economy and helps pay for its public services. Yet tourism in Florida is essentially a twentieth century development.

Throughout its long history as a Spanish possession, broken only briefly between 1763 and 1783 by a term under British rule, Florida remained an unpromising wilderness of swamp and forest, aptly described as 'a colony without colonists'. Little changed following its transfer to the United States in 1819, and for most of the nineteenth century Florida continued to be seen as a negative area by-passed by the mainstream of American life.

By 1890, however, the first recognition of the state's potential as a tourist area appeared when a wealthy industrialist, Henry Flagler, extended the Florida East Coast Railroad to West Palm Beach, where he built a sumptuous hotel to welcome his guests. These included American millionaires like the Astors, the Rockefellers and the Vanderbilts who, attracted by the mild climate, spent every winter in Florida. The state thus established its initial reputation as a high-class winter holiday destination.

Miami: Florida's First Tourist Honeypot

In 1896 Flagler's railway was extended southward to Miami, then merely an army outpost and fruit-growing centre. But in 1897 the opening of the first hotel, the Royal Palm, signalled the 'discovery' of Miami and the beginning of large-scale tourism to the area. Miami also quickly developed as a home port for cruise ships taking holiday-makers to more exotic locations even further south. Population increased rapidly from under 5000 in 1900 to 42 750 (city) by 1920 and to 100 000 (metropolitan area) by the mid-1920s.

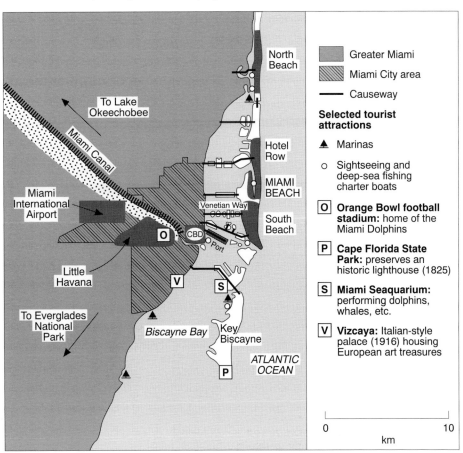

Figure 11.4.2
Miami and Miami Beach

While Miami's tourist industry originated in the area that became the city's CBD, nowadays it is concentrated almost entirely on the Miami Beach peninsula, separated from the mainland by Biscayne Bay (Figure 11.4.2). This separation has allowed Miami Beach to develop quite differently from the rest of the metropolitan area.

Miami Beach is an entirely artificial creation which needed huge initial investments. Before its transformation in the early twentieth century much of the peninsula consisted of worthless sandbars and mangrove swamps. The latter's black, oozy mire had to be reclaimed with fill dredged from the navigation channel of Biscayne Bay and with topsoil hauled from the Everglades.

Since its creation, Miami Beach has developed northward in three distinct resort zones:

South Beach

Originally a coconut plantation, by 1920 this area had become the new focal point of tourism in the Miami region. The explosion of tourism in South Beach resulted from the opening of Venetian Way, the first of the artificial causeways giving access by car – which after 1910 superseded the railway as the major stimulus of growth. By the 1930s 'SoBe' was the fashionable place to be, as 'snowbirds' from the northern states and Canada travelled south to enjoy the sun and escape the harsh winter back home.

Over the years, however, the area's relatively small, low-rise (2–4 storey) hotels have mostly been converted into flats for senior citizens, many of whom choose to spend their retirement in Florida. Thus South Beach is now predominantly a residential area rather than a tourist resort, though its art deco architecture comprises a national historic district which attracts many visitors.

Hotel Row

This area developed with the resurgence of the tourist industry after the Second World War, the main boost coming with the growth of air travel. Hotel Row is characterised by imposing, high-rise hotels which form a 'concrete canyon' running for kilometres along the seaward side of the Miami Beach peninsula (Figure 11.4.3). Originally built for the luxury trade, many of the hotels now cater for package tourists, especially in summer – Florida's low season. The beach sand suffers badly from long-shore drifting and is seldom used as a recreational resource any more.

Figure 11.4.3
Hotel Row, Miami Beach

1 (a) Which popular holiday destination on the eastern side of the Atlantic occupies an approximately equivalent latitudinal position to Florida?

(b) Compare that destination's climatic statistics with Miami's (Figure 11.4.4) and account for any significant differences.

2 Account for Florida's original holiday image as a select, high-class *winter* resort and explain how this image has changed during the twentieth century.

3 (a) With reference to Figures 11.4.1 and 11.4.4, suggest why summer is the low season for tourism in Florida.

(b) What special climatic hazard affects Florida in late summer and autumn?

4 What factors have permitted Miami's visitors to take longer holidays in recent decades?

5 Which of Miami Beach's three tourist areas would best suit (i) a family seeking a relatively low-cost summer holiday; (ii) young singles wanting an action-packed vacation among their own age group?

6 For the benefit of fly-drive package tourists, compile a 2-week motoring itinerary of Florida that would include a broad sample of the varied attractions that the state has to offer its visitors. Start and finish the itinerary at Miami, adding wherever possible brief descriptive notes about the places of interest selected (Figure 11.4.1).

Motel Row

A more recent surge of uncontrolled development is a ribbon of large, often ornate motels built in the late 1950s and 1960s at North Beach. This area abounds with sporting facilities and pulsates with nightlife. Interspersed with the motels are high-rise, modern condominiums. Reflecting the current problems of Miami, great emphasis is placed on security in these buildings, which are virtually mini-fortresses.

Beyond North Beach, Miami's coastal development is now almost continuous, with seafront communities stretching northward all the way through Fort Lauderdale to Palm Beach and further north still. Popular beach resorts along this coast include Surfside, Sunny Isles, Bal Harbour and Hollywood. The whole comprises a kind of subtropical seaside megalopolis whose 1990 population totalled 3 193 000 (Miami: city 359 000, metropolitan area 1 937 000).

Modern trends in Miami's tourism

- Relatively speaking, modern tourism in the Miami region is in decline, due to:
  (i) the spectacular rise of alternative destinations in Florida, especially Orlando;
  (ii) relative to its size, Miami has become America's most crime-ridden city. In 1991, 18 256 crimes were committed per 100 000 population, and of these, 4191 were violent crimes;
- Miami's visitors are getting older; the younger set are moving on to trendier places;
- hotels in Miami Beach continue to be converted into condominiums for senior citizens;
- the winter is still high season in Miami, with most visitors continuing to originate in the north eastern states and Canada;
- the tourist season has been progressively extended into the summer months when lower rates attract lower-income Americans and foreigners on package holidays;
- all visitors are tending to stay longer;

Figure 11.4.4
Miami: climatic data

	J	F	M	A	M	J	J	A	S	O	N	D	Annual
Temperature	19.4	19.9	21.4	23.4	25.3	27.1	27.7	28.0	27.4	25.5	22.4	20.1	24.0°C
Precipitation	51	48	58	98	163	187	171	177	241	208	71	43	1516 mm

- more people are flying to Miami, then renting a car;
- Miami's tourist industry is now heavily dependent on Hispanic employees;
- Miami's port, the 'Cruise Capital of the World', is still the home of some 20 world-class cruise ships currently handling around three million passengers a year.

Orlando: Florida's latest tourist honeypot

Just over a quarter of a century ago central Florida consisted almost entirely of rolling expanses of citrus groves, cattle ranches, dense woodlands, swamps and streams. Today this part of Florida is synonymous with fantasy and fun; it has more hotels, theme parks and family attractions than any other tourist destination in the world.

It began in 1971 near Orlando, with the opening of Disney World's Magic Kingdom, a collection of fantasy theme areas ranging from Adventureland with its jungle cruise and Frontierland with its steamboat ride, via trips to the moon from Space Mountain or 20 000 leagues under the sea in Jules Verne's *Nautilus*, along Main Street USA to Cinderella's castle, and into Fantasyland to meet the world's most famous mouse and his pals (Figure 11.4.5).

These are but a few of the treats and thrills awaiting visitors to the Magic Kingdom – which itself is just one element in the 100 km² Disney World. This also includes the EPCOT centre and Disney–MGM studios together with hotels, campsites, first-class golf courses, a marina, a nature conservation area and a 12 000-place car park – all on reclaimed swamp, lake and forest.

Figure 11.4.5
Disney World's Magic Kingdom

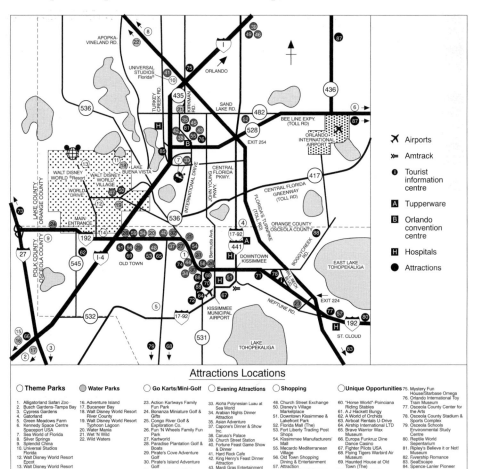

Figure 11.4.6
The clustering of attractions in Greater Orlando

Disney's Magic Kingdom was soon followed by Sea World, Wet 'n Wild, Mystery Fun House, Water Mania, Medieval Times, Universal Studios, other Disney ventures such as EPCOT, Disney-MGM Studios and Pleasure Island, and the very latest arrival, Splendid China, which presents replicas of over 60 famous Chinese sights, including the Great Wall of China! In all, some 50 theme parks and other attractions of every description now dot the central Florida landscape (Figure 11.4.6).

To match the explosion of visitor attractions, Orlando not only had to build a new international airport but also had to provide some 75 000 rooms in a wide variety of accommodation ranging from budget-grade motels and self-catering villas to luxury hotel suites. This volume of accommodation is little less than staggering, given the short period in which it was developed. Indeed, Greater Orlando might well share Disney World's claim to be 'the most popular holiday destination in the world'.

Figure 11.4.7
The growth of Orlando's population, 1970–1990

	Population (1000)		
	1970	1990	Growth %
City	99	165	
Metropolitan area	453	1225	

Note: Total US population grew 22.3% over the same period

QUESTIONS

1 (a) Complete Figure 11.4.7 and comment on the entire data.
  (b) Suggest who has benefited from such rapid growth and who has been disadvantaged.
  (c) Comment on the likely spatial changes that have resulted.

2 (a) With reference to Figure 11.4.6, which elements of the pre-Magic Kingdom era are still evident in the modern landscape of Greater Orlando?
  (b) What are your personal views about the transformation that has occurred during the past 25 years or so?

3 The attractions shown on Figure 11.4.6 clearly demonstrate the principle of similar enterprises clustering together (aggregation). Outline the economic advantage accruing from this phenomenon, and suggest some of the likely problems or limitations.

4 Florida's Disney World (like California's Disneyland before it) has obviously been a huge success financially.
  (a) Suggest why Disneyland Paris (formerly Eurodisney) appears to be unable to match its American counterparts in attracting the crowds and making profits.
  (b) What advice would you offer the managers attempting to improve the situation at Disneyland Paris?

Shadow across the Sunshine State

Despite the undoubted success of Florida's tourist industry, a serious problem has emerged to cast its shadow across future tourism throughout the state.

Florida's crime wave

Florida is America's most crime-afflicted state. In 1991, 8547 crimes of all types were committed per 100 000 population, and the crime rate has risen inexorably over 25 years; whereas in 1968 the chances of being involved in a serious crime were 1 in 30, by 1993 the odds had fallen to 1 in less than 8. Nowadays more than 35 000 visitors are robbed, raped or attacked every year, and in 1993, 22 tourists were murdered. These statistics received massive publicity in Europe and elsewhere; the immediate results were not unexpected:

- many people cancelled holidays already booked;
- future bookings plummeted 20 per cent;
- inquiries about holidays in Florida dropped from 60 to 10 a day;
- on the New York Stock Exchange, Disney shares fell in value.

By the end of 1993 a travel industry survey showed that British tourists rated the entire USA 13th in a list of most-feared holiday destinations, ahead of Third World countries such as Peru, India, China and Thailand. Of the 500 questioned, 21 per cent said that fear of crime would prevent them visiting the United States, with over half expressing particular concern about Florida. Figure 11.4.8 attempts to put this important issue into proper perspective.

Not without risk

AROUND 40m tourists have gone to Florida in the past year. Of these, 22 have been shot to death. Some, frighteningly, were targeted as tourists; they were seen leaving the airport, juggling luggage and maps, and with shiny new rental cars. Others, like the British man and his girlfriend shot near Tallahassee this week, would have been attacked whether they were tourists or not. They got into harm's way.

Florida has done what a state reasonably can: tried to disguise its rental cars, by removing number-plates beginning with the tell-tale letters Y and Z, laid on extra policemen, issued tourists with safety tips. It has now sent another 800 armed state employees out to guard the roads, has laid on 24-hour security for rest stops, and is asking the federal government for an extra $4m for security. In an effort to blank the horrors out of people's minds, to stop them associating palm trees and Mickey Mouse with blood, all advertising for the state of Florida has been stopped.

Yet the problem is actually quite small: to repeat, 22 people out of 40m, in a state which has America's highest rate of violent crime (1,184 incidents per 100,000 people, according to the FBI's latest report). It would not rank at all, but for a tourism industry worth an annual $30 billion. What Florida is now trying to do is to guarantee that visitors will never come to harm. Face up to it: no power on earth can do that.

The Economist
18 September 1993

Figure 11.4.8
Not without risk

1 How can Florida's tourists (especially new arrivals) be easily identified by local criminals, and why are tourists favourite targets of the criminals?

2 Why did British Airways, in response to the crime problem, reschedule its three weekly flights from Heathrow to Miami to arrive earlier than 7 pm?

3 'If I was a tourist in Florida I'd get myself on the next plane home.' US Secretary of State, 1993.

'Visitors to Florida are in no more danger than they would be in any other major holiday destination anywhere in the world.' Director of Marketing (Europe), Florida Division of Tourism, 1994.

'People are over 1000 times more likely to die from an illness or accident on holiday than they are from an act of violence.' Home and Overseas Travel Insurers, 1993.

(a) With reference to the above statements and to Figure 11.4.8, write your own appraisal of the Florida crime factor.

(b) Draw up a list of safety tips that you would issue to tourists departing for a holiday in Florida.

People – Environment Interactions

12.1 Continental Overview

It is hardly surprising that an area the size of North America is affected by a range of natural hazards, some of which can reach catastrophic proportions. However, the people of the continent are fortunate in two ways. Firstly, the overall hazard threat is significantly less than in a number of other comparably sized areas around the world. And secondly, the US and Canada have had sufficient wealth to invest heavily over a considerable period of time in hazard prevention and defence.

It is arguable which hazards are most feared by North Americans. Earthquakes and volcanoes, which can have a devastating impact in other parts of the world result in a relatively low loss of life in the US and Canada. Volcanic eruptions are few and far between and only occur in the most sparsely populated parts of the continent. Although California is one of the most susceptible areas of the world to earthquakes, strict building regulations have reduced the impact of this frightening hazard to a level envied by other countries. However the fear of the 'Big One' always lurks in the hazard zones when high technology and heavy investment might be totally overwhelmed by the forces of nature.

Correctly or not, the impact of hazards is measured in terms of cost as much as it is in loss of human life. The only data relating to hazards presented in the US Statistical Abstract is about tornadoes, floods and tropical storms (Figure 12.1.1). When Hurricane Andrew hit Florida and the Gulf Coast in August 1992, it was the most expensive natural disaster in American history with a total cost of over $25 billion (Section 12.3).

Because of the unpredictable nature of natural hazards, the financial impact varies considerably from year to year (Figure 12.1.2). The National Hurricane Centre says that states along the Atlantic and Gulf coasts should prepare for more frequent and intense hurricanes, while the United States Geological Survey warns that there is a 65 per cent chance of a catastrophic earthquake in the next 25 years. The costs will be astronomical, according to Applied

Figure 12.1.1
Tornadoes, floods and tropical storms, 1984–1994

Hazard	1984	1985	1986	1987	1988	1989	1990	1991	1992	1993	1994 prel.
Tornadoes, number	907	684	764	656	702	856	1133	1132	1303	1173	1082
Lives lost, total	122	94	15	59	32	50	53	39	39	33	69
Most in a single tornado	16	18	3	30	5	21	29	13	10	7	22
Property loss of $500 000 and over	125	69	75	38	48	60	91	64	(NA)	72	83
Floods: Lives lost	126	304	80	82	29	81	147	63	87	101	72
Property loss ($ million)	4000	3000	4000	1490	114	415	2058	1416	800	16 400	1224
North Atlantic tropical storms and hurricanes:											
Number reaching US coast	12	11	6	7	12	11	14	8	6	8	7
Hurricanes only	1	6	2	1	1	3	–	1	1	1	–
Lives lost in US	4	30	9	–	6	56	13	15	24	9	38
Property loss (1990 $ million)	77	4457	18	8	9	7840	57	1500	25 000	57	973

– Represents zero.　NA Not available.

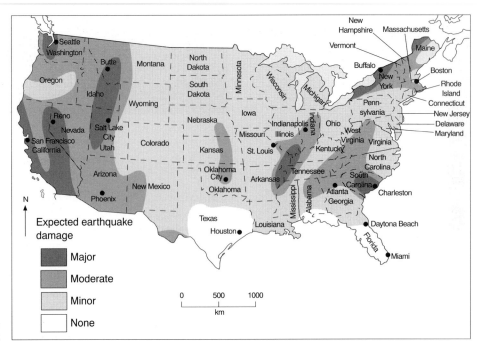

Figure 12.1.2
USA: Seismic risk

Insurance Research, based in Boston. The firm predicts that, by the year 2020, the total bill from earthquake damage, primarily in California and Washington state will be $245 billion, and total hurricane expenses along the Atlantic and Gulf coasts, as well as Hawaii, will reach $281 billion. That is why state and local officials, as well as insurance companies, are urging Congress to change the current federal disaster programme from one that emphasises relief under almost any circumstances to one that encourages prevention through techniques such as better building code enforcement and incentives to build away from coastlines, fault lines and floodplains.

The impact of natural hazards in Canada is less catastrophic, but nevertheless extremely serious at times. Figure 12.1.4 shows where the main problems lie. The most expensive natural catastrophe in Canadian history, resulting in losses of over $400 million, was a violent hailstorm that struck Calgary on 7 September 1991.

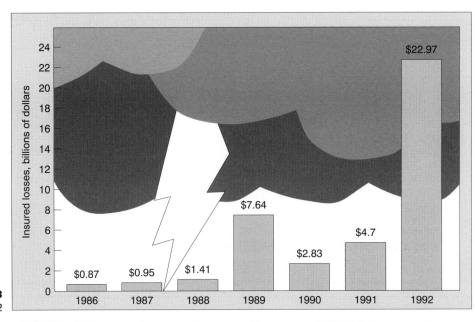

Figure 12.1.3
Canada: The cost of catastrophe, 1986–1992

QUESTIONS

1. Produce an annotated map of North America showing the areas affected by major natural hazards.
2. Why is Canada less affected by major hazard incidence than the US?
3. Explain the pattern of seismic risk on Figure 12.1.2
4. Why does the cost of catastrophe vary so much over time? Explain the annual variations shown in Figure 12.1.3.

NATURAL HAZARDS IN CANADA

FLOODS AND DROUGHTS

Most floods are caused by sudden and intense downpours. The most severe flood in Canadian history occurred on October 14–15, 1954, when Hurricane Hazel brought 214 mm of rain to the Toronto region in just 72 hours. The Don and Humber rivers and the Etobicoke Creek flooded, killing 80 and causing the equivalent of over $100 million in damage in today's dollars. Spring snowmelt has also been the cause of many of Canada's worst floods. For instance, in 1950, the Red River in Manitoba rose 10 metres above normal, submerging one-fifth of Winnipeg and forcing the evacuation of 100,000 people.

Too little water has been a recurring problem on the Prairies since the soil was first tilled. The most serious drought came in the 1930s. Between 1933 and 1937, the region received only 60% of its normal rainfall. The drought of the 1980s equalled that of the 1930s in duration and intensity, reducing grain exports by $4 billion in 1988 alone.

STORMY WEATHER

No region of Canada is secure from the fury of winter blasts and blizzards. Intense winter storms are frequently accompanied by numbing cold, ice or heavy snow. For the residents of Newfoundland, the blizzard of 1959 was one of the worst on record. The storm, which struck on February 16, took six lives, left 70,000 Newfoundlanders without power, crippled telephone service, and blocked highways and roads with drifts 5 metres high.

In the summer months, thunderstorms are frequent afternoon occurrences across much of the southern part of the country. Southwestern Ontario holds the distinction of having the highest average number of thunderstorm days at 34. This also is the region most frequently hit by tornadoes (an average of 21 each year). On May 31, 1985, eight tornadoes moved across southern Ontario, causing $100 million in property damage and killing 12. Tornadoes also occur in the West. On July 31, 1987, one hit Edmonton, Alberta, killing 27, injuring over 200, leaving 400 homeless and causing more than $250 million in damage.

In terms of property alone, the most expensive natural catastrophe in Canadian history was a violent hailstorm that struck Calgary on September 7, 1991. Insurance companies eventually paid about $400 million to repair over 65,000 cars, 60,000 homes and businesses, and a number of aircraft.

Canada Year Book, 1994

Figure 12.1.4
Natural hazards in Canada

12.2 Mississippi Basin Flood Hazard

The Mississippi River drainage basin is the third largest in the world and the largest in North America. The basin includes all or parts of 31 American states and two Canadian provinces, and funnels over 40 per cent of the continent's natural drainage into the Mississippi. Its headwaters can be divided into two areas: one originates at the continental divide in the Rocky Mountains, the other in the Great Lakes lowlands. Both areas can accumulate large winter snowfalls.

In mid-July 1993, following two months of unusually heavy rain (mainly cyclonic but supplemented by summer convection storms typical of the continental interior), the Midwest suffered severe flooding as the Mississippi–Missouri river system broke its banks along an 800-km stretch.

Figure 12.2.1
The Mississippi River drainage basin

In the early hours of Monday, 19 July, the Missouri River at St Louis rose to its highest recorded level, more than 14 metres above normal, almost topping the 15-metre wall that protects the city's 900 000 inhabitants. In Iowa, one of the worst-affected states, more than 50 000 people were made homeless and over 250 000 were left without clean drinking water. In all, more than 40 000 km² of farmland and hundreds of communities were inundated, with the loss of 41 lives. It was not until October that the flood waters finally receded. The final cost of damage exceeded $25 billion, rivalling Hurricane Andrew as the most expensive natural disaster in US history.

Taming the Mississippi: the work of the US Army Corps of Engineers

Over the course of millions of years the meandering Mississippi has carved out a floodplain up to 200 km wide, in which frequent flooding has deposited deep layers of fertile alluvium. Occasionally, however, the floods have been devastating in human terms. In 1927, for instance, an area almost the size of Scotland was inundated, over 400 lives were lost and 637 000 people made homeless.

Following that disaster, it was realised that flood control, previously considered a local responsibility, could only be made effective at a national level. Under the 1936 Flood Control Act, Congress took responsibility for nationwide flood protection. In the Mississippi basin the federal government, through the US Army Corps of Engineers, began a long-term project to provide a major flood protection system along much of the Mississippi and its tributaries.

This work has consisted mainly of heightening and strengthening the earthen embankments known as levees, behind which the valuable towns, cities, farms and factories are protected from the river. Many of these levees were originally constructed as early as the 1880s, but they were only three metres high and 16 metres wide at the base. With the aid of modern earth-moving equipment the present-day levees have been raised to a height of 10 metres and extended to 96 metres wide. In all, there are now over 4500 km of levees along the Mississippi

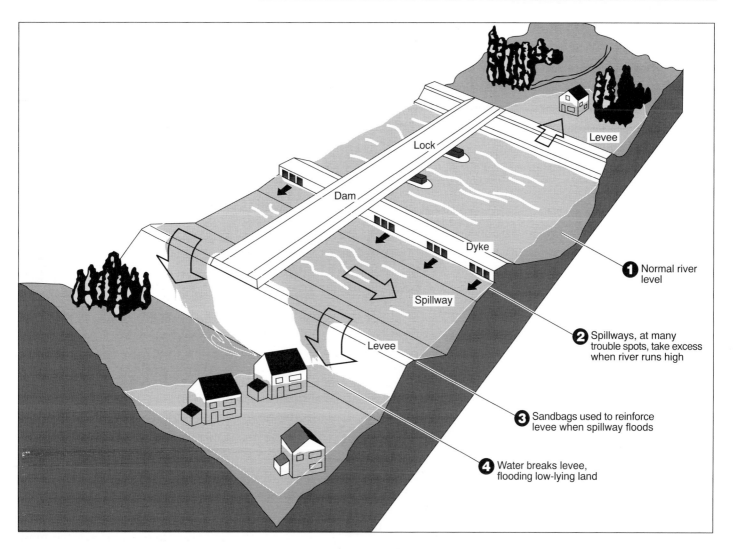

Labels on figure:
Lock
Levee
Dam
Dyke
Spillway
Levee

1 Normal river level

2 Spillways, at many trouble spots, take excess when river runs high

3 Sandbags used to reinforce levee when spillway floods

4 Water breaks levee, flooding low-lying land

Figure 12.2.2
Flood control on the Mississippi

system. Supplemented by parallel spillways (Figure 12.2.2), and by reservoirs constructed on many of the tributaries to delay the passage of floodwaters into the Mississippi, the levees are designed to withstand 1:100-year flood levels.

A secondary objective of the project was to improve navigability. In the early nineteenth century the Mississippi system was the principal means of transporting goods over long distances within the USA. However, it was handicapped by a continuously shifting shallow, meandering channel; indeed, the river was so treacherous that the average paddle-steamer survived only 18 months of river journeys before disaster struck.

The Corps of Engineers, which has had responsibility for improving river navigation for over 100 years, undertook to establish a modern 2.75-metre channel by building thousands of stone wing-dykes which collect sediment and divert the current to the opposite bank, where the river excavates a deeper channel. As the river shifts its course, the dykes are constantly built and rebuilt to maintain the channel.

The Corps also decided to shorten the river by excavating a series of cut-offs through major meander necks, effectively straightening the Mississippi's course and reducing its length by 270 km. This not only shortened the shipping route but by speeding water downstream it lowered flood levels. Unfortunately, the river kept trying to meander again, so almost 1000 km of revetments (overlapping concrete mattresses) have been laid across the soft river bed as far as the deep-water channel, stabilising it and protecting the vulnerable river banks from erosion.

These engineering works have undoubtedly contributed greatly to the revival of the Mississippi system's water-borne commerce, which is now more than three times greater than that on the St Lawrence–Great Lakes. However, the whole project has cost $10 billion and, once started, the work can never end. The river is a dynamic system and rebels against its artificial straitjacket. Consequently it takes $180 million a year just to maintain the control structures and rectify the problems the river continually produces.

The great Mississippi flood debate: environmentalists v engineers

The flood of 1993 highlighted a question that had been developing for some time: does the Corps' engineering works on the Mississippi cause, rather than prevent, flooding? The debate can be summarised thus:

The environmentalists' accusations

- river engineering has turned small floods into big ones, and big ones into catastrophes;
- levees force the water upwards because it cannot spread out across the floodplain;
- the water is forced to back up, causing flooding upstream, despite the reservoirs located there;
- the added pressure and greater speed of the water breaks down the levees, causing more violent floods than in a natural situation;
- the same amount of water comes through the system each year, but water levels get progressively higher.

The engineers' response

- floods are caused by heavy rainfall, not engineering works;
- past flood levels have not been measured accurately and have been over-estimated by up to one third. Sophisticated current-meters and other equipment now measure the river's flow very precisely;
- the levees do raise water levels but the upstream reservoirs hold back some water and thus achieve a balance of levels. Most years the reservoirs more than offset the effects of the levees;
- navigation works have little or no effect on big floods like those of 1993;
- the Corps is required to build to the 1:100-year flood level. Raising the levees would give greater protection but at greater cost. The 15-metre levee at St Louis is designed to stop all but a 1:500-year flood. Similar protection could be provided by the engineers along the whole river system but the taxpayer must be prepared to pay for it.

QUESTIONS

1 What other factor, apart from heavy rainfall, contributes to the flood hazard in the Mississippi Basin in spring and early summer?

2 Referring to Figure 12.2.1, suggest what functions are served by the locks and dams on the upper Mississippi, and the major dams on the Missouri.

3 Write a concise account of the flood control system on the Mississippi, as depicted in Figure 12.2.2.

4 Draw appropriately annotated diagrams to illustrate the effects of levees and reservoirs on the Mississippi–Missouri system at high water, according to (i) the environmentalists; (ii) the engineers.

5 Whose arguments – the environmentalists' or the engineers' – do you find the more persuasive, and why?

Floodplain urbanisation and the 'levee effect' cycle

About one-sixth of all US urban land lies within the 1:100-year flood plain, over half of all floodplain land has been developed, and urban areas are expanding on to floodplains at a rate of two per cent per annum. Not surprisingly, urban communities, once established on floodplains, demand flood protection. As soon as embankments are built, the 'levee effect' cycle comes into operation: the floodplain is perceived to be safe for development, land values rise, further development occurs, and improved flood protection is demanded. Since the real cost of levee-building falls on the federal government (i.e. the general tax-payer) rather than on the floodplain residents themselves, there is little incentive to break the cycle.

Ironically, floodplain urbanisation increases the frequency and magnitude of flooding in several ways:

- impermeable surfaces such as roofs, roads and car parks greatly increase runoff. Urbanisation may increase small floods ten-fold, while a 1:100-year event may be doubled in size by extensive artificial paving of the basin area;
- smooth surfaces combined with urban drains and sewers deliver water much more rapidly to the river channel, speeding the onset of the flood and reducing the time lag between storm rainfall and peak flow by perhaps 50 per cent;
- bridge supports and riverside facilities serving floodplain communities reduce the carrying capacity of the river channel. Such works on the Mississippi have reduced the channel's capacity by one-third since 1837.

The National Flood Insurance Programme and land use planning

When floods occur, they affect poorer people disproportionately because many floodplain residents are on low incomes and cannot afford to buy land in safer areas. Often they cannot afford flood insurance either: more than 80 per cent of the 1993 flood victims were uninsured.

The National Flood Insurance Programme (NFIP) was introduced by the federal government in 1968 owing to the reluctance of private insurance companies to risk catastrophic losses. It not only provides low-cost, variable insurance (dependent upon the risk of flooding in different locations) but, more importantly, aims to control development within over 17 000 participating floodplain communities.

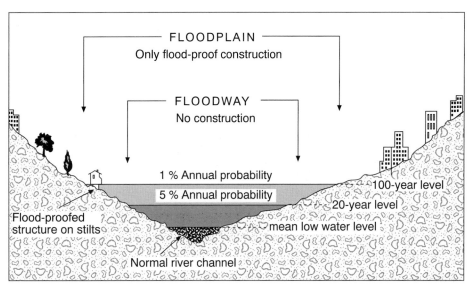

Figure 12.2.3
The delineation of hazard planning zones on a hypothetical floodplain

Basically, to obtain NFIP insurance, no new construction is permitted in the 1:20-year floodway zone, while existing residential development in the rest of the floodplain must be raised to at least the 1:100-year level and be flood-proofed (Figure 12.2.3). All new property-owners within the 1:100-year flood-plain must buy insurance at normal commercial rates.

Such moves towards land use planning within floodplain areas have gathered strength in recent years and become entwined with radical proposals by environmentalists to turn back the clock on floodplain development. They call for all engineering works on the Mississippi to be removed and for the river to revert to its natural state. People would have to learn to live with the resulting floods or get off the floodplain. All farmland would again become natural wetlands and riverine forests, while many flood-prone communities would be relocated to safer sites.

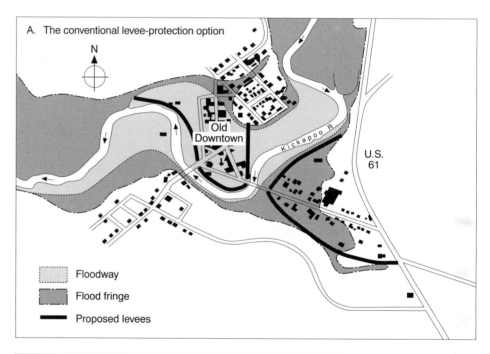

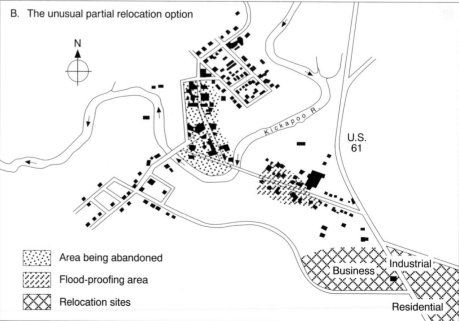

Figure 12.2.4
Adjustments to the flood hazard at Soldiers Grove, Wisconsin

Several communities along the Mississippi were wiped out by the floods of 1993. The residents of one such community, Valmeyer, Illinois, decided to move to a 205 ha tract on top of a river bluff, some 115 metres higher than the original site and thus completely safe from future flooding.

The case of Soldiers Grove, Wisconsin, a small community on the Kickapoo tributary of the Missouri River, supports the wisdom of Valmeyer's residents in choosing to relocate away from the floodplain. Following a series of floods in the 1970s, the Corps of Engineers proposed to protect the CBD, sited within a meander on the floodway, and other property by building two levees and an upstream reservoir. In 1978 yet another flood caused the townspeople to relocate the entire CBD, believing that the benefits of such a move would outweigh those of the levee-building scheme (Figure 12.2.4). The costs of publicly acquiring, evacuating and demolishing all buildings in the floodway and of flood-proofing all properties in the floodplain are estimated to have been about the same as the proposed levees. If so, Soldiers Grove has gained from its relocation option. The levees would have protected the community from most floods but would not have offered the opportunities provided by relocation. For example, with compensation payments, businesses could build more modern premises incorporating cost-saving energy conservation and solar energy facilities at a more attractive site along a major highway.

While even partial urban relocations, such as Soldiers Grove, have been rare in the past, the experience of the devastating flood of 1993 may persuade increasing numbers of floodplain communities to follow Valmeyer's radical example in the future.

QUESTIONS

1 Suggest why, despite the obvious risks, settlement was initially attracted to floodplain land.

2 Explain why modern urban development continues to expand on to floodplains, and how this expansion contributes to the flood hazard.

3 (a) Distinguish between the floodway and the floodplain in terms of the flood hazard (Figure 12.2.3).
   (b) Referring to Figure 12.2.3, comment on the terms and conditions of the government's flood insurance scheme (NFIP).

4 (a) Summarise the advantages and disadvantages of relocating the Soldiers Grove CBD (Old Downtown) (Figure 12.2.4).
   (b) Why were the town's three other areas (to the north east, south east and south west of the CBD) able to remain in their original locations?

5 Why is the environmentalists' call for the removal of all flood-prone communities from the Mississippi Basin unlikely to be heeded? What would you recommend as a reasonable compromise?

12.3 Hurricanes in the South Eastern States

General background

Hurricanes are tropical cyclones with wind velocity of Force 12 (121 km/h or 75 mph) on the Beaufort Scale (Figure 12.3.1). These are essentially very low pressure systems which draw in great quantities of warm moist air from tropical seas. They develop along the Inter-tropical Convergence Zone (ITCZ) where the trade winds of both hemispheres meet in the low-pressure Equatorial Trough. The ITCZ changes position with the seasons; between July and October it is located at approximately 15°N. At this time, optimum conditions for hurricane formation over the North Atlantic Ocean are most likely to occur, including:

- a warm water surface of at least 27°C;
- a large sea area to supply the overlying air with large quantities of water vapour;

- a pronounced vertical instability in the air mass;
- little or no vertical wind shear;
- sufficient distance from the Equator for the Coriolis force to deflect the unstable air mass, which begins an anticlockwise spiral.

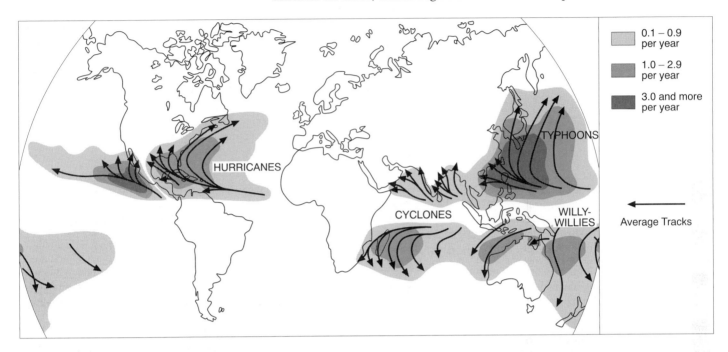

Figure 12.3.1
Tropical cyclones: terminology, location and average annual frequency

The eye of the hurricane forms above a relatively cool area of sea. Around the eye, winds spiral around and upward, drawing in increasing quantities of moist air. This forms a thickening wall of cloud revolving around the cooler, tranquil eye. The input of warm, moist air maintains the hurricane's growth for several days (Figure 12.3.2).

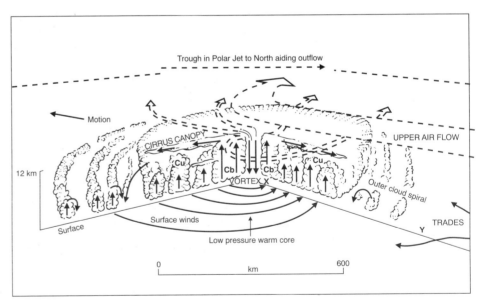

Figure 12.3.2
Circulation and movement associated with a typical hurricane

During the air's ascent, large-scale condensation releases enormous amounts of previously latent energy. Near the tropopause some of the ascending air is spun outwards as an extending canopy of cirrus cloud. The rest subsides at the core of the system, becoming warm and dry through adiabatic expansion. Such adiabatic warming causes further convection and draws in even more moisture from the ocean. The whole whirling system continues to develop, with latent heat as the main source of energy. This mechanism, termed the 'heat engine', pumps out thousands of megatonnes of energy every day.

The Azores High, a semi-permanent zone of high pressure which dominates the North Atlantic during summer, guides hurricanes on the easterly trade winds which flow around its southern margin. Thus a typical hurricane moves along a curving track from the mid-Atlantic, through the West Indies and into the southern states of the USA (Figure 12.3.1).

Once over land, hurricanes quickly lose their energy. By the time they reach the Midwest or North East a week or so later, hurricanes have usually diminished to tropical storms (wind speed 62–119 km/h). Occasionally the remnants of hurricanes cross the Atlantic on the westerly air flow and bring high winds and above-average precipitation to the British Isles and adjacent parts of continental Europe (e.g. Hurricane Lili, 28 October 1996).

QUESTIONS

1 (a) With reference to Figure 12.3.1, explain the initial westerly track of all tropical cyclones and their subsequent poleward and easterly 'recurvature'.
  (b) Why do northern hemisphere tropical cyclones cause more deaths and property damage than their southern hemisphere counterparts?

2 Explain in your own words (i) the 'heat engine' mechanism of hurricane development; (ii) the anticlockwise rotation of hurricanes.

3 (a) Account for the formation and location of the three main cloud types on Figure 12.3.2.

  (b) Why do warm, calm conditions characterise the eye of the hurricane?
  (c) Describe and explain the likely variations in ocean surface conditions along line X–Y.

4 (a) Studies have shown that more intense hurricanes (category 3 or above) strike Florida and the US east coast during wet years in the Sahel region of Africa (1940s–1960s; 1988–present) than in dry years (1970–1987). Attempt to explain this correlation.
  (b) Explain how global warming (greenhouse effect) could produce more intense hurricane activity in the future.

Hurricane Andrew, 16–28 August 1992

Figure 12.3.3
Satellite photograph showing Hurricane Andrew, passing over Florida

Andrew was a relatively small but ferocious Cape Verde hurricane that wrought unprecedented devastation along a path through the north western Bahamas, the southern Florida peninsula and south central Louisiana (Figures 12.3.3 and 12.3.4). With a central pressure of only 922 mb, violent winds up to 320 km/h, and a 5-metre storm tide, Andrew was almost a category 5 hurricane on the Saffir-Simpson scale, so rare that only one or two occur every century. The final cost of damage exceeded $25 billion, making Andrew the most expensive natural disaster in US history.

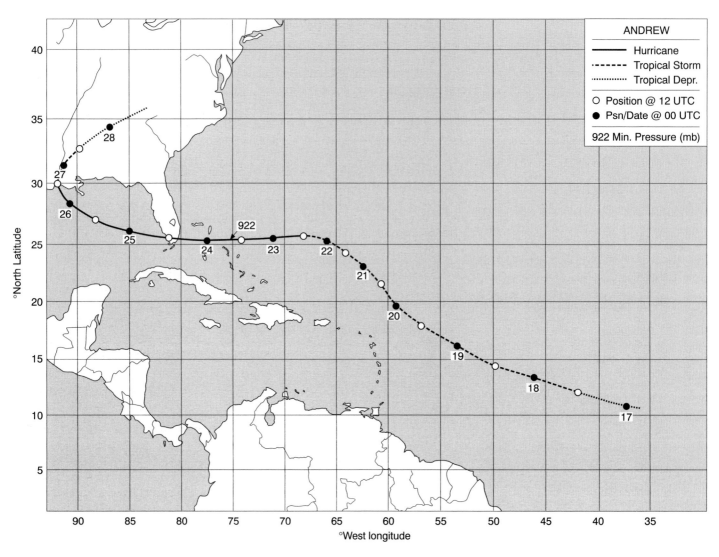

Figure 12.3.4
Best track positions for Hurricane Andrew

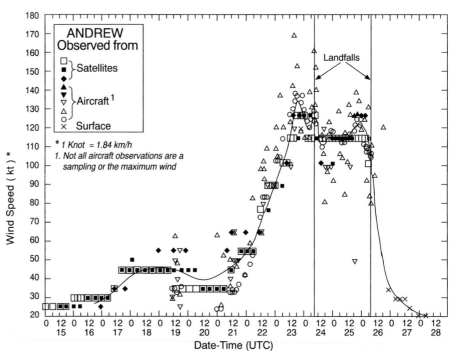

Figure 12.3.5
Best track maximum sustained wind speed
curve for Hurricane Andrew

Andrew's origins and development

14 August 1992 A tropical wave crossed from the west coast of Africa to the tropical North Atlantic Ocean.

15 August Moving at about 40 km/h, the wave passed to the south of the Cape Verde Islands.

16 August Convection increased and narrow spirals of cloud developed. At 1800 hours (UTC) the tropical wave became a tropical depression (wind speed up to 61 km/h).

17 August The depression deepened and became Andrew, the first Atlantic tropical storm (wind speed 62–119 km/h) of the 1992 hurricane season (July–October) (Figure 12.3.5).

18–19 August Steering currents carried Andrew on a north westerly course away from the West Indies and towards Florida. Strong south-westerly vertical wind shear and surface high pressure to the north limited convection within the system.

20 August While Andrew retained a vigorous circulation aloft, it was still classified as a tropical storm with 80 km/h surface winds and extremely high central pressure of 1015 mb (Figure 12.3.6).

21 August The vertical wind shear over Andrew decreased and a ridge of high pressure extended from the US south east coast into the Atlantic with its axis just north of Andrew. The tropical storm turned westward, accelerated to nearly 30 km/h and quickly intensified.

22 August Andrew reached hurricane strength (over 119 km/h) during the morning. An eye formed and the rate of strengthening increased.

23 August The ridge of high pressure to the north held steady, driving Andrew nearly due west over the north west Bahamas.

24 August Andrew weakened as it left the Bahamas but rapidly reintensified as it moved over the Straits of Florida. The hurricane's eye diameter decreased and convection strengthened up to the time Andrew crashed ashore near Homestead Air Force Base, Florida at 5.05 am local time (Figure 12.3.7). Andrew continued nearly due west across the extreme southern portion of the Florida peninsula.

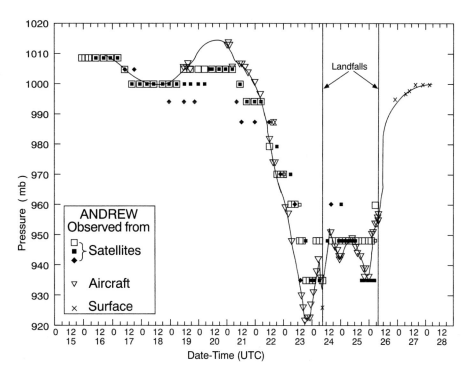

Figure 12.3.6
Best track central pressure curve for Hurricane Andrew

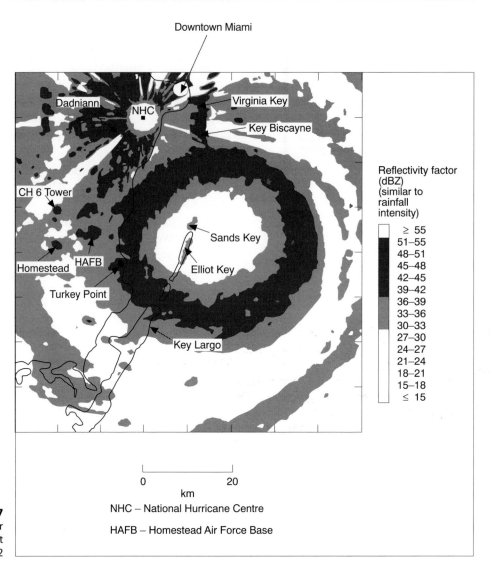

Downtown Miami

Dadniann

NHC

Virginia Key

Key Biscayne

CH 6 Tower

Sands Key

Homestead

HAFB

Elliot Key

Turkey Point

Key Largo

Reflectivity factor
(dBZ)
(similar to
rainfall
intensity)

≥ 55
51–55
48–51
45–48
42–45
39–42
36–39
33–36
30–33
27–30
24–27
21–24
18–21
15–18
≤ 15

0   20
km

NHC – National Hurricane Centre

HAFB – Homestead Air Force Base

Figure 12.3.7
Hurricane Andrew: pattern of radar reflectivity near Miami at 0830 UTC, 24 August 1992

Although pressure rose to about 950 mb, Andrew was still a major hurricane when its eyewall passed over Florida's Gulf coast. As Andrew tracked over the Gulf of Mexico it again intensified and continued to move rapidly as it turned toward the west northwest.

25 August As Andrew reached the north central Gulf of Mexico, the high pressure system to its north east weakened. With the change of steering currents, Andrew turned toward the north west and its forward speed dropped to about 15 km/h.

26 August The hurricane thus struck not New Orleans, as originally forecast, but a sparsely populated section of the south central Louisiana coast with category 3 intensity at about 0830 UTC.

27 August About 10 hours after landfall, Andrew again weakened rapidly to tropical storm strength, and then to depression status 12 hours later. During this phase, Andrew moved north and then accelerated toward the north east. It continued to produce very heavy rain along its track, as well as several destructive tornadoes in Louisiana, Mississippi, Alabama and Georgia.

28 August By midday, Andrew had begun to merge with a frontal system over the mid-Atlantic states of Pennsylvania, New Jersey and New York, thus losing its separate identity.

QUESTIONS

1 (a) Study Figures 12.3.3 and 12.3.4. At approximately what time on what date was the satellite photograph taken?
   (b) From an atlas map of the area shown in Figure 12.3.3, calculate the approximate diameter of the main hurricane system.

2 (a) Referring to Figure 12.3.4, for how many days did Andrew exist as a hurricane?
   (b) How was Andrew able to reintensify and maintain its hurricane status between leaving Florida and making landfall in Louisiana?
   (c) Why did Andrew rapidly degenerate into a tropical storm after landfall in Louisiana?

3 (a) Describe and explain the relationship between the graphs of wind speed and pressure (Figures 12.3.5 and 12.3.6).
   (b) Comment on the methods and intensity of observations recorded.

4 Referring to Figures 12.3.2 and 12.3.7, rearrange the following statements to describe accurately the sequence of weather likely to be experienced on the Atlantic coast of Florida as Hurricane Andrew approached and passed overhead:

- 10 to 45-minute rainless interval with calm or light breezes.
- Wind freshens and rain begins to fall; heavy cloudiness prevails.
- Dark, heavy mass of spiralling cloud becomes visible.
- Rainfall gradually decreases over two to three days, and winds fall light.
- Heavy rainfall occurs patchily until the leading cloud-wall (near the eye) passes over.
- A broad trail of cirrus cloud becomes visible.
- Heaviest downpours and strongest gusts from the opposite direction as the second cloud-wall passes over.

Andrew's aftermath

Hurricane Andrew left in its wake a 30 to 55 km wide swathe of damage totalling $25 billion. The toll of human misery was immense. Nearly two million people were evacuated from vulnerable coastal areas in the United States, and around 220 000 were left homeless. Storm-related deaths numbered 54.

In Florida alone, 63 000 homes were destroyed and another 100 000 were severely damaged, in an area likened to a nuclear battlefield. Hardest hit were the rural and suburban areas about 30 km south of downtown Miami – Homestead, Florida City, Kendall – most of which were totally destroyed. South Miami and parts of Coconut Grove and Coral Gables (including the National Hurricane Centre) were also badly damaged. Along with piles of debris which were once houses, Andrew left a surreal scene of devastation: schools, shopping centres, petrol stations and churches demolished; cars smashed and overturned; boats damaged and blown ashore; crumpled aircraft; and a severely scarred landscape of uprooted trees (Figure 12.3.8).

Figure 12.3.8
Damage caused by Hurricane Andrew

Weathering A Crisis

When Hurricane Andrew ripped through Florida and the Gulf Coast in August 1992, it was the most expensive natural disaster in American history, costing insurance companies $15.5 billion and the federal government another $10 billion. According to the Florida insurance department, as much as 40 percent of Andrew's damage could have been prevented if building code standards had been enforced. Instead, widespread deficiencies were found in thousands of construction sites in Florida, where wind resistance standards are 10 mph tougher than those in any other state.

Code enforcement had been lax because state officials fell behind during the 1980s construction boom and have never caught up. Dade County, for example, approved 20,000 building permits in 1991, but it has only 60 building inspectors.

American City and Country
January 1994

Figure 12.3.9
Weathering a crisis

Over 20 000 US troops were deployed to the stricken area to deter looting, help clean up the debris and build temporary tent cities. Even so, tens of thousands of people went for days without food, water, medical care or shelter in the 35°C heat before help arrived, while millions of homes and businesses lost power supplies.

Animals also fell victim to the storm. Thousands of cats and dogs, not allowed into the emergency shelters, were left on the streets to fend for themselves. One homeowner woke up to find a shark, washed over in the storm, floating in his swimming pool; others found fish in their TV sets. Over 2000 monkeys and baboons, used by the University of Miami for research, escaped when their cages were destroyed; hundreds of the animals, rumoured to be infected with the AIDS virus, were shot by police and frightened, armed residents. Most animals at Miami's Metrozoo escaped injury, but 300 rare tropical birds flew away to freedom and, perhaps, their deaths.

When images of the hurricane hit the news media, foreign aid flowed in from around the world as if southern Florida – one of the most sophisticated regions in the United States – were a Third World country. Canada, Japan and Taiwan all sent relief. Even Russia offered to send workers and machinery to help with the clean-up.

Devastating as it undoubtedly was, Hurricane Andrew had three redeeming features:

- While the human death toll was shocking, the loss of life was relatively small compared with that caused by past hurricanes of similar magnitude; previously, storm-surge flooding was the greatest threat, over 90 per cent of hurricane fatalities in the United States being due to drowning.
- Andrew helped an area often beset with ethnic and racial conflict to become a more cohesive community. South Florida residents, whether white, black, Cuban, Haitian, Guatemalan or all-American, worked together to rebuild the devastated areas.
- The damage done to Miami's vital tourist industry was minimal. Had Andrew's diameter been larger, or the hurricane made landfall just a few kilometres to the north, Miami and its beach resorts would probably have been wiped out. New Orleans in Louisiana similarly escaped because of a last-minute change in Andrew's track.

12.4 Tornadoes: Violent Twisters of the Great Plains

Tornadoes occur in many countries, including Britain, but more are recorded in the United States than anywhere else in the world. They are rapidly rotating columns of air with strong winds and violent downdraughts. Within the USA, tornadoes are quite common throughout the Midwest and the eastern side of the country generally (Figure 12.4.1). They also strike the south eastern states in winter, especially during severe thunderstorms, and can develop from decaying hurricanes. However, they are most frequent and destructive in 'Tornado Alley' on the Great Plains. Here the conditions that spawn tornadoes are ideally met. Complex depressions develop near the eastern Rocky Mountains, drawing in warm, moist air masses from the Gulf of Mexico. These interact with cool air from the north west, forming a squall line along which 'supercell' thunderstorms begin to form.

Tornado development

A supercell thunderstorm erupts when warm, moist air breaks through an overlying stable layer and moves upward through cool, dry air (Figure 12.4.2). In the northern hemisphere, the updraft is tilted to the north east and rotates anticlockwise when viewed from above. The warm air parcels decelerate in the

QUESTIONS

1 With reference to Figure 12.4.1, describe and account for (i) the pattern of tornado occurrences in the United States; (ii) the north easterly direction of tornado paths.

2 Explain why most tornadoes in the US occur in late spring and early summer. (June, with an annual average of 155 tornadoes, is the peak month).

Figure 12.4.1
Average annual occurrence of tornadoes in the US

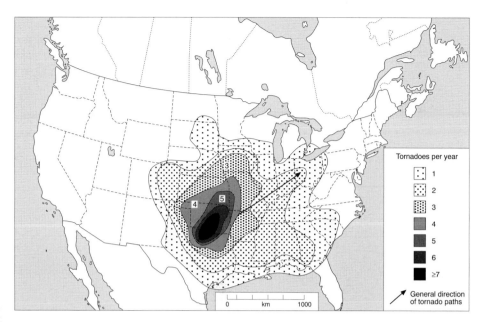

Figure 12.4.2
Anatomy of a tornadic storm with 'twister' at ground level

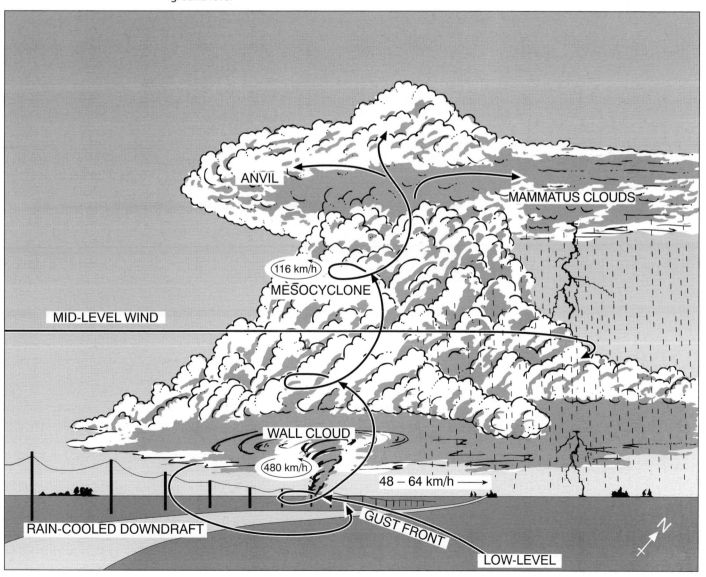

stratosphere, fall back down and spread sideways in the anvil. In the north east part of the storm, rain falls out of the tilted updraft into mid-level dry air, cooling it and causing it to sink. The supercell's rotation pulls some of the rain and the cool air around to the south west side of the storm.

Near the ground, warm air and rain-cooled air meet in a turbulent boundary called the gust front. Lowered wall clouds and tornadoes tend to form along this line, near a cusp marking the storm's centre of rotation. Here pressure may be only 10 per cent of that of the surrounding air. Tornado paths along the ground are unpredictable but are generally aligned from south west to north east.

Destructive power

Figure 12.4.3
Tornado destruction in Texas

Tornadoes are capable of immense, if short-lived, devastation. The damage inflicted on buildings requires wind speeds of up to 440 km/h. Windward walls of buildings (generally to the south west) almost always fall inward, implying that structures are most often damaged by the wind's brute force, not by the sudden drop in atmospheric pressure, as previously thought. Consequently, residents of Tornado Alley are no longer advised to open windows to reduce the pressure inside; many people were injured by flying glass as they rushed to open their windows. Nor are residents now advised to hide in the south west corner of the house, where they would be in the most danger of walls falling in on them. Instead, residents are urged to shelter in a central cupboard or storage area because of the added protection of interior walls.

The strength of tornadoes can also be gauged by their ability to move heavy objects. Trees, cars and people are hoisted skywards, aircraft flipped over, and 70-tonne railway coaches derailed. Roof fragments can travel tens of kilometres and light debris appears to be lifted several kilometres into the main storm, falling to earth up to 325 kilometres away. Most of the debris falls to the left of the tornado's path, often in well-defined bands according to weight.

Tornadoes can take a heavy toll of human life. Historically, the worst year was 1925 when the Tri-State Tornado killed 689 people along a 350-km path through Missouri, Illinois and Indiana. More recently, 361 tornado-related deaths occurred in 1974, while in April 1991, 22 people died when a tornado swept through Kansas and Oklahoma. Moreover, for every death that occurs, about 20 other people are seriously injured. One remarkable escape received wide publicity in November 1995 when a midnight tornado flattened the home of the Wells family in rural Prairie County, Arkansas. Seven-month-old Joshua Wells was ripped from his cot by 320-km/h winds and hurled 400 metres into a rice field. Miraculously, Joshua survived with barely a scratch, but his parents were killed by the freak storm.

QUESTIONS

1 With reference to Figure 12.4.2, explain why (i) the windward walls of buildings in Tornado Alley are usually to the south-west; (ii) residents used to be advised to open windows to reduce pressure inside their houses before the arrival of a tornado; (iii) most of the debris falls to the left of a tornado's path.

2 Your chances of surviving a tornado can be improved by seeking the best shelter. Consider the following situations, then select and explain the safer option from each pair.
   (i) Inside a car on the open road/Inside a house in the middle of a field.
   (ii) Lying down on a building's flat roof/Sitting in a basement room.
   (iii) Sheltering in a south west facing room/Sheltering in a north east facing room.
   (iv) Sheltering in a large upstairs room/Sheltering in a small ground-floor room or cupboard.
   (v) Running across open countryside parallel to the tornado path/Lying in a ditch transverse to the tornado path.

3 Which children's story and Hollywood film is linked with Kansas and Tornado Alley?

12.5 The Active Edge of North America: Plate Tectonics and Volcanoes

Plate Tectonics

Following earlier debates about continental drift and studies of the ocean basins, by the late 1960s the theory of plate tectonics had been formulated. This widely accepted theory postulates that the Earth's crust, between 5 and 70 km thick, consists of seven major plates and about a dozen smaller ones (Figure 12.5.1) which together comprise the lithosphere. These rigid lithospheric plates rest on and move over an underlying layer of partially molten rock known as the asthenosphere (Figure 12.5.2).

Since the continents are composed principally of granitic rocks, they are lighter than the basaltic ocean floor and so 'float' higher in the mantle than does the oceanic crust. The boundary between the crust and the mantle is marked by the Mohorovičić Discontinuity (moho).

It should be noted that the lithospheric plates are much thicker than the oceanic or continental crust. Moreover, their boundaries do not usually coincide with those of the oceans and continents. Some plates carry oceans, some continents, some both. For example, the Pacific Plate is entirely oceanic, but the northern section of the American Plate carries the western half of the North Atlantic Ocean as well as the North American continent.

Huge convection currents in the asthenosphere, generated by heat from radioactive decay, act as the titanic engine that drives the lithospheric plates across the Earth's surface. The plates interact along their boundaries, diverging, converging, or slipping past each other along transform faults. At the plate boundaries occur major forces responsible for many large-scale surface features; such processes include volcanism, earthquakes and orogeny.

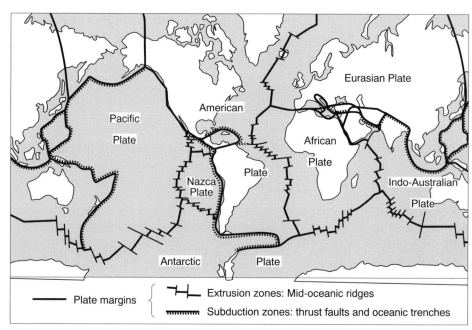

Figure 12.5.1
Tectonic plates of the world

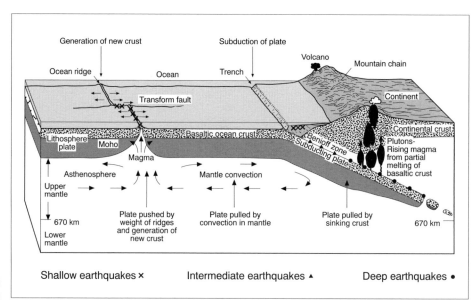

Figure 12.5.2
Plate tectonic mechanisms

Interaction at the Pacific–North American plate boundary

At oceanic spreading centres in the Pacific plate, magma wells up from the mantle to create new basaltic ocean crust. The sea floor is forced to spread laterally away from such divergence centres, and part of the Pacific Plate moves slowly (perhaps 2 cm a year on average) but inexorably towards the western edge of North America. The creation of new sea floor is accompanied by the formation of ocean ridges, volcanic activity and shallow earthquakes as the oceanic crust successively tears, heals and tears apart again.

The excess oceanic crust thus produced is subducted (drawn down) where the Pacific Plate converges with the North American Plate and descends below it. The greater buoyancy of the continental crust ensures that the continental crust survives while the ocean floor continuously renews itself (Figure 12.5.2).

Since subduction involves the descent into the mantle of a mass of cold rock some 100 km thick, numerous deep earthquakes occur along the plane of compression, usually between 300 and 700 km below the surface. Moreover, since the descending oceanic plate contains large quantities of water (which acts as a flux) and is progressively heated as it is drawn down, at a depth of 100 km or more the subducted basaltic crust and oceanic sediments melt to become an andesitic (acidic) magma. Because it is lighter than the surrounding rock, the magma rises through the continental crust in the form of dome-shaped plutons.

One hundred million years ago the Sierra Nevada range formed the western edge of North America. Subsequent erosion has revealed many granitic domes, ancient plutons once buried 1.5 km deep and measuring from 1.5 km to dozens of kilometres across. Some of the most spectacular plutons form the dramatic landscapes of Yosemite National Park (Figure 12.5.3).

Sixty million years ago the Atlantic Ocean was widening more rapidly than the Pacific. This caused North America to drift far to the west, overriding the Pacific Plate before it could be subducted into the mantle. Instead it scraped along under the North American Plate, the friction shortening and crumpling the continental plate to form the Rocky Mountains, a process that lasted 20 million years. Today, the Rockies are not only the continental watershed but also the boundary between the generally static continent to the east and the active continent to the west.

Fifteen million years ago the Pacific seafloor plate slid northward, creating a transform fault in California. Ever since, part of the western edge of the continent has been dragged northward by movement along the San Andreas Fault, notorious for its destructive deep earthquakes. This kind of plate boundary is also known as a strike-slip fault or fracture zone. Because crust is neither produced nor consumed, it is regarded as 'neutral', and since only horizontal movement is involved, no volcanism is present.

The coast ranges that extend along the western margin of North America from California to southern Alaska are also believed to be an unusual product of plate tectonics. These low mountains are geological misfits relative to the surrounding terrain and are thought to have originated in the ancient Pacific as oceanic plateaus, island arcs or other thick pieces of oceanic crust which would not subduct. Instead, they were scraped off the descending Pacific Plate and remain lodged against the edge of the continent, complicating the processes of subduction and mountain building.

The Cascade Range: subduction zone volcanoes

The subduction that formed the Western Cordillera of North America (i.e. all the ranges from the Rockies to the Pacific coast) largely ceased around ten million years ago. However, continuing volcanic eruptions in the Cascade Range are testimony to residual subductive activity along the Pacific–North American plate boundary. Here plutons have risen explosively to the surface.

The Cascade Range forms the link between the Sierra Nevada to the south, through Oregon and Washington states, to the Coast Mountains of British Columbia to the north. Throughout their 100 km extent the Cascades contain many majestic snow-capped volcanic cones exceeding 3000 metres. Such summits include northern California's Lassen Peak (3187 m; first recorded eruption 1650, most recent 1914) and Mt Shasta (4392 m; first and last 1786), while Mt Hood (3424 m; 1800, 1907) is the highest point in Oregon and Mt Rainier (4392 m; 1825, 1882) is the highest in Washington and the Cascade Range (Figure 12.5.5). Most of the volcanoes are extinct or dormant but some have been active in the recent past. Mt Baker (3285 m; 1820, 1870) steamed ominously in 1975 and Mt St Helens (originally 2950 m; first recorded eruption 1500) erupted violently in 1980 and has been active on a number of subsequent occasions.

Figure 12.5.3
Half Dome, an ancient pluton, Yosemite National Park, California

QUESTIONS

1 Using the words and phrases provided below, complete the following statements summarising the 'laws of plate tectonics':
- the Earth's surface is divided into rigid plates, which form the . . . . ;
- the plates are created at the . . . . , which are called . . . . ;
- the plates move apart from one another without being . . . . They slide on a . . . . called the . . . . ;
- the plates are destroyed at the . . . . , called . . . . , where they plunge into the . . . . However, only the . . . . of the plates are swallowed up in this process;
- the light . . . . move with the plates that carry them, but they are not . . . . ;
- bordering the plates are ridges, trenches and . . . . In general, plate boundaries do not . . . . with the borders between the oceans and the continents;
- the energy of the . . . . is . . . . at the plate boundaries, either . . . . (earthquakes or the formation of . . . .) or . . . . (volcanoes or the formation of . . . .).

List of words and phrases for Q1

accretion zones	dissipated	mountains
asthenosphere	Earth's interior	oceanic parts
coincide	lithosphere	oceanic ridges
continents	mantle	oceanic trenches

deformed	mechanically	plutonic rock
subduction zones	submersible	thermally
transform faults	viscous substrate	

Based on Allegre, *The Behavior of the Earth*

2 The Atacama Trench, as predicted by plate tectonic theory, appears off the Pacific coast of South America but has no North American equivalent. Referring to the account and Figures 12.5.1 and 12.5.2, explain the absence of such an off-shore feature between California and Alaska.

3 What additional evidence on Figure 12.5.4 supports (i) your explanation for Q2 above; (ii) the belief that the continents are generally of older origin than the ocean floors?

4 Referring to Figure 12.5.2, answer the following questions:
   (a) At approximately what angle does the oceanic plate descend beneath the continental plate?
   (b) What name is given to the seismic zone where subduction occurs?
   (c) At what distances below the surface do deep earthquakes begin and cease?

5 Using Figure 12.5.2 as a guide, draw annotated diagrams to illustrate the formation of (i) the Rocky Mountains; (ii) the Coast Ranges.

6 Account for the origin and present appearance of Half Dome (Figure 12.5.3).

Figure 12.5.4
The age of the sea floor

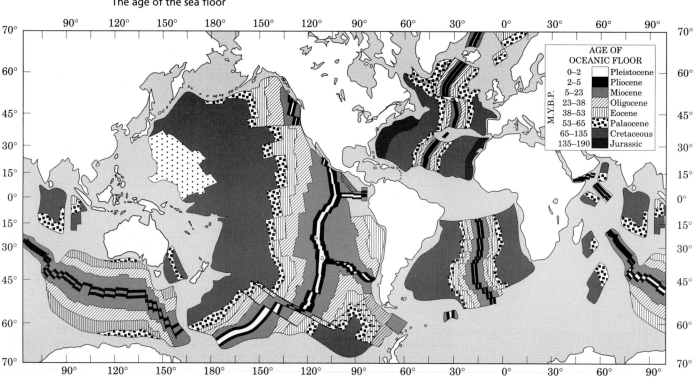

Figure 12.5.5
Mt Rainier, Cascade Range, Washington

Explosive volcanism: the eruption of Mt St Helens, 1980

Before 1980 Mt St Helens was a steep volcanic cone capped by snowfields and small glaciers. It had erupted during at least 20 different phases over the past 4500 years, the last active period being between 1831 and 1857. In 1978 US government scientists wrote a report predicting that a major eruption was likely before the end of the century. On 18 May 1980 the prediction became reality: Mt St Helens erupted in violent and spectacular fashion (Figure 12.5.6).

On 20 March the first tremors of an earthquake swarm occurred, culminating in the great eruption two months later. By the end of March a series of small-scale emissions of ash had begun, and by the end of April a large bulge had appeared on the volcano's northern slopes, caused by a rising column of magma. The sequence of explosive events of 18 May 1980 is summarised in Figure 12.5.7.

Figure 12.5.6
The eruption of Mt St Helens, 18 May 1980

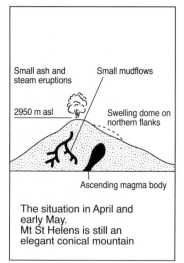

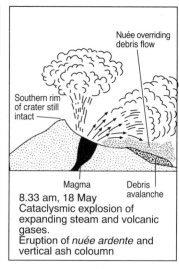

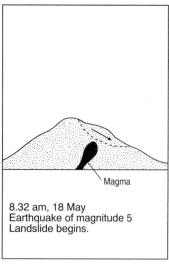

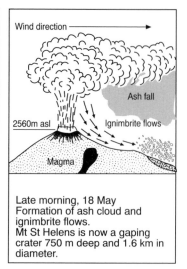

Figure 12.5.7
Stages in the eruption of Mt St Helens, May 1980

The principal phenomena and effects of the eruption may be summarised as follows:

Debris avalanche or landslide

This 2 km³ mass of cold material including rock, ice, snow and soil rushed downhill at 75 metres per second, leaving deposits up to 100 metres thick. Entering the valleys of the local drainage system, it produced torrential mudflows which destroyed houses, bridges, roads and many thousands of commercially valuable trees. The mudflows also clogged streams, a reservoir and the navigation channel of the Columbia River (Figure 12.5.8).

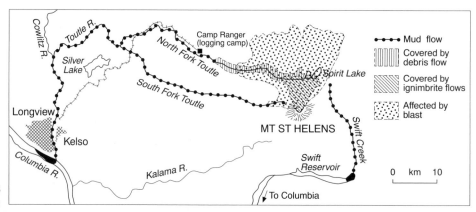

Figure 12.5.8
Local effects of the Mt St Helens eruption

Nuée ardente ('glowing cloud')

This was a searing lateral blast of hot dust and rock torn from the core of the volcano and suspended in incandescent gases. Reaching speeds of up to 250 km/h, it flattened all 10 million trees within the 550 km² blast area and damaged thousands more around it. The nuée caused nearly all the 66 human fatalities resulting from the eruption. Most of the victims were sightseers who ignored warnings to stay clear of the danger zone around the volcano, which normally would have been virtually devoid of people.

Flows of ignimbrite ('fire-cloud rock')

These were similar to the nuée ardente except that the material suspended in the gas was mainly pumice fragments derived from the magma itself, whose temperature during eruption was around 700°C. Falling from the eruption column, these fragments left an apron of deposits up to 40 metres thick.

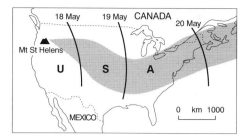

Figure 12.5.9
Ground track of the ash cloud following the 18 May eruption

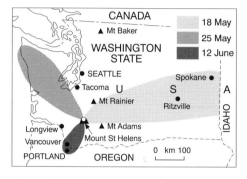

Figure 12.5.10
Areas affected by ash falls from the eruptions of 18 May, 25 May and 12 June

Ash cloud

This boiled quickly up into the atmosphere and spread out sideways, blown by the wind. At an altitude of 17 km the cloud consisted mainly of fine ash particles, but at 20 km it became an aerosol of sulphates condensed from the sulphurous gases of the eruption. The cloud obscured the sun, turning day into night along its path across the United States (Figure 12.5.9), while the *tephra* (ash fall) damaged homes, vehicles and crops along a wide swathe. At Ritzville (Figure 12.5.10), for example, the ash – described as 'like grey talcum powder' – was seven cm thick. The long-term climatic effects of these outpourings remain unclear.

Mt St Helens produced five smaller eruptions during the remainder of 1980. Since then a dome of highly viscous dacitic lava has grown intermittently within the crater. This dome measures about 250 metres high and one km in diameter and is estimated to contain 75 million cubic metres of extruded material.

Comparisons with the past

Although spectacular and devastating, the 1980 eruption of Mt St Helens was only moderate in terms of the volume of new magma involved. Nearby, former volcano Mt Mazama rose to about 3650 metres until a cataclysmic explosion blew away the top of the mountain some 6600 years ago. An estimated 70 km³ of debris – 40 times more than that ejected by Mt St Helens – was distributed over a 10-state area.

The cavity that resulted could no longer support the remainder of the cone, which collapsed into the void of the magma chamber. The resulting caldera now contains Crater Lake, 10 km in diameter, the deepest (590 metres) and perhaps the loveliest of all American lakes. Subsequent volcanic activity within the caldera has produced Wizard Island, a cinder cone rising from the floor of the cavity to a height of 233 metres above the lake's surface, which itself is 1885 metres above sea level (Figure 12.5.11).

Figure 12.5.11
Crater Lake and Wizard Island

Even the massive eruption that destroyed Mt Mazama is eclipsed by Long Valley Caldera in California. When that volcano blew its top 700 000 years ago it ejected 600 times as much material as did Mt St Helens, the ash fall extending from Hudson Bay in the north, to Nova Scotia in the east, to Cuba and the Yucatan Peninsula in the south – an event of truly continental proportions.

QUESTIONS

1 (a) What caused the 'earthquake swarm' prior to the eruption of Mt St Helens? (Figure 12.5.7).

 (b) Why is volcanism invariably accompanied by seismic activity (earthquakes) but not vice-versa?

2 (a) The eruption of Mt St Helens is said to have been a simultaneous combination of Plinian- and Pelean-type events. Explain this statement in detail.

 (b) Describe the effects of the eruption on the volcano's profile (Figure 12.5.7).

3 (a) Explain in your own words what happened to the great volume of material ejected by the explosion (Figures 12.5.7–12.5.10).

 (b) How and why did the areas affected by ash fall from the later eruptions differ from that of the initial eruption? (Figures 12.5.9 and 12.5.10).

 (c) How might such ash clouds affect climate?

4 Mt St Helens is the most active volcano in the continental United States. What evidence is there that it is likely to remain so?

5 Describe the probable eruptive history of Mt Rainier (Figure 12.5.5).

6 With reference to Figure 12.5.11, explain the difference between a caldera and an ordinary crater.

7 Explain why the scale of volcanic eruptions in the Cascades appears to have diminished within the last million years or so.

Intra-plate volcanism: the Hawaiian hot spot

Most volcanoes are concentrated along plate boundaries but about five per cent of the world total occur within plates. The Hawaiian volcanoes are good examples of such intra-plate activity. They occur near the centre of the northern section of the Pacific Plate but dramatically indicate its past movements.

Here a linear volcanic chain has active volcanoes – Mauna Loa and Kilauea – at its south eastern end with a major continuation of reefs, coral atolls and islets – the Hawaiian Ridge – to the north west of the Hawaiian Islands (Figure 12.5.12). Then occurs a dog-leg bend into the submerged Emperor Seamounts which continue northward to the edge of the Pacific Plate (Figure 12.5.13).

Figure 12.5.12
The Hawaiian Islands

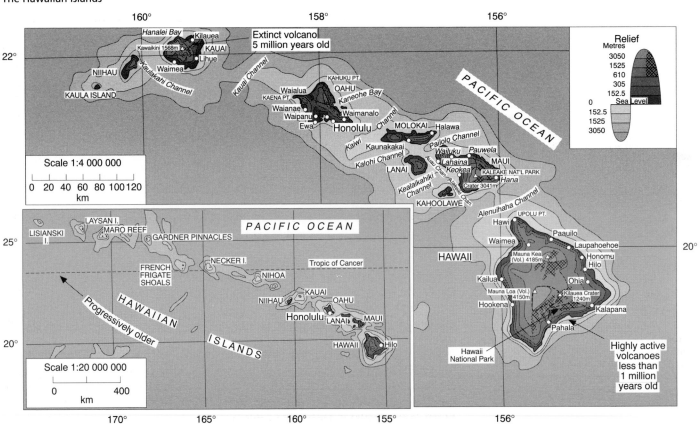

Figure 12.5.13
The andesite line and the products of the Hawaiian hot spot

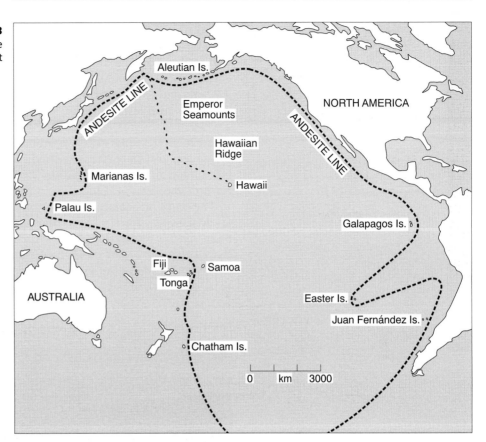

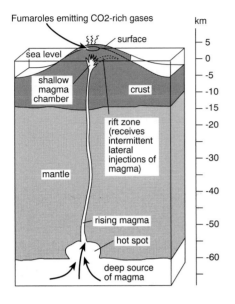

Figure 12.5.14
Cross-section of an active volcano formed above the Hawaiian hot spot

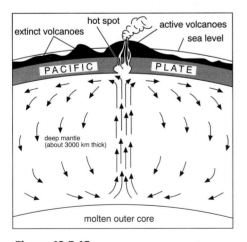

Figure 12.5.15
Stationary hot spot and moving plate: the life and death of Hawaiian volcanoes

This chain of progressively older, extinct volcanoes is thought to be the product of a 'hot spot', a magma-generating centre, located at a deep fracture within the plate far below the lithosphere, which has maintained its position for at least 80 million years. Above the Hawaiian hot spot a volcano formed at the plate's surface (Figure 12.5.14). As the plate moved on – at an average rate of approximately 8–10 cm a year – the volcano became extinct, then eroded, and eventually sank below the surface of the sea (Figure 12.5.15). Meanwhile a new volcano appeared above the hot spot, and so the process has continued for millions of years.

The basic lava produced by such volcanoes always occurs inside the 'andesite line' separating the rocks of the true ocean basin of the Pacific from those of greater acidity lying on the continental side of the line. Thus intermediate rocks such as andesite, dacite and rhyolite are always found outside the andesite line (Figure 12.5.13).

Effusive volcanism: the eruption of Mauna Loa, 25 March 1984

Mauna Loa (4170 metres asl, but rising over 9000 metres above the seafloor that surrounds the Hawaiian Ridge) is estimated to have a volume of 40 000 km³, making it the world's largest volcano. It is an excellent example of a convex, gently sloping shield volcano built up by countless effusive eruptions (Figure 12.5.16). It is still very active, erupting every 3–4 years on average, with fountains and streams of incandescent fluid lava of basaltic character issuing from the summit and from two rift zones on its flanks.

After a year of increased earthquake activity, Mauna Loa erupted on 25 March 1984. The recorded sequence of events was as follows:

1.25 am A fissure split the summit caldera (3–5 km in diameter and formed by prehistoric subsidence). Lava, spurting from this fissure at a temperature of 1140°C, accumulated within the caldera.

QUESTIONS

1 (a) In which compass direction did the Pacific Plate move before 40 million years ago? (Figure 12.5.13)

(b) What change of direction occurred approximately 40 million years ago?

2 Account for the different types of lava produced by volcanoes on either side of the andesite line (Figure 12.5.13).

3 List and account for the principal contrasts between Mt St Helens and Mauna Loa, using the following headings: location, origins, magma/lava types, eruptive characteristics, resultant landforms, probable future activity/extinction, threat to local human communities.

Dawn A new line of lava fountains formed at an elevation of 3800 metres but within two hours the fissure had descended to 3450 metres, the new lava fountains being 50 metres high and 2 km long. The lava flow descended the south east flank of the mountain.

4.00 pm The lava fountains dwindled but new earthquakes heralded the opening of new fissures lower down.

4.40 pm Final vents opened at 2900 metres asl – about seven km distant from those at 3450 metres asl. A vigorous outflow of lava ensued. Although the fountains were now only 20 metres high, the outpouring amounted to approximately 0.5 million m³ per hour. Within 24 hours the lava advanced 12 km north east towards the town of Hilo.

The eruptions continued for another 10 days but the rate of advance progressively slowed – from 6 km on the second day, to 4 km on the third day, and to 3 km on the fourth day. By 15 April the eruption was over. The longest lava flows extended 27 km, stopping 10 km from the outskirts of Hilo. The new lava covered 48 km², but no one was killed or injured, and the only damage was the cutting of power lines and the blocking of a few jeep tracks.

Figure 12.5.16
Hawaiian shield volcanoes: lava on Mauna Loa in the foreground, with Mauna Kea visible in the background

12.6 Earthquakes in California: on the Edge of Disaster

One very important consequence of California's location on the active edge of the continent is the risk of major earthquakes along the junction of the North American and Pacific tectonic plates. The San Andreas fault, which for 1050 km of its length runs through California, roughly north to south, is the most famous and best-studied section of this plate junction (Figure 12.6.1). Several damaging earthquakes have occurred along or close to the fault in the twentieth century. The most devastating was that which destroyed much of San Francisco in 1906. More recently, the two biggest quakes have rocked the same city in October 1989 (magnitude 6.9 on the Richter Scale; epicentre at Santa Cruz) and Los Angeles in January 1994 (magnitude 6.6; epicentre at Northridge). The latter occurred on one of the many transverse and subsidiary faults which criss-cross the 'Golden State' (Figure 12.6.2).

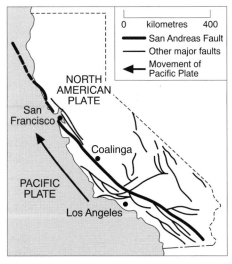

Figure 12.6.1
California: the San Andreas fault and tectonic plates

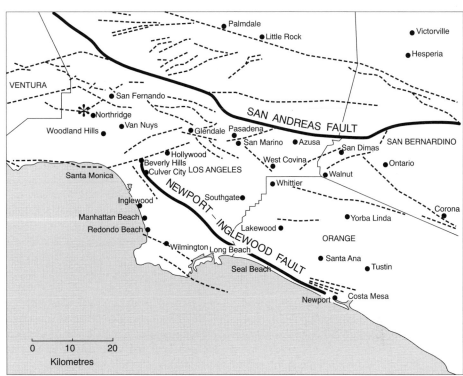

Figure 12.6.2
Fault lines in Southern California

The Northridge Earthquake, January 1994

Figure 12.6.3
24 dead in Los Angeles quake

24 dead in Los Angeles quake*

It had been the city's worst earthquake for 150 years, but switchboard operators in Los Angeles still urged their callers to have a nice day.

Outside, at least 24 people lay dead, hospitals were inundated with 'a tidal wave' of wounded, freeways were disappearing in clouds of concrete dust, homes were burning and thousands had been left homeless.

Yesterday, in 40 cataclysmic seconds, the irresistible power of nature pushed the Santa Monica mountains upwards, squeezed the lush San Fernando valley together and left a trail of chaos in its wake.

Giant freeways buckled — including Amer-

ica's busiest, the Santa Monica — huge buildings collapsed, skyscrapers swayed, landslips swept away scores of luxury homes, mobile home parks blazed and gas mains exploded leaving islands of flame in streets flooded from burst water pipes.

At Northridge, a northern suburb of Los Angeles that was the quake's epicentre, 14 people died when a three-storey building was reduced to two levels. Another 18 were trapped as the ground floor of the 130-apartment complex disappeared beneath the top two storeys.

'It was like a bomb going off,' said Eric Pearson, a third-floor res-

ident. 'One big jolt lifted the building off its foundation, pulled it over and slammed it down within a three-second period. When the shaking was over, the building tilted over diagonally so it fused all the doors shut.'

A police motorcyclist died when he failed to see the freeway ahead had collapsed and soared over the edge of the abyss.

Crushed

An electricity power station supplying much of the city was put out of action by water escaping from a fractured aqueduct. A train carrying sulphuric acid was derailed, filling the air with lethal fumes.

Even worse, the city was bracing itself for further disasters ahead. Geologists at California Institute of Technology said they had recorded 30 tremors after the main shock, which registered 6.6 on the Richter scale.

They predicted that these could continue for five weeks and could not rule out a more violent earthquake.

Yet many were thankful that it struck at 4.31am on a public holiday — yesterday was Martin Luther King Day. Had it been virtually any other time, the toll would have been catastrophic as packed freeways collapsed.

The 1989 quake that devastated San Francisco, registering 7.1 on the Richter scale, killed 67 when it struck in the late afternoon peak period.

Yesterday's quake shook the homes of the stars in Malibu and the Hollywood Hills and was felt in San Diego, 125 miles south and Las Vegas, 275 miles east.

It knocked out power across much of the metropolitan area and, because of the interdependence of power grids, cuts were reported as far north as Portland, Oregon and Seattle in Washington State.

Fires raged either side of the Golden State Freeway. The road's intersection with the Antelope Freeway collapsed, crushing cars — at least one person died there.

Along Rodeo Drive, renowned as the world's most opulent shopping street, many shop windows exploded on to the pavement. Hollywood Boulevard's Walk of Fame glittered with shattered glass.

Unease

California's Governor Pete Wilson declared a state of emergency. President Clinton freed the way for federal aid by declaring the region a disaster area.

Geologists are advising people in threatened areas to take all essential belongings from their homes today because it might be their last chance.

The underlying anxiety is all the greater because yesterday's earthquake was not on the notorious San Andreas fault, which experts say is '13 months pregnant' and overdue for 'The Big One'.

Daily Mail, 18 January 1994

* The final death toll was 51.

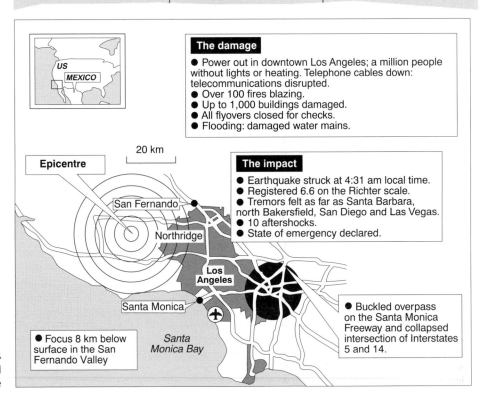

The damage

● Power out in downtown Los Angeles; a million people without lights or heating. Telephone cables down: telecommunications disrupted.
● Over 100 fires blazing.
● Up to 1,000 buildings damaged.
● All flyovers closed for checks.
● Flooding: damaged water mains.

The impact

● Earthquake struck at 4:31 am local time.
● Registered 6.6 on the Richter scale.
● Tremors felt as far as Santa Barbara, north Bakersfield, San Diego and Las Vegas.
● 10 aftershocks.
● State of emergency declared.

● Buckled overpass on the Santa Monica Freeway and collapsed intersection of Interstates 5 and 14.

● Focus 8 km below surface in the San Fernando Valley

20 km

Epicentre

San Fernando

Northridge

Los Angeles

Santa Monica

Santa Monica Bay

Figure 12.6.4
The Northridge earthquake: impact and damage

Figure 12.6.5
The Simi Freeway, Los Angeles, severed by the Northridge earthquake

Figure 12.6.6
The Mercalli Scale of earthquake intensity

Scale	Intensity	Description of effect	Maximum acceleration (mm/sec²)	Corresponding Richter scale	Approximate energy released in explosive equivalent
I	Instrumental	detected only on seismographs	<10		1 lb (0.45 kg) TNT
II	Feeble	some people feel it	<25		
III	Slight	felt by people resting; like a large truck rumbling by	<50	<4.2	
IV	Moderate	felt by people walking; loose objects rattle on shelves	<100		
V	Slightly strong	sleepers awake; church bells ring	<250	<4.8	
VI	Strong	trees sway; suspended objects swing; objects fall off shelves	<500	<5.4	small atom bomb, 20 000 tonnes TNT (20 kilotonnes) hydrogen bomb, 1 megatonne
VII	Very strong	mild alarm; walls crack; plaster falls	<1000	<6.1	
VIII	Destructive	moving cars uncontrollable; chimneys fall and masonry fractures; poorly constructed buildings damaged	<2500		
XI	Ruinous	some houses collapse; ground cracks; pipes break open	<5000	<6.9	
X	Disastrous	ground cracks profusely; many buildings destroyed; liquefaction and landslides widespread	<7500	<7.3	
XI	Very disastrous	most buildings and bridges collapse; roads, railways, pipes, and cables destroyed; general triggering of other hazards	<9800	<8.1	60 000 1-megatonne bombs
XII	Catastrophic	total destruction; trees driven from ground; ground rises and falls in waves	>9800	>8.1	

Source: After Bryant (1990).

Note The Mercalli Scale is a semi-quantitative measure of earthquake intensity devised by an Italian scientist in 1902 and modified in 1931. It depends upon a descriptive list of mainly visual phenomena of increasing devastation gauged at the seismic measuring station. For this reason it is less accurate than the Richter Scale, which has largely replaced it for scientific purposes.

Q U E S T I O N S

1 Distinguish between the earthquake's focus and its epicentre (Figure 12.6.4).

2 (a) Charles Richter, who devised the Richter Scale, stated: 'Earthquakes don't kill people, buildings do.' Elaborate this assertion.

(b) Suggest the safest places to shelter inside a building during an earthquake.

3 The death toll resulting from the Northridge earthquake was relatively small because of its timing (Figure 12.6.3). Describe a worst-case scenario when a similar earthquake would probably cause much greater loss of human lives.

4 (a) Referring to the evidence in Figures 12.6.3–12.6.6, where would you place the Northridge earthquake on the Mercalli Scale?

(b) Explain any difference between your Mercalli rating and the official 6.6 rating on the Richter Scale (Figure 12.6.6).

5 (a) Referring to Figure 12.6.6, explain the 'maximum acceleration' data.

(b) Suggest the factors affecting the 'intensity' of earthquakes (Figure 12.6.6).

Living on the fault line

At the San Andreas fault, the edge of the North American plate (on which most of the continent sits) and the edge of the Pacific plate (which carries most of the California coastline) attempt to move past each other at an average speed of four cm/year – about the same rate as fingernails grow (Figure 12.6.7). Where they succeed, in so-called 'creeping zones', numerous small tremors occur; these do little damage and can often be detected only by the most sensitive seismographs. In contrast, some sections of the fault record no movement in any direction for years at a time. In some places the rocks have been locked together for centuries. Such inactive sections are known as 'lock zones', 'seismic gaps' or 'asperities' (Figure 12.6.8).

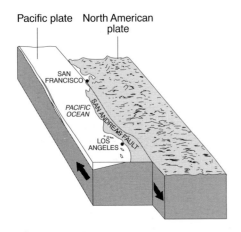

Figure 12.6.7
Tectonic movement along the San Andreas fault

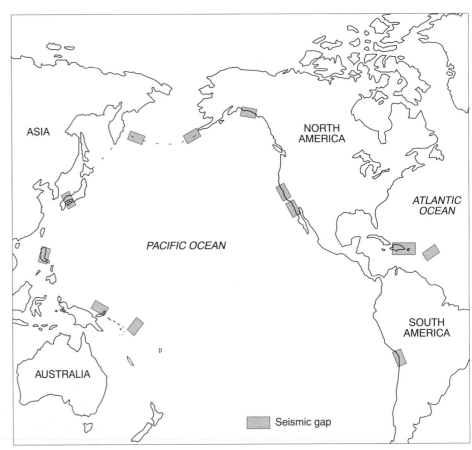

Figure 12.6.8
Major seismic gaps or asperities around the Pacific plate

Eventually, when the rocks can no longer stand the enormous stresses, the two plates force their way past each other in a violent release of accumulated energy usually causing earthquakes that register at least 7 on the Richter Scale (Figure 12.6.9). The 1989 Santa Cruz quake and its many after-shocks, which occurred in one of the asperities along the San Andreas fault, almost completely filled a seismic gap, altering the probability of future earthquakes in that section of the fault (Figure 12.6.10).

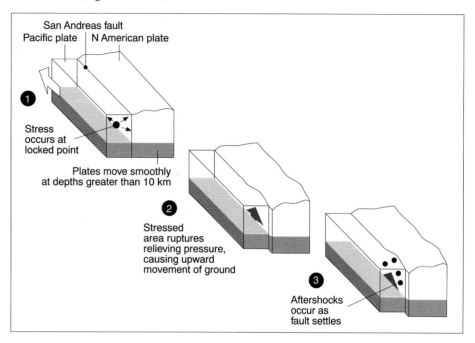

Figure 12.6.9
Seismic activity along the San Andreas fault

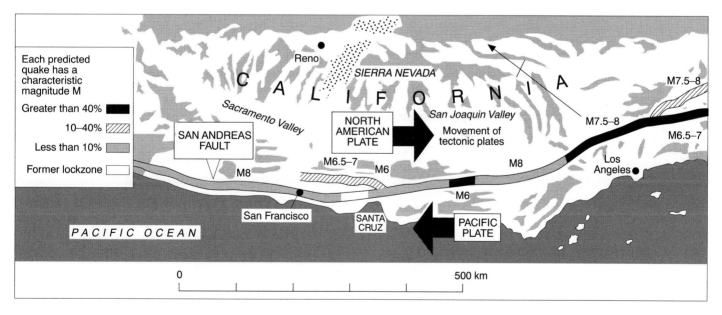

Figure 12.6.10
Earthquake probability along the San Andreas fault, to c. 2020

Of great concern, however, is the situation around Los Angeles, where other seismic gaps must inevitably produce what seismologists call 'The Big One' at some point in the near future. The 1994 Northridge earthquake occurred on a transverse fault that had remained inactive for over 200 years. This should be seen as a warning of the havoc to come when the San Andreas fault itself unleashes pent-up energy equivalent to 60 000 hydrogen bombs.

1 How are seismologists able to identify the location and extent of seismic gaps? (Figure 12.6.8).

2 Suggest the likely sequence and scale of seismic activity along sections of the San Andreas fault that are termed 'intermediate zones', relative to creeping zones and lock zones.

3 (a) Referring to Figure 12.6.9, explain (i) why the two plates move smoothly at depths greater than 10 km; (ii) the causes of aftershocks.

(b) How do aftershocks affect people at the surface?

4 (a) With reference to Figure 12.6.10, how would you now rate the Santa Cruz area in terms of earthquake probability?

(b) Which area of California is currently at greatest risk from earthquakes?

5 Los Angeles lies on the Pacific plate and San Francisco on the North American plate. Referring to Figure 12.6.10, calculate the number of years it will take, at the current rate of plate movement, for the two cities to share the same latitude.

6 Study the details of the devastation outlined in Figure 12.6.11. How would you, as a member of the disaster-relief planning team, prioritise action to alleviate suffering and repair damage following a major earthquake along the Newport-Inglewood fault zone?

7 (a) How has the California state government attempted to prepare for the 'Big One'?

(b) In what sense are the structural engineers 'competing against time'?

The 'Big One': The Ultimate Nightmare

As every Southern Californian has heard time and again, the terrible quake known as the 'Big One' will emanate from the San Andreas fault, the 1050-km-long rip in the Earth that runs the length of the state and whose southern tip skirts the eastern edge of the City of Angels. But now, geologists and disaster experts believe even greater devastation could be wreaked by a smaller quake in the Newport-Inglewood fault zone, a series of geological fissures that traverse greater Los Angeles from Newport Beach to Beverly Hills. (See Figure 12.6.2.)

Just how bad the devastation would be is the subject of a study by the California Division of Mines and Geology, which published a scenario in 1988 for a Newport-Inglewood quake of magnitude 7 on the Richter scale. The scenario's worst-case predictions are grim indeed. The quake would rock an area that is home to 9 million people, triggering landslides that would smash posh residential areas. Long stretches of freeway would buckle. The heaving earth would rupture water, power and natural-gas lines, igniting numerous gas and chemical fires. Damaged refineries and sewage facilities would dump oil and raw waste onto fabled beaches and into the sea.

While the Golden State cannot know when or where its next quake will strike, disaster planners have been using the 1988 scenario and four others like it to get ready. The state now boasts one of the most stringent building codes in the world and has in place well-coordinated disaster-relief plans for attending to the wounded and repairing damaged infrastructure, says Fred Turner, a structural engineer with the state's Seismic Safety Commission. The most difficult task remaining is shoring up existing structures to make them more earthquake resistant, a multibillion-dollar investment that the economically depressed state can ill afford right now. 'On the other hand,' warns Turner, 'We are competing against time.'

Anticipating the worst
In the worst-case scenario, these facilities would be hardest hit:

Disabled hospitals
All four of the acute-care hospitals that lie within the fault zone would probably become non-functional, putting 1,900 beds out of service. Hospitals as far as 25 miles away would also be vulnerable, with a total of 34 per cent of the area's 14,500 beds probably being knocked out.

Fractured highways
More than 130 miles of highway and 350 bridges lie near enough to the Newport-Inglewood fault zone to be severely shaken. At least 16 arteries would probably be closed for at least 24 hours, landslides and liquefaction being major causes of blockages. A 10 km stretch of the San Diego Freeway, which lies only 4 km from the fault, would close indefinitely.

Damaged airports
Four of the five principal public airports in L.A. and Orange counties, which together serve 60 million passengers a year, could suffer damaged runways and buildings, although all but one would probably stay open.

Dangerous harbours
The Los Angeles-Long Beach Harbour would light up with a major fire from leaking oil facilities. Ships would suffer damage from smashing against their moorings in every harbour along the coast from Santa Monica to Newport Beach.

Free-flowing sewage
The quake would rupture numerous water mains, and water treatment plants would be severely disabled. Residents would go without water for several days in some areas. Raw sewage would dump directly into streams, inundate neighbourhoods, and foul the region's famous shoreline, in some cases for several weeks.

Flaming gas lines
Gas line breaks would ignite fires along the fault zone and in adjacent areas.

Spilled oil
Damage to refineries and storage facilities would spill petroleum into the sea and would fuel fires inland.

The heaving earth
Along the Newport-Inglewood fault zone, which straddles Los Angeles and Orange counties, the ground would shake violently enough to move houses off their foundations, topple walls and masonry and sever underground pipes.

Landslides: Though the shaking would diminish in intensity farther away from the fault, landslides would pose a danger in narrow canyons and on the densely populated hillsides.

Liquefaction: Large pockets of greater L.A. extending from San Fernando Valley to Newport Beach are in danger of liquefaction – a geological term meaning that apparently solid soil turns into something like quicksand.

U.S. News & World Report,
31 January 1994

Figure 12.6.11
The big one

13 National and International Relations

13.1 Regional Disparities and Developments

The period since 1945 has witnessed a growing awareness of inter-regional contrasts by both the US and Canadian governments, due primarily to the heightened perception of economic variations by the general population and a developing realisation of political power by disadvantaged areas and ethnic groups.

The era of regional planning developed first in western Europe but later spread to the rest of the developed world, and in more recent times many developing nations have implemented regional policies, albeit with widely varying degrees of success. In comparison with many European nations, North America reacted slowly to regional inequalities. Although previous policies can be recognised as having specific regional benefits, it was not until the 1960s that both the USA and Canada introduced comprehensive systems of regional development planning.

Identifying problem regions

Although many criteria have been used to identify area distress, most analyses focus on per capita income and levels of unemployment, as both factors have been reported consistently over a long period of time.

While significant income variations are apparent in both the USA and Canada, the general trend in the two countries has been one of convergence, at least until the last decade or so. Figure 13.1.1 shows the shifting per capita income balance between the nine census divisions of the US. The recent resurgence of New England and the Middle Atlantic contrasts markedly with the third region in the traditional 'Manufacturing Belt', the East North Central, which recently dipped below the national average for the first time. Clearly New England and the Middle Atlantic have enjoyed greater success in attracting new growth industries to replace the large number of jobs lost in traditional manufacturing.

Figure 13.1.1
US: per capita income by division, 1900–1995

Division	Percentage of US average					
	1900	1940	1970	1980	1990	1995
New England	135	127	108	106	118	116
Middle Atlantic	143	132	114	108	116	115
East North Central	107	112	105	101	98	101
West North Central	98	81	94	94	94	95
South Atlantic	51	77	90	93	98	98
East South Central	50	49	74	78	79	83
West South Central	61	64	85	94	85	87
Mountain	140	87	90	95	89	92
Pacific	163	135	111	115	103	103
Mean deviation (%)	37	28	12	8	11	9

Per capita incomes in the Pacific region have fallen in a relative sense in recent years while the East South Central remains the poorest region in the nation. Throughout the whole time period all three divisions of the South have recorded per capita personal incomes below the national average but have steadily moved towards the norm, with the exception of the West South Central between 1980 and 1993. Figure 13.1.2 shows in greater detail the clear patterns of relative poverty and wealth across the USA. The state with the highest per capita income is Connecticut ($28 110) while Mississippi ranks bottom of the fifty-state list ($14 894).

Figure 13.1.2
USA: personal income per capita (1995 $US)

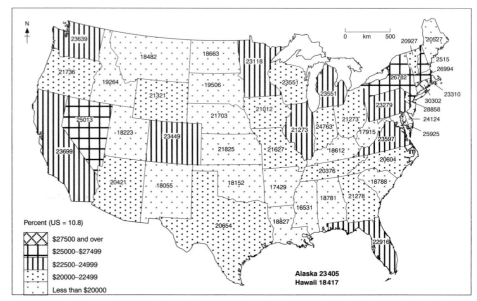

Regional per capita income trends for Canada are indicated in Figure 13.1.3. The most noticeable development has been the income improvement in both the Atlantic provinces and Quebec, at a fairly steady rate over the period analysed. Taking into account both per capita income and total populations the nuclei of wealth in the country are in Ontario and the western provinces of Alberta and British Columbia. The mean deviation data, with the limitations noted, show steady convergence throughout the whole time period, thus differing from the USA in recent years.

Figure 13.1.3
Canada: per capita personal income by province, 1950–1994

Province	Percentage of Canadian average					
	1950	1960	1970	1980	1990*	1994*
Newfoundland	51	56	63	64	79	79
Prince Edward Island	55	57	67	71	77	86
Nova Scotia	74	76	78	79	86	86
New Brunswick	70	68	72	72	82	84
Quebec	86	87	89	95	91	91
Ontario	121	118	118	107	110	110
Manitoba	101	99	93	90	91	92
Saskatchewan	83	89	72	91	86	87
Alberta	101	100	99	112	101	102
British Columbia	125	115	109	111	105	105
Yukon and N.W. Territories	n.a.	106	95	103	n.a.	n.a.
Mean deviation (%)	23	19	18	16	12	11

*1990 and 1994 data = Average Family Income

In 1993 over 10 million North Americans were unemployed. The 'good times' of the 1950s, 1960s and early 1970s, when economic growth was relatively steady and very few people were without a job, have long gone. Not unexpectedly the continental figure hides huge regional variations. Unemployment is clearly an important indicator of regional distress, but while there is a definite correlation between per capita income and unemployment, it is far from conclusive, particularly in the USA. A number of affluent states suffer from unemployment rates at a higher level than might be expected due to the contraction of key industries (structural unemployment) or from a level of immigration considerably higher than the job opportunities available. Also, regions where farming is more important tend to suffer from disguised unemployment on the land. However, a relationship that may not appear significant at the state level may become very obvious with more detailed analysis at the county level.

Regional trends in Canada from 1974 (Figure 13.1.4) show that the Atlantic provinces and Quebec have consistently recorded jobless rates above the rest of the country. The relatively low unemployment rates of the Prairie provinces, the most important agricultural area in the country, is partly explained in terms of disguised unemployment on the land. The pattern of unemployment in the USA has generally been much more complex (Figure 13.1.5), certainly at the census division level, for reasons already discussed. At this point it may be interesting to compare income and unemployment data for the USA with a range of other indicators (Figure 13.1.6) which reflect differences in the quality of life.

Figure 13.1.4
Canada: unemployment rate[1] by province, 1974–1995 (%)

	1974	1979	1984	1989	1995
Newfoundland	13.0	15.1	20.2	15.8	18.3
Prince Edward Island	[2]	11.2	12.8	14.1	14.7
Nova Scotia	6.8	10.1	13.0	9.9	12.1
New Brunswick	7.5	11.1	14.8	12.5	11.5
Quebec	6.6	9.6	12.8	9.3	11.3
Ontario	4.4	6.5	9.0	5.1	8.7
Manitoba	3.6	5.3	8.4	7.5	7.5
Saskatchewan	2.8	4.2	8.0	7.4	6.9
Alberta	3.5	3.9	11.1	7.2	7.8
British Columbia	6.2	7.6	14.7	9.1	9.0

[1]Annual averages.
[2]Estimates less than 4000.

Figure 13.1.5
US: unemployment rates by region, 1980–1995

	1980	1985	1990	1995
New England	6.3	4.6	5.6	5.4
Middle Atlantic	7.5	6.7	5.2	6.2
East North Central	9.2	8.6	5.7	4.8
West North Central	5.3	6.0	4.1	3.9
South Atlantic	6.8	6.7	5.3	5.1
East South Central	7.9	9.2	6.3	5.7
West South Central	6.1	8.6	6.2	5.9
Mountain	6.3	7.2	5.3	4.9
Pacific	7.5	7.9	5.1	7.3

Regional development policy in the USA

The first comprehensive federal programme to assist areas in economic distress was initiated by the 1961 Area Redevelopment Act, but previous legislation can be identified as having specific regional effects. By far the most prominent of such actions was the establishment of the Tennessee Valley Authority in 1933

Census Region	A	B	C	D	E	F	G
New England	6.5	4136	84.1	9.6	2608	8.7	240.4
Middle Atlantic	8.4	4417	91.2	8.8	2444	11.9	257.3
East North Central	9.3	4951	79.4	11.0	2078	11.0	331.7
West North Central	8.1	4496	59.0	12.1	1855	8.8	339.3
South Atlantic	9.6	6257	79.1	13.1	1998	10.9	287.9
East South Central	9.7	4624	57.5	13.2	1559	15.9	384.6
West South Central	8.3	5866	76.6	12.9	1763	15.0	568.2
Mountain	7.4	6097	72.7	18.0	1875	9.7	322.4
Pacific	6.8	6172	91.5	12.7	2076	10.0	259.7

A – Infant Mortality Rate 1993
B – Total Crime Rate per 100 000 pop. 1994
C – % population living in metropolitan areas 1994
D – Suicides per 100 000 population 1993

E – Per capita federal income tax 1993
F – Food Stamp Program – % households participating 1993
G – Total per capita energy consumption [M.BTU] 1993

which, although a landmark in the history of regional assistance, was bitterly opposed by many politicians.

In the 1950s the poorer regions became more and more vociferous in their demands for federal assistance and the balance of Congressional opinion shifted slowly in favour of regional aid. However the ensuing policies were criticised both by those who regarded them as lightweight and by a consistent lobby which sought the abandonment of such federal intervention. The main legislation introducing federal assistance to distressed areas were the Area Redevelopment Act 1961, the Public Works and Economic Development Act 1965 and the Appalachian Regional Development Act of the same year.

The Area Redevelopment Act 1961

The Area Redevelopment Administration (ARA) was established to co-ordinate assistance to specified areas of intense poverty and unemployment. Such aid was primarily in the form of low-interest, long-term loans to extend local investment for projects which had difficulty in obtaining the necessary extra finance from regular sources. Federal help was also directed towards infrastructure projects and retraining schemes, usually in the form of loans, as a means of increasing the attractiveness of recipient regions to new industry.

Localities which requested and qualified for assistance were designated Redevelopment Areas. The county formed the primary unit of organisation. Each Redevelopment Area was required to produce and submit an Overall Development Programme to the ARA, highlighting sectors requiring new investment. By the end of the scheme in 1965, 1120 Redevelopment Areas had been established, incorporating more than one-third of all counties in the USA. Political compromise, to gain sufficient support for the scheme to get off the ground, resulted in inadequate funds spread over far too large an area. The total investment in the four year period was just over $300 million. The ARA estimated that 71 000 new direct jobs were created by the programme which was replaced by the 1965 Public Works and Economic Development Act. The main value of the 1961 Act was as a pilot scheme for the more effective 1965 Act, although the former legislation did have a reasonable degree of success in certain regions.

The Public Works and Economic Development Act 1965

The Economic Development Administration (EDA) replaced the ARA, with increased funding for more help in the form of direct grants. The criteria for federal assistance under the Act were:

- substantial unemployment;
- persistent unemployment;

- low median family income;
- high out-migration;
- Indian reservations;
- prospect of a sudden rise in unemployment;
- at least one eligible area in every state.

In an important contribution to the literature on the subject, Benjamin Chinitz used the eligibility criteria to identify seven types of regional distress [Figure 7]. Chinitz argued that the difficulties suffered by the regional types are often so different that lasting solutions are only possible with the application of remedial plans specifically tailored to the problems evident in each region and that such flexibility was clearly lacking in US regional development policy.

Eligible areas were still required to submit an Overall Economic Development Programme which was to be revised annually. A new development unit, the Economic Development District (EDD) was established. This had to contain at least two counties and a 'growth centre' which would generate spread effects to the wider region. The vast majority of EDDs contained between five and ten counties with the most intense concentrations in Appalachia, the South, the Upper Great Lakes region, the Northwest and the southern Mountain region.

Initially, the number of counties eligible was approximately 900 but in 1971 the eligibility criteria was relaxed because of political compromise and soon 1800 counties were included. A commensurate increase in funding was not forthcoming and this greatly limited the potential benefits of the programme. Lack of adequate investment and poor selection restricted many growth centres to a role well below that originally envisaged. Public works schemes were an essential part of the programme but much debate ensued concerning the allocation of EDA funds, particularly between infrastructural and cultural projects.

The Act also introduced an even larger planning unit, the Economic Development Region, to cover multi-state areas. It was felt that the 'Title V' Regional Commissions set up to oversee the EDRs would provide more co-ordinated planning over wider regions, irrespective of state boundaries. The commissions in fact had little impact, mainly because they controlled such limited funds.

The whole regional programme survived a number of concerted attempts by the powerful opposition lobby to ditch it but was finally terminated by the Reagan administration in 1981. The only exception was assistance to the Appalachian region.

Appalachia

The Appalachian region has received by far the largest injection of federal funding through the two unrelated and contrasting programmes of the Tennessee Valley Authority and the Appalachian Regional Development Act. Both schemes are unique in the history of US regional economic development.

The nature of Appalachia's economic problems stems from a combination of the region's physical landscape and people's inability or unwillingness to recognise the limitations of such an environment. The economy of the region from the beginning has been heavily over-dependent on mining, forestry and agriculture and was thus subject to the periodic fluctuations characteristic of such activities.

Although agricultural potential differs widely in a region with such a diversity of physical landscape, modern farming in general is very limited in extent and there has been much abandonment of farmland. Indiscriminate and unscientific methods of farming by early settlers frequently resulted in appalling soil erosion. Appalachia remained a region of semi-subsistence agriculture well into the twentieth century.

Widespread timber clearance, both to make land available for agriculture and for direct commercial exploitation, exacerbated the situation in this fragile highland ecosystem. In the latter part of the nineteenth century the region was penetrated by railroads, and its luxuriant stands of both hardwoods and softwoods were exploited by a multitude of timber companies with virtually no thought to the future regeneration of this valuable resource. Limited financial understanding led many early residents to sell valuable timber and mineral rights for relatively small sums and the exploiting companies did virtually nothing to enhance the well-being of local communities. In addition, the communities' apparent lack of concern for their own futures inhibited the development of education and other public services. The unstable politics of Appalachia created by divided loyalties in the Civil War, the Prohibition era and the coal union wars of the 1930s further hindered economic development.

From the 1870s to the early 1900s, as the railway network expanded, the coal companies moved into the region and remained to impart a devastating physical and social legacy. Although deep mines were at first dominant, the location of much high quality coal at or very near to the surface encouraged strip mining after 1945. This laid bare huge tracts of land, ripping away topsoil and vegetation, disrupting drainage systems, polluting rivers, and destroying animal life. The local rural population and poor European immigrants were brought together to work in the newly constructed mining camps and, although conditions were frequently poor, mining did provide a much-needed source of employment. However, this was only for as long as the market for coal remained high.

The economic depression of the 1930s and the erosion of coal's traditional markets by new sources of energy proved catastrophic for many Appalachian communities. Mechanisation and rationalisation of the industry in an attempt to increase competitiveness in the 1950s created further unemployment in a region where alternative job opportunities were virtually negligible. As the nation's major coal producing region, Appalachia bore the brunt of the industry's decline, which was halted temporarily by the oil crises of the 1970s. The role of the timber and coal companies was predominantly one of exploitation with very low levels of investment in industrial and social infrastructure. Such a lack of investment resulted in national economic expansion generally by-passing Appalachia. The region was relatively inaccessible and the people were characterised by poor health, low education and skill levels and low incomes.

Appalachia clearly lacked spontaneous growth centres, with few urban areas in the region capable of providing the services, concentrated labour force and other external economies needed to support growth. Such a situation has been particularly evident in central Appalachia. It also became clear that the in-migration of manufacturing firms did not necessarily result in considerable economic growth. The character of manufacturing activity in some rural places works against long-lasting economic advancement. Many incoming manufacturing units were branch plants generating little intra-regional exchange of revenue and resources. Location in Appalachia was often primarily to take advantage of lower wage rates and the incentives offered by state and federal programmes.

The Tennessee Valley Authority: formative years

The Tennessee Valley Authority (TVA) established in 1933, was the first attempt at regional development in the USA incorporating federal, state and local government. The TVA region (Figure 13.1.7) covers 104 000 km² in seven Appalachian states and has become a model for similar projects throughout the world.

The original objective of the TVA was to plan, conserve and develop the Tennessee River basin, the fifth largest in the USA. The valley had become a focal point of poverty, even by Appalachian standards, due to a history of indiscriminate cultivation, poor farm management and totally unco-ordinated and short-sighted timber harvesting. The following facts go some way to illustrate the poverty in the area in 1933:

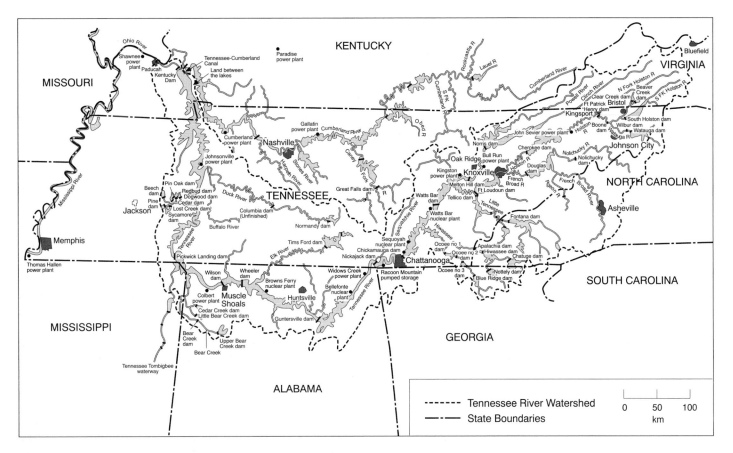

Figure 13.1.7
The Tennessee Valley Authority area

- average income and average farm production per person was only two-fifths of the national average;
- only three out of every one hundred farms had electricity;
- the basin's population suffered from a high incidence of malaria and vitamin deficiency diseases.

Farmers in the region were generally in debt and their families under-nourished. Manufacturing industry was negligible, as was other alternative employment. Problems of soil erosion were exacerbated by steep slopes and heavy rainfall, resulting in very variable and declining yields. Unchecked runoff frequently choked the Tennessee River causing flooding, and the low summer flows made navigation extremely difficult. The urban areas with their limited size and lethargic lifestyles characterised the general mood of depression emanating from their surroundings. In 1933 the region had a population of over two million, one eighth of whom were black with most of the rest being 'poor whites'.

The Tennessee Valley Act of 1933 assigned six important tasks to the Authority. These were:

- to control flooding;
- to improve navigation;
- to enhance the use of land in the basin;
- to develop electric power facilities;
- to reforest the basin where necessary;
- to upgrade the socio-economic conditions of the population.

The TVA has been generally recognised as a significant success although as one might expect of a scheme of such size it has not escaped criticism. However much of the argument levelled against the TVA can be identified as politically rather than economically motivated, as the concept of intervention of this kind is the antithesis of many Americans' views on the role of government.

By the late 1960s the Tennessee Valley region had undergone a fundamental change. Employment was no longer dominated by agriculture and now had a larger ratio of manufacturing workers than the national average. Per capita incomes had risen to 70 per cent of the US level and the previous huge outmigration had been largely stemmed. An urban-industrial society had emerged in the Valley.

The TVA today

The TVA is a unique federal corporation that supplies electricity to over 7 million people and works to develop the region's other resources. TVA facilities generating electric power include 11 coal-fired plants, two nuclear plants, 29 hydroelectric dams, four combustion-turbine units, and a pumped storage plant. Together, they provide 25 500 MW of capacity. Eight Corps of Engineers dams and four ALCOA (Aluminium Company of America) dams also contribute to the TVA power system which pays its own way through electricity sales. In fact, nearly 98 per cent of TVA's total funding comes from power sales of more than $5 billion a year. Congressional appropriations, $140 million in 1994, are the primary source of funding for the non-power programmes. Other funding for the non-power programmes, $18 million in 1994, is provided through proceeds from timber and land sales, user fees, and sales of non-power assets. TVA's non-power programmes focus on three key areas – the environment, the river, and economic development.

The Appalachian Regional Development Act 1965

While the TVA did much to regenerate that part of Appalachia within its jurisdiction, socio-economic conditions in the rest of the region improved little. The immense poverty and structural problems of the area captured the attention of John F Kennedy during his crucial West Virginia presidential primary contest in 1960 and led to the establishment of the President's Appalachian Regional Commission in 1963 to enquire into the problem. The Commission reported a year later, followed by the passage of the Appalachian Regional Development Act (ARDA) in 1965, just before the Public Works and Economic Development Act.

The Appalachian region as defined by ARDA stretches from north eastern Mississippi to southern New York (Figure 13.1.8). The region is primarily rural, with 356 out of the 397 counties so classified. Containing about 10 per cent of the country's population, Appalachia illustrates considerable economic diversity. In 1965 Maryland's designated counties had an average per capita income more than double that of Kentucky's.

The highway programme has been the focal point of ARDA investment, accounting for 80 per cent of total funding in earlier years. However, by 1975 funds allocated to other projects such as improving water and sewage facilities, access roads, housing loans, health schemes, vocational education, child development programmes, and environmental improvement projects amounted to 45 per cent of total expenditure.

In the 1970s, partly in response to criticism concerning its regional role, the ARC reoriented funding so that the poorest regions would benefit most. The major basic division of Appalachia was into north, central and south. The region-wide discussions of needs and solutions which took place following the 1975 amendments to the ARDA focused on a series of 'Questions for Appalachia' related to these main divisions. Northern Appalachia was characterised mainly as being part of the original 'industrial hearthland', a region of coal mines and steel mines centred on Pittsburgh and with many medium-sized cities often having a limited service role for the isolated areas between. The railway network and services had deteriorated and insufficient modern highways had been built.

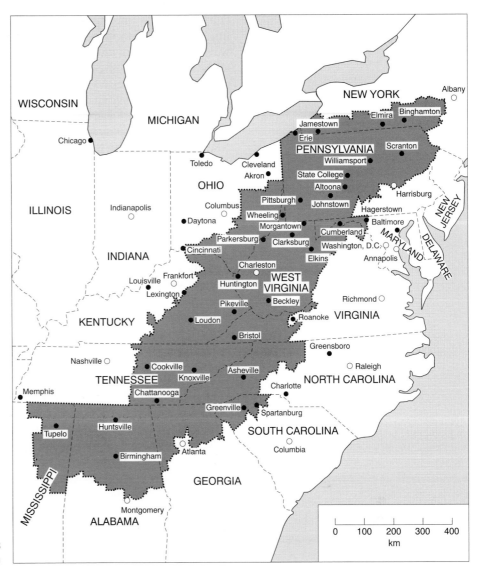

Figure 13.1.8
The Appalachian development region

The area suffered from declining basic manufacturing, slow replacement by newer industries, unattractive cities and an abused environment. The major needs here were seen as providing replacement jobs in the short term and an improved environment in the longer term.

Central Appalachia had for long been a region dominated by coal mining and with a very low level of general investment. The decline of mining resulted in acute problems for people who had always been poor and had not had access to average opportunities in education, health care and employment. Outmigration had been massive between 1950 and 1970 and many counties lost more than half their population. The region required better internal communications for the delivery of basic services, and access to more diversified employment, together with a massive effort in housing improvement. Much of the region required development from scratch, and was, on any count, the poorest section.

Southern Appalachia was undergoing rapid growth and industrialisation, including the multiplier effects of TVA, but suffered chiefly from poor access to services and poor housing, particularly in rural areas but also in some towns. It was also dominated by relatively low-wage industries and required diversification for more balanced growth, together with the provision of training for higher-level skills.

Because of their rural character, their relative poverty and their low tax bases, many Appalachian states and communities had previously found it difficult to raise the matching share required in many programmes before federal funds could be granted. Although these communities were eligible in all other ways for grants for the construction of basic public facilities, before the existence of ARC they often could not take advantage of a number of federal programmes. In response to this problem, Congress designed a unique feature of Appalachian legislation: the supplemental grant programme. Under this scheme, the federal share in grant programmes may be raised from the usual 50 per cent to as much as 80 per cent of the cost of construction, so that a state or community can participate by putting up as little as 20 per cent as its matching share. The Appalachian states have used supplemental grants to construct many types of public facilities, including vocational education schools, sewage treatment plants, recreational facilities, libraries and airports.

The ARC can point to a list of important achievements. For example:

- in the mid-1960s, infant mortality in much of the region was twice the national average. It now equals the national average and is significantly higher in only a few remote counties;
- a network of modern clinics has put primary health care within a 30-minute drive of every Appalachian, a goal that had been achieved by 1985;
- the ratio of active physicians grew from 90 per 100 000 people in 1965 to 132 thirty years later;
- throughout the region, communities once plagued by chronic water shortages now benefit from dramatic improvements in water supply;
- in 1965 one in three Appalachians lived in poverty, more than twice as many as in the rest of the nation. By 1990, this number had been cut in half and Appalachia's poverty rate was less than three points above the national average. Since 1965 per capita income has risen by nearly five percentage points to 83 per cent of the national average;
- ARC has been responsible for the rehabilitation and construction of more than 14 000 housing units;
- in 1965 Appalachia had an enormous 'education deficit'; only 32 out of every 100 people over 25 had finished high school. Now 68.4 per cent of adults over 25 have graduated from high school;
- the previous very high rate of outmigration has declined markedly;
- the 4870 km Appalachian Development Highway System is now two-thirds complete.

QUESTIONS

1 With reference to Figure 13.1.1 describe the trends in regional per capita income in the USA since 1900.

2 Analyse the regional data for Canada presented in Figure 13.1.3.

3 Assess the relationship between per capita income and unemployment in (i) Canada; (ii) the USA.

4 (a) From Figure 13.1.6 select what you believe to be the five best measures of socio-economic development. Justify your selection.

(b) Rank each region for the criteria selected and comment on any variations illustrated.

(c) Now add the ranking figures for each region to reach a composite quality of life index. Re-rank your aggregated figures to show the regions' rank order from 1 to 9.

(d) To what extent does your composite quality of life ranking compare to the rankings for per capita income and unemployment?

Regional development policy in Canada

Prior to the 1950s no explicit federal regional development policy was pursued in Canada, although certain programmes such as the Prairie Farm Rehabilitation Act had firm regional implications. The first direct effort to compensate for regional disparities was the Equalisation Programme established in 1957.

Phase one: Equalisation Programme

The Depression years blatantly exposed the economic weakness of the poorer provinces and led to the concept advocated by the Rowell-Sirois Commission in 1939 that the federal system should enable every province to provide services of average Canadian standards to its population without having to impose heavier than average tax burdens. Equalisation remains an integral part of the federal system. However, it is not a regional development programme in the true sense, in that payments are not conditional on funds being used for development purposes. Equalisation payments are calculated on the basis of a formula that compares the revenue-raising capacities of the provinces. Thus in 1992–3 Ontario, Alberta and British Columbia received no Equalisation payments; the other seven provinces were allocated varying amounts ranging from C$450 per head in Saskatchewan to C$1500 per head in Newfoundland. While the introduction of the Equalisation Programme eased the financial burden on the poorer provinces it did nothing directly to tackle the structural weaknesses which caused regional imbalance. Consequently a number of individual measures were introduced in the 1960s to meet such needs.

Phase two

This second phase of development policy was initiated by a New Products Programme for surplus manpower areas which commenced in 1960 to help areas of high unemployment and slow economic growth. The scheme permitted firms to obtain double the normal rate of capital cost allowances on most of the assets it acquired to produce products which were new to designated areas.

The Agricultural and Rural Development Act (ARDA) of 1961 was designed to alleviate the high incidence of low incomes in rural areas through federal-provincial programmes to increase small farmers' output and productivity. In 1966 a Fund for Rural Economic Development (FRED) was set up to provide comprehensive rural development schemes in areas characterised by widespread low incomes and major problems of adjustment, but considered to have development potential. Under FRED, agreements were signed with four provinces, for five separate plans covering the interlake region of Manitoba, the Gaspé region in Quebec, the Mactaquac and northeast areas of New Brunswick, and all of Prince Edward Island.

A totally area-specific scheme was established by the Atlantic Development Board (ADB) which was set up in 1962 with the objective of improving the economic structure of the Atlantic provinces. A similar agency was established in Quebec in the form of the Eastern Quebec Development Board.

The Area Development Incentives Act (ADIA) aimed to help areas of chronic high unemployment. The Act used accelerated capital cost allowances, income tax exemptions and cash grants as an inducement to manufacturing industry to locate in worst affected areas. The scheme, although achieving certain successes, was generally criticised for lack of co-ordination and long-term planning.

Phase three: Department of Regional Economic Expansion (DREE)

The third phase of regional development policy in Canada was initiated in 1969 by the establishment of the Department of Regional Economic Expansion (DREE). While earlier efforts had characteristically concentrated on worst-affected regions, DREE proposed to centre activity on areas which had the potential for significant economic growth. As well as continuing the work of earlier schemes, DREE embarked on two new projects: the 'Special Areas' programme and a new package of industrial incentives under the Regional Development Incentives Act (RDIA).

The Special Areas scheme helped to upgrade existing and potential growth centres, rendering them more attractive locations while RDIA provided a direct inducement to industry to locate in designated regions, particularly in the Special Areas.

The Special Areas programme, which was seen as experimental, lasted for only three years and a major policy review in 1972 witnessed the emergence of a new level of federal-provincial co-operation with the introduction of General Development Agreements (GDAs) and their subsidiary agreements. These ten-year programmes covered a wide range of development projects from construction of industrial and social infrastructure to financial assistance for resource development, secondary industry and tourist projects.

Phase four: Department of Regional Industrial Expansion (DRIE)

During the 1970s DREE's regional development approach was increasingly considered to be too restricted in scope and in 1982 a new strategic approach was announced which can be classed as phase four in the evolution of regional development policy. A new Department of Regional Industrial Expansion (DRIE) was set up, merging the regional programmes of the existing DREE with the industry, small firms and tourist components of the Department of Industry, Trade and Commerce. The revised federal-provincial agreements became known as 'Economic and Regional Development Agreements'. Reflecting the severe cyclical problems affecting the economy in the early 1980s, the new approach shifted away from the disadvantaged regions for a time to place greater emphasis on nation-wide assistance. Under DRIE, the programme of incentives made available financial assistance for firms throughout Canada, ranging from investment grants for new plants and plant modernisation to financial and technical assistance for product development, research and marketing. The eligibility of a region for assistance was determined by its 'Development Index' which was based on a variety of socio-economic characteristics.

Phase five: new policy directions

By the mid-1980s there was growing recognition that, despite a variety of efforts over the previous 25 years, unacceptable levels of regional disparity continued to exist. In 1986 these concerns led to a fundamental restructuring of regional development policy, which laid the foundation for the current structure of regional development in Canada. Significantly there was a decentralisation away from Ottawa to give regional agencies the primary responsibility for development within their local area. Greater flexibility in terms of support and wider consultation and participation at the local level were also promoted.

In 1987, the new policy resulted in the creation of two major regional development agencies: the Atlantic Canada Opportunities Agency (ACOA), to develop and implement programmes contributing to the long-term economic development of the Atlantic provinces; and Western Economic Diversification (WD), to develop and diversify the Western economy to make it less vulnerable to international economic developments and fluctuating commodity prices. A new 'flagship' economic development department – the Department of Industry, Science and Technology – was also created to promote the effective integration of advanced technology and competitive industrial capacity.

The Atlantic Provinces: a problem region

The four provinces of New Brunswick, Newfoundland, Nova Scotia and Prince Edward Island make up the region of Atlantic Canada. Just over 2.3 million people live in an area covering half a million km². The Atlantic provinces have long been regarded as the major problem region in Canada, characterised by slow economic growth, heavy reliance on primary industries, low per capita incomes and persistently high unemployment rates. In an attempt to rectify such relative deprivation the region has figured prominently in Canadian regional development programmes. Although this injection of federal funding has resulted in significant improvements covering many aspects of the regional economy, the Atlantic provinces still lag behind the rest of the nation according to most socio-economic indicators.

Development in the Atlantic provinces has been hindered by a number of factors, particularly the paucity of natural resources, the low level of manufacturing industry and capital investment, and the scattered nature of rural settlement. Although the provinces are the most rural of Canada's regions, the generally infertile soils and cool summers have restricted agricultural improvement. A consequence of such a lack of agricultural potential is the highest percentage of rural non-farm population in the country.

The principal nuclei of the region are the ocean-facing seaports of St John in New Brunswick, Halifax in Nova Scotia and St John's in Newfoundland. However, these major urban areas are of limited size, with 1995 populations of 129 000, 343 000 and 177 000 respectively. These urban nuclei account for the lowest percentage of population for any region in the country. Such a small and dispersed population provides a very limited attraction for industries attempting to achieve a reasonable level of economies of scale. Low capital intensity in the private sector and poor public services have often been cited as disincentives to new industry.

The four provinces have benefited from a range of region-specific programmes and the highest levels of assistance under the centrally administered policies. With the introduction of a decentralised approach in 1987, the Atlantic Canada Opportunities Agency (ACOA) was the first of the regional agencies to be created, with a funding allocation of $1.05 billion over a five year period. ACOA's overall objective is to foster the long-term economic renewal of Atlantic Canada through:

- a self-sustaining entrepreneurial climate;
- more successful small and medium-sized businesses;
- more lasting employment opportunities;
- increased earned income;
- national policies and programmes that reflect the aspirations and opportunities of Atlantic Canada;
- an expanding competitive economy.

To further these aims ACOA operates two main programmes: Action and Co-operation, and Advocacy and Co-ordination. The former comprises a package of assistance (grants, loans, interest-rate subsidies, loan insurance and equity support) for small and medium-sized enterprises, as well as the provision of information, advice and consultancy, and access to technology. Action Programme assistance may provide up to 50 per cent of investment costs. Eligible projects encompass both commercial and non-commercial operations involving activities such as innovation, business studies, capital investment, supplier development, market development and business support.

Under the Co-operation Programme, ACOA and each of the provincial governments enter into a series of cost-shared, federal-provincial co-operation agreements which fund infrastructure initiatives targeting the business environment for entrepreneurship, market and trade development, innovation and technology transfer, human resource development and a sustainable environment. Agreements are also signed between ACOA and non-governmental organisations. Between June 1987 and August 1992, agreements worth almost $600 million were signed for projects ranging from agriculture development, forestry and fisheries, minerals, energy and industrial development projects to transport, tourism and cultural initiatives (Figure 13.1.9).

QUESTIONS

1 Outline the characteristics that combine to make the Atlantic Provinces Canada's major problem region.

2 Summarise the evolution of regional development policy in Canada.

3 How does ACOA operate to improve socio-economic conditions in the Atlantic provinces?

Figure 13.1.9
The Atlantic provinces co-operation programme initiatives, June 1987– August 1992

	Number of initiatives	Federal share $ million	Provincial share $ million	Total cost $ million
Newfoundland	18	265.1	125.2	390.3
Prince Edward Island	13	95.2	59.8	155.0
Nova Scotia	18	333.4	365.8	699.2
New Brunswick	20	367.4	202.3	569.7
Pan-Atlantic	3	10.0	4.0	14.0
Total	72	1071.1	757.1	1828.2

The Advocacy and Co-ordination functions involve advocating the region's interests and co-ordinating federal economic programmes in Atlantic Canada to ensure maximum impact. Activities targeted by ACOA include fisheries, offshore mining, tourism, transportation, trade policy, environment and shipbuilding.

13.2 The North American Free Trade Agreement (NAFTA)

NAFTA came into effect on 1 January 1994 and will eliminate most tariffs and other restrictions on free trade and investment between Canada, the United States and Mexico by the year 2003; remaining tariffs will be removed by 2008 (Figure 13.2.1).

The creation of NAFTA was hastened by:

- the ever-increasing economic challenge from Asia (especially Japan) and Western Europe;
- the completion of the internal market of the European Union (EU) and the establishment of the European Economic Area (EEA) in 1993;
- a concern that the Uruguay Round of GATT talks on free trade might fail, commercially disadvantaging those countries left outside trade blocs.

The first significant move towards a North American trade bloc was the signing of the Canada–United States Automotive Products Trade Agreement (Auto Pact) in 1965; in 1988 the two countries extended this to the comprehensive Canada–United States Free Trade Agreement (FTA or CUSTA). Two years later Mexico formally requested a free trade deal with its northern neighbours, and after another four years of keen debate (mainly in the United States) NAFTA was approved by the governments of all three countries. In effect, the terms of the 1988 FTA were extended to include Mexico, whose economy was then less than five per cent the size of those of the United States and Canada combined (Figure 13.2.4). This is the equivalent of adding the Dutch economy to the EU.

NAFTA is not only an important milestone in the development of global free trade but also an interesting economic combination of a relatively poor Third World country with two of the world's richest nations. Indeed, never before have developing and industrialised countries opened their markets to each other so completely. With around 390 million consumers and a combined GDP of over $7.6 trillion (1994), the NAFTA countries vie with the EEA (the EU and Iceland, Norway and Liechtenstein) to become the world's biggest trade bloc.

It should be emphasised, however, that NAFTA differs from the EU's Maastricht Treaty in a number of ways:

- it is limited to trade issues;
- it does not attempt to redistribute wealth to poorer regions, either within or beyond its boundaries;

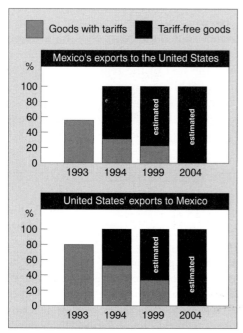

Figure 13.2.1
NAFTA tariff elimination

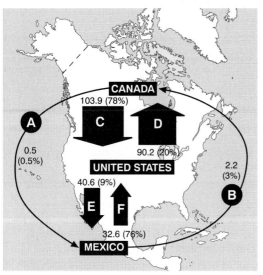

Figure 13.2.2
Value of intra-North American exports
($ billion) and as % of total exports, (1992)

- it has no ambition to establish a common currency;
- it does not seek to achieve a political union or federation;
- it does not aim to establish a customs union with common external tariffs;
- it retains all existing border controls;
- it does not permit free movement of labour.

Like the EU, NAFTA has a provision for more countries to be included, subject to the approval of all the original signatories. Chile – despite its geographical location – seems likely to become NAFTA's next member.

NAFTA's terms

These may be summarised thus:

- all tariffs on goods qualifying as North American will be phased out within ten years, though special rules will apply in key sectors such as energy, agriculture, textiles and clothing. Canada's ban on large-scale transfers of water remains unaffected by NAFTA;
- trade in services will also be facilitated, e.g. transportation, finance and investment;
- other provisions give relief or protection to 'sensitive industries' (e.g. some agricultural products, like US sugar, will be protected for 15 years), and technical and environmental standards;
- each country's cultural, health and social programmes remain unaffected.

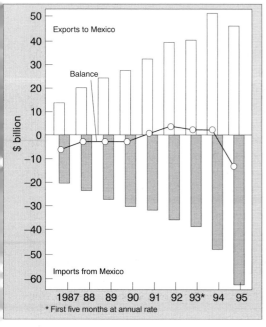

Figure 13.2.3
US trade with Mexico, 1987–1995

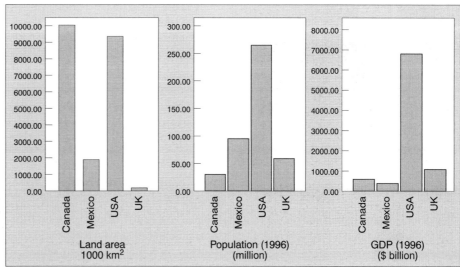

Figure 13.2.4
The NAFTA countries and the UK: comparisons of size

Largely to assuage American concerns about Mexico, special 'supplemental agreements' or 'side deals' were added to NAFTA. These side deals provide for annual reviews of 'import surges' or dumping of products and, if necessary, a penalty: a resumption of pre-NAFTA tariffs, usually for three years. Also, three-country commissions monitor the enforcement of NAFTA-related national laws, while any two signatories can investigate suspected breaches of regulations concerning environmental standards, health and safety in the workplace, minimum wages and child labour. Fines of up to $20 million can be imposed, but governments would pay.

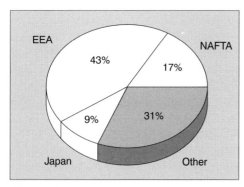

Figure 13.2.5
NAFTA countries' share of world trade

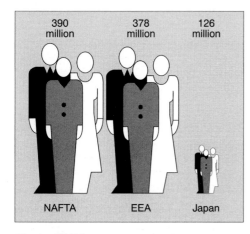

Figure 13.2.6
NAFTA's population: comparison with the EEA and Japan

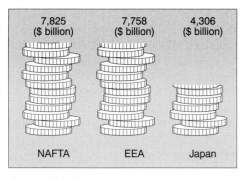

Figure 13.2.7
NAFTA's GDP: comparison with the EEA and Japan

NAFTA's benefits

These are potentially huge. NAFTA promises to secure the economic future of all three partners by creating jobs, improving efficiency and raising incomes. In the first five years the Agreement is expected to generate an additional $15 billion a year for the participants, thus recouping the combined government outlay of $75 billion.

These benefits will accrue because free trade encourages each member country to concentrate on what, in economic terms, it does best. As Mexico was the most protectionist country of the three, NAFTA will have the biggest impact there, and its economy will gain the fastest. Indeed, Mexico's annual growth is expected to be boosted by 0.5 per cent in the first three years and by 1 per cent thereafter.

Although Mexicans already spend more on American products than Germans, Japanese or Canadians, as they grow richer they will buy even more goods and services from both their northern neighbours – especially cars, telecommunications equipment and financial services; consequently, the United States and Canada will in turn grow richer, so all three partners gain from NAFTA.

Increasing prosperity is expected to encourage Mexico's political stability. Since one objective of US foreign policy is to secure stable democracies throughout the western hemisphere, the hope is that the whole of Latin America will follow Mexico's example and eventually join NAFTA – though the name would have to change!

A more immediate benefit of NAFTA to the United States is that it should reduce, if not stop, the flow of illegal Mexican immigrants once better opportunities and conditions are established in their homeland.

Concerns about NAFTA

Two major concerns – jobs and the environment – continue to be voiced by NAFTA's opponents (mostly Americans), who claim that the Agreement's economic benefits will be outweighed by negative outcomes in these two important areas.

Job losses in the United States

Critics of NAFTA argue that because Mexican workers are poorly paid – their wages are only about one-eighth those of their American counterparts – American companies and investment will relocate south of the border, use Mexico as a platform from which to flood US markets with cheap goods, and thus cause factory closures and mass unemployment in the United States.

The highest estimate of job losses (5.9 million) was made in 1992 by US presidential candidate Ross Perot, who warned of a 'giant sucking sound' as jobs moved to Mexico. More realistic analyses suggest that the United States may suffer a loss of up to 500 000 jobs over the decade to 2003, but these losses should be set against both the new jobs that NAFTA will generate in the USA and the normal monthly variation in employment of around 200 000.

As recently as 1987, manufacturing wages were 13 times higher in the United States than in Mexico, but the gap between Mexican and American wages has been narrowing quickly. When fringe benefits are taken into account, Mexicans do even better. Including subsidised food and housing and the standard Christmas bonus of one month's pay, these benefits add up to 62 per cent of base pay in Mexico; fringe benefits in the United States amount to only eight per cent of base pay. Consequently, average aggregate labour costs in Mexico (i.e. wages plus benefits) are only 35 per cent lower than in the US and 40 per cent lower than in Canada.

QUESTIONS

1 Describe and account for the pre-NAFTA pattern of intra-North American trade (Figure 13.2.2), and US trade with Mexico (Figure 13.2.3).

2 To whose exports (Figure 13.2.1) does NAFTA give the greater protection up to 2004, and why?

3 Discuss the proposition that the United Kingdom would be better off in a NAFTA-type European grouping rather than the present European Union.

4 Referring to Figures 13.2.4–13.2.7, write a brief review of NAFTA's main characteristics.

Moreover, labour costs comprise only about 15 per cent of total manufacturing costs, and when US companies take account of all the other costs of closing down existing plants and opening new ones in Mexico, the real advantages of relocating to benefit from cheap labour begin to evaporate. Indeed, in the decade before NAFTA only 96 000 American jobs were transferred to Mexico under the *maquiladoras* system, which transformed the Mexican export economy (Figure 13.2.8). In reality, most American jobs will survive simply because American workers are much more productive than Mexicans; and for those who are made redundant, NAFTA provides a substantial fund for retraining.

Mexico's maquiladoras and the problem of environmental pollution

CIUDAD JUAREZ, on the border between Mexico and the United States, is one of the places that environmentalists think of when they argue that NAFTA is bad for people's health in both countries.

In one generation, under a customs regime that resembles free trade, the place has grown from a sleepy frontier town to a city of well over 1m people. Living conditions in the shanty-towns that sprawl up the hillsides to the east and the high plains to the south are as bad as anywhere in Mexico. A third of the city's inhabitants have no piped water. The two-thirds that do send all their sewage to ditches of stinking black ooze that parallel the Rio Grande for nearly 30 kilometres.

The untreated sewage mingles with industrial effluent, including heavy metals and solvents. Several times a year residents stumble across illegal dumps of solid toxic waste. A pall of brown smog often envelops both Juarez and El Paso, its American neighbour across the river.

Migrants have flocked to Juarez to seize the jobs offered by the *maquiladoras*—assembly plants that import components and export finished goods to the United States free of either country's customs duties. Since 1969 more than 2,000 of these plants have sprung up along the border, paying wages about one-sixth of American ones. Juarez is Mexico's *maquiladora* capital. Its 450 plants employ 150,000 people. Their owners include big companies like Ford, General Motors and Toshiba, as well as a host of smaller, mainly American firms.

For years these companies could do much as they liked. But the biggest growth industry now on this part of the border is environmentalism. Eager to get NAFTA in place, President Carlos Salinas has repeatedly told the Americans that Mexico does not want polluting factories. At American insistence, a separate accord linked to NAFTA allows for heavy fines—even possible trade sanctions—if national environmental laws are not enforced.

Environmentalists praise Mexico's 1988 pollution-control law, but say that enforcement has been corrupt or non-existent. Mexico's newly-appointed federal attorney-general for the environment, says that is changing. His staff of inspectors has more than doubled to 415. They carried out 16,000 inspections in his first 14 months, compared with 4,500 over the previous four years, he says. They found minor offences in 70% of cases, and serious infringements in 15%. On the border, they temporarily shut down parts or all of 62 plants, including 28 *maquiladoras*, to force them to clean up. Municipal authorities have been given new environmental powers. Ciudad Juarez is starting to act firmly to enforce water standards.

Work is due to start soon on two sewage-treatment plants, a much-delayed $55m project that should open by October 1994, according to the city's ecology director. He has warned companies to stop discharging untreated waste by April 1994. After that, he will start fining them.

Figure 13.2.8
Mexico's *maquiladoras* and the problem of environmental pollution

Figure 13.2.9
Some Americans remain sceptical about NAFTA's benefits

QUESTIONS

1 Referring to Figure 13.2.8, explain the origins of, and rationale for, the *maquiladoras* along the US–Mexican border.

2 Summarise the environmental problems associated with the *maquiladoras* and the attempts to clean up this area. What role will NAFTA play?

Disposal of hazardous waste poses tougher problems. By law, the *maquiladoras* must ship toxic waste back over the border. Some are now doing so, but officials admit that perhaps 30% stays in Mexico. It has only one commercial toxic-waste disposal site, in far-off Monterrey, so much is dumped illegally. Mexico has plans for new sites. But, with charges for disposal half those in the United States, demand will long outstrip available sites.

Many big *maquiladoras* appointed environmental managers and began investing in water-treatment plants and vapour-collectors some years ago. Most of the problems come from smaller companies. The authorities temporarily shut down Presto Lock, an American padlock maker, after they found it had dumped effluent, including cyanide and heavy metals, into the drains.

Some environmentalists remain sceptical. The instinct of many Mexican officials is still to cover up problems, and staff are short of laboratories. Many companies are yet to be inspected; more than half have not filed the environmental inventory the law requires. Environmentalists also say conditions will improve only marginally until a planned border clean-up fund, financed by the countries on either side, is set up.

Estimates of the cost of controlling pollution and bringing infrastructure up to scratch range from $6 billion to $16 billion for the whole border. So far the only pledge is for $1.8 billion from the World Bank. Though the Mexican government is supposed to match this, its money will go to environmental improvements throughout the country, not just on the border.

It is not only the Mexican side of the border that needs the money. El Paso is among America's poorest cities, and 60,000 of its 770,000 people live in shanty towns amid third-world misery. For public-health purposes, it forms a single metropolis with Juarez. Authorities and environmentalists from both cities are seeking federal funds for a joint project to manage air quality.

The NAFTA debate has focused unprecedented attention on the border region. It has also brought an environmental awareness that is starting to take deep root. Both Juarez and El Paso hope that NAFTA may provide the money they need to tackle problems long ignored.

Bordering on the intolerable
Monterrey

The Economist,
16 October 1993

Environmental damage along the US–Mexican border

NAFTA's critics also complain that Mexican manufacturers have lower costs because they are not burdened by the USA's responsible but expensive environmental standards. Even when Mexico has passed legislation to protect the environment, the laws are either poorly enforced or ignored completely. Such a 'polluter's paradise' will further encourage American companies to take advantage of lower costs south of the border.

The response to this charge is that Mexico's environmental problems are caused by poverty and poor sanitation. As the national economy is boosted by NAFTA, Mexicans themselves will begin to demand better environmental and safety standards, and old polluting factories will be replaced by clean technology from abroad.

Furthermore, World Bank studies have shown that the savings to be gained by First World manufacturers from lax Third World environmental laws are marginal, being outweighed by the costs of relocation and the less certain political conditions. Even cement-making, the USA's 'dirtiest' industry, would not gain by moving to Mexico.

In any case, NAFTA provides for the United States and Mexico to spend more than $700 million over three years to clean up the border area polluted by the *maquiladoras* (Figure 13.2.8), and calls for a larger fund ($6–$16 billion) for longer-term environmental protection and infrastructure development in the same area.

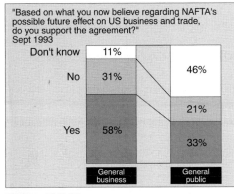

Figure 13.2.10
Results of NAFTA survey in the USA,
September 1993

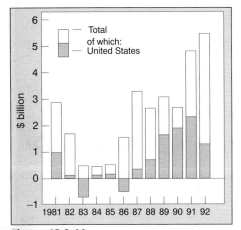

Figure 13.2.11
Foreign direct investment flows into Mexico,
1981–1992

NAFTA's risks and challenges

Despite American and, to a lesser extent, Canadian concerns, on balance it seems that Mexico is taking the biggest risk by joining NAFTA. Already, under the *maquiladoras* system, some Mexican industries (e.g. cement, cars, glass, steel, petrochemicals and some sectors of agriculture) have been forced to restructure, and these are likely to increase their exports under NAFTA. But others (e.g. wood products, paper, textiles, leather goods, toys and electrical goods) have suffered competition from Asia where wages are even lower than in Mexico.

In the context of NAFTA, Mexico's disadvantages include lower efficiency, higher interest rates, higher energy costs and higher freight charges. Removal of trade barriers has exposed Mexican firms to American and Canadian competitors that are generally much wealthier, more competitive, more technologically sophisticated and with better-educated and more highly-skilled workforces. Moreover, Mexico's northern neighbours have combined economies nearly 20 times bigger than its own. Even so, Mexico entered NAFTA with confidence and optimism, believing that free trade can help their economy grow within a decade to the size of Canada's at the start of the NAFTA era. The 30 per cent devaluation of the Mexican peso and other checks on the country's economic growth in 1995 have caused concern within NAFTA. Only time will tell if Mexico has made the right decision in joining this trade bloc.

In contrast, Canada and the United States are two of the world's most successful economies because they have always embraced change – spatial, social and economic – more readily than most other countries. The changes now required by NAFTA – if accepted – can only result in more and better jobs, and higher incomes for Canadians and Americans. The fact that NAFTA also benefits Mexico economically and politically should be regarded as a bonus.

QUESTIONS

1 Account for the reluctance of many Americans to accept NAFTA as beneficial to them (Figure 13.2.9), and the different responses made by business people and the general public (Figure 13.2.10).

2 Outline and explain the changes that might be expected in the flow of foreign investment into Mexico in the wake of NAFTA (Figure 13.2.11).

3 As a locational adviser to an established American company manufacturing high-quality clothing primarily for the US market, you are required to draw up a report to improve the company's competitive position, enhance its share of the market, and raise profit margins. Present detailed arguments for and against the following scenarios: (i) remaining in the United States; (ii) relocating in Mexico; (iii) relocating in Taiwan. Which option would you recommend to the company, and why?

4 'The NAFTA countries will not prosper at each other's expense; trade, production and investment are not zero-sum games.' Elaborate this statement of free trade principles.

13.3 North America, APEC and the Pacific Rim

At the end of the Second World War, having twice come to the aid of Europe in the twentieth century, the United States emerged as a global superpower and a major influence on the Western democracies and their relations with the Soviet Union and the communist bloc. During the 40-year Cold War, in its attempts to contain the spread of communism, the United States (sometimes with, sometimes without Canada) entered into numerous military treaties, pacts and agreements around the world, perhaps the most successful being NATO.

At the same time, having also become one of the world's foremost economic powers, the United States took the lead in establishing and developing such important post-war organisations as the IMF, World Bank, OECD and G7. With its programme of aid to the Developing World and its links to a wide variety of countries through trade treaties and on-going defence commitments, the United States remains a truly global power in every respect, though for reasons of history, commerce and personal links, its principal interests have traditionally inclined towards Europe. As the year 2000 approaches, however, it is increasingly being said that the Euro–American century is closing and the Asian–American century is opening, with an accompanying shift of world economic power from the Atlantic to the Pacific (Figure 13.3.1).

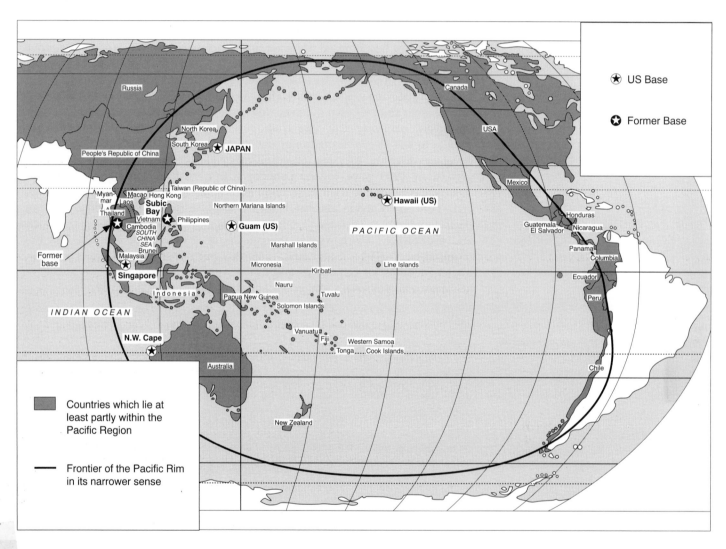

Figure 13.3.1
The Pacific Rim: the world's next centre of power?

Asia–Pacific Economic Co-operation forum (APEC)

The organisation that embodies the increasing importance of the countries of the so-called 'Pacific Rim' is the Asia–Pacific Economic Co-operation forum (APEC). Founded in 1989 following an initial proposal by Australia, APEC's membership has grown as shown in Figure 13.3.2.

Figure 13.3.2
The growth of APEC

1989	Australia, Brunei[1], Canada, Indonesia[1], Japan, Malaysia[1], New Zealand, Philippines[1], Singapore[1,2], South Korea[2], Thailand[1], United States.
1991	China, Hong Kong[2], Taiwan[2].
1993	Mexico, Papua New Guinea.
1994	Chile

[1]ASEAN members [2]Newly industrialised countries (NICs), the 'Four little tigers'.

With such a diverse membership, APEC (which met for the first time only in 1993 in Seattle, USA) has been criticised for being little more than a talking shop, with scarcely anything in common other than a Pacific shoreline. Indeed, APEC's initial aims were unclear, with some Asian members (particularly Malaysia) pushing for an exclusively East Asian economic group – in effect, an expanded ASEAN.

In addition, some of the smaller South-East Asian countries feared that their interests would be brushed aside by the organisation's giants: the United States, Japan and China. To allay these fears, APEC has located its headquarters in Singapore and has agreed to hold every second annual meeting in a South-East Asian country.

APEC's most enthusiastic supporters include the United States, Canada, Australia and New Zealand, who were concerned at the prospect of viewing a booming Asia from distant edges of the Pacific. With East Asian economies growing two to three times faster than those of the United States or the European Union, Asia will probably account for nearly a quarter of world GDP by the year 2000. Moreover, APEC's 18 members already account for a growing proportion of world trade – 46 per cent of all world exports in 1993, with 70 per cent of their exports going to other APEC countries.

At its 1994 meeting in Indonesia, APEC gave itself direction and substance by agreeing to transform the Pacific Rim from a mere geographical expression into the world's largest free trade area. The organisation's members committed themselves to start bringing down their trading barriers in 2000; the developed countries will remove all their restrictions on intra-APEC trade and investment by 2010, and the developing countries by 2020.

While the developed countries can be readily identified, the developing countries remain undefined. By 2010 South Korea and even China may be classified as developed, but Malaysia insists on retaining its developing status and concessions for as long as possible, stipulating that the 2020 target date is not binding on any signatory. In fact, the 15 or 25-year time scale of APEC's free trade declaration can be criticised for lacking urgency. If all East Asian countries halved their existing tariffs immediately and encouraged more foreign investment, 90 per cent of the resulting growth in world GDP would occur in the APEC region while the GDP of the ASEAN countries would rise by 5 per cent.

Figure 13.3.3
Selected APEC countries: GDP and defence spending, 1992

	GDP* 1992, $bn	GDP, 1985–92, annual average increase, %	Defence spending, 1992, % of GDP
United States	5710	2.0	5.3
Japan	2430	4.4	1.0
China	2140	7.6	5.0
Indonesia	540	6.6	1.4
S Korea	400	9.6	3.8
Thailand	320	10.1	2.7
Taiwan	290	7.7	4.8
Malaysia	140	8.3	4.8
Hong Kong	110	6.6	0.4
Singapore	50	7.8	5.4

*at purchasing power parities

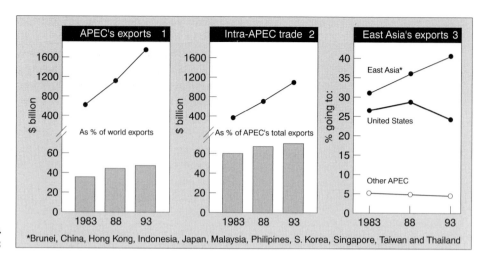

Figure 13.3.4
Changing patterns of APEC trade, 1983–1993

*Brunei, China, Hong Kong, Indonesia, Japan, Malaysia, Philipines, S. Korea, Singapore, Taiwan and Thailand

1 Summarise the reasons for the formation of NATO and explain why it may be said that the organisation 'won' the Cold War.

2 Eight of the APEC countries listed in Figure 13.3.2 also belong to a worldwide organisation of which Britain is the senior member. Name the organisation and the eight APEC countries.

3 Identify the five APEC members that can currently be described as 'developed'. Outline their economic characteristics and those of the NIC's (Figures 13.3.2 and 13.3.3).

4 With reference to Figure 13.3.4, use the graphical information to write a summary of APEC's changing patterns of trade.

5 (a) Suggest why APEC's developing countries have been allowed 25 years to remove their trade barriers, but developed countries have only 15 years.
(b) Which developed country probably stands to gain most from APEC's leisurely free-trade deadlines, and why?

Japan and the US trade deficit

While the United States' membership of APEC guarantees the continuation of American political and military involvement in the Asia–Pacific region, the organisation's two other great powers – Japan and China, who are expected to vie for economic leadership in the coming century – currently pose contrasting challenges to the United States (Figure 13.3.5).

Japan is currently Asia's dominant power, but this status is based on economic strength rather than military might, since a pacifist constitution was imposed by the United States after the Second World War.

Japan's remarkable post-war economic recovery has been achieved in a variety of ways:

Figure 13.3.5
APEC's trade warriors

- considerable financial aid from the United States ($4.6 billion, 1946–65);
- production of arms and equipment for the United Nations allies in the Korean War, 1950–1;
- exploitation of the growing North American and West European post-war economies and those countries' liberalisation of trade;
- protection of Japan's own industries by a complex system of structural, cultural and legal barriers, including spurious product-safety rules affecting imports. (Japan's imports of manufactured goods amount to only 3.1 per cent of its GDP; the equivalent figure for other G7 countries is 7.4 per cent);
- direction of industrial strategy by the all-powerful Ministry of International Trade and Industry (MITI), which has helped Japan to achieve technological leadership in targeted industries. In recent years, Japanese companies have filed more patent applications than the USA and the UK combined;

- increasingly heavy investment in labour-saving robotics and other automation;
- movement of assembly, then full-scale production to low-cost countries in Asia or into major export markets as other Asian countries began to compete, adopting Japanese production methods;
- investment in a highly competitive education system, including more than 500 universities;
- development of an extreme work ethic and company loyalty – though the government has recently introduced more statutory holidays and tried to reduce working hours.

Japan's success has generated not only the world's second largest economy and high standards of living for most Japanese, but a huge, apparently permanent trade surplus, most notably with the United States. In addition, Japan has built up a large reserve of foreign currency, much of which has been invested in the United States and Europe.

Figure 13.3.6
Japan's trade surplus, 1993–1995

	With US ($ billions)	With World ($ billions)	US as % of World
1993	50	131	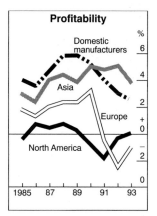
1994	56	132	
1995	59	139	

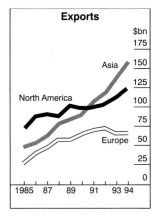

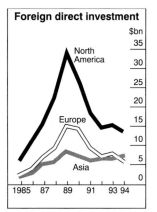

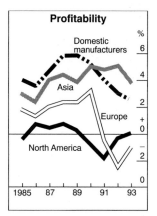

Figure 13.3.7
Japan's exports and foreign investment, 1985–1994

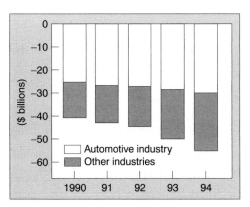

Figure 13.3.8
The US trade deficit with Japan, 1990–1994

Car wars

American concerns about access to Japanese markets have focused on several key economic sectors, including high technology, telecommunications and medical equipment, insurance and, above all, cars and car parts. While broad agreement has been reached on other products, two years of talks failed to resolve the dispute over the trade in cars, which is responsible for a large percentage of the US deficit with Japan (Figure 13.3.8) and nearly 25 per cent of America's global deficit.

The US car industry bore the brunt of the Japanese export onslaught in the 1980s, and by 1991 the Japanese share of the car market in the United States had peaked at 29 per cent. By 1994 Japan still had a 25 per cent share of the US car market, but American cars accounted for only 2 per cent of the Japanese market – despite improvements in quality and supplying some models with right-hand steering (like the British, the Japanese drive on the left-hand side of the road).

The United States government demanded changes in two areas:

- Japanese car dealers should put more American cars on sale. (Around 40 per cent of dealers are owned by Japanese car manufacturers, and less than 2 per cent have franchise agreements with US car makers).
- Japanese car manufacturers should buy more parts from the United States. (In 1994 Japanese car firms bought $20 billion-worth of US car parts, but 85 per cent of these were used in their US-based factories).

With Japan unable or unwilling to accede to these demands, in May 1995 the US government formally complained to the newly-established World Trade Organisation (WTO) that Japan's market was closed to foreign cars and car parts. Moreover, the United States would retaliate with a 100 per cent tariff on imports of Japanese luxury cars. Although this market totalled about $6 billion, no Japanese manufacturer would lose more than 3 per cent of its worldwide sales. Even so, Japan's response was to ask the United States for 'urgent consultations' at the WTO – the first stage of a formal complaint.

The two countries eventually stepped back from this confrontation and agreed to try and settle their differences on an informal bilateral basis. It remains to be seen, however, whether this truce in the long-running US–Japanese 'car wars' can be made permanent.

The United States and China – superpower of the future

China has one of the world's largest economies but at the same time it is one of the world's poorest developing countries, with a GDP per capita far lower than Japan's – even allowing for the difference in cost of living (Figure 13.3.9). Nevertheless, with a population numbering 1.2 billion and a rapidly expanding economy (more than 9 per cent a year over the last decade), China intends to become the dominant power in Asia within a generation and a world power within another.

For the United States the big question is whether or not to treat China, a communist dictatorship with an abysmal record on human rights, as another Soviet Union that must be opposed and contained, if necessary at huge economic and military cost. Currently, the hope is that China's need for peace and stability to achieve its economic ambitions will ensure that it remains an enthusiastic trading partner rather than becoming a military rival complete with a nuclear capability.

Two aspects of this uncertainty illustrate the nature of current US–China relations:

China's commercial piracy

Although China has succeeded in marking up a $30 billion trade surplus with the United States (Figure 13.3.10), so far this has not caused the same outrage as Japan's surplus. However, the US government calculates that, in China, the

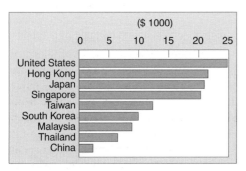

Figure 13.3.9
Selected APEC countries: GDP per person at PPP, 1993

1 (a) Complete the table (Figure 13.3.6).
   (b) Explain, in your own words, how Japan has succeeded in accumulating such huge international trade surpluses.

2 Explain the likely effects on Japan's trade surplus of a steep fall in the value of the yen against the US dollar and other foreign currencies (as happened in 1994–95).

3 (a) Comment on the salient trends in Japan's exports and foreign investment since 1985 (Figure 13.3.7).
   (b) Which region seems likely to be the most important to Japan towards and beyond the year 2000?

4 Calculate for each year the percentage of the total US deficit for which the automotive industry (cars and car parts) is responsible (Figure 13.3.8).

5 Argue the case for and against the American action over the car trade deficit with Japan, taking into account the following factors: (i) the existence of APEC and WTO, to which both the United States and Japan belong; (ii) the inability of US manufacturers to compete successfully in the luxury car market with Japanese and European car makers; (iii) US dealers in Japanese luxury cars employ thousands of Americans; (iv) thousands of Americans already own Japanese luxury cars.

form of commercial piracy known as intellectual property theft (IPT) accounts for the loss of $3 billion-worth of American business. Around 98 per cent of computer software produced in China is said to be pirated from American originals. Similarly, 90 per cent of musical recordings made in China are 'stolen' from American companies and artists. American complaints have focused on around two dozen compact-disc factories in booming south China. These were allegedly turning out copies of popular artists, including Pearl Jam, Madonna – and even Barry Manilow!

Faced with the threat of the United States imposing 100 per cent tariffs on a range of their exports, and of losing their most-favoured nation (MFN) status, the Chinese ostentatiously closed several (but not all) of the offending factories in 1995. The United States resolved to keep up the anti-IPT pressure on China, unilaterally and through APEC.

So far, car exports have not contributed to China's trade surplus with the United States. Indeed, Chinese car production remains little more than embryonic. The potential market is huge, but only about 200 000 cars a year are made in China, with another 300 000 imported – despite duties of over 200 per cent. The main US concern is to gain an investment foothold in the Chinese car industry by setting up a joint-venture factory. However, in 1994 the Chinese government imposed a two-year ban on new foreign joint-venture car plants, to allow the country's 160 vehicle producers to consolidate into six or seven big groups. Only when this has been accomplished will China allow foreign companies to participate in the large-scale expansion of its car-making industry.

China pushes south

China's growing economic and military strength has increased its confidence in foreign affairs. It has already succeeded in negotiating the return to Chinese sovereignty of Hong Kong (British since 1842) and of Macao (Portuguese since 1577) in 1997 and 1999 respectively. More problematic is Taiwan, whose anti-communist government held China's seat at the United Nations until 1971 and still regards itself as the rightful ruler of the whole of China. Beijing's view is that Taiwan must in due course be reunited with the rest of China. The worry is that this may be attempted by invasion from the mainland, in which case the United States, which has protected and supported Taiwan since 1949, could become involved in a direct military confrontation with China.

China has also recently asserted its claims in the South China Sea, 80 per cent of which it regards as its territorial waters (Figure 13.3.11). Its claims extend as far south as the Spratly Islands, a 965-km string of rocks, islets and coral reefs that were charted and named by the British Admiralty in the mid-nineteenth century. Regarding the islands as worthless, Britain never claimed them. Today, however, the Spratlys and the surrounding seas are thought to contain oil and other valuable mineral deposits; consequently they have become another regional flashpoint.

Two countries, China and Taiwan, now lay claim to all the islands, while four others – Brunei, Malaysia, the Philippines and Vietnam – claim parts of the group. In 1974 the Paracel Islands, also claimed by Vietnam, were forcibly annexed by China. In 1988 the same two countries came to blows over the Spratlys but, in an ASEAN-sponsored declaration, agreed to settle their dispute peaceably. In 1994 Indonesia discovered that China had laid claim to its Natuna gas field, but the Indonesians' request for clarification has not received a response. Moreover, in 1995 the Philippines found Chinese military structures on Mischief Reef, 130 nautical miles (240 km) off the Philippines' Palowa Island and well within its 200 mile (370 km) exclusive economic zone (EEZ).

China's creeping assertiveness in the region has rung alarm bells, since the United States has defence pacts with the Philippines and Thailand, and 70 per cent of all Japan's oil imports pass through the waters now claimed by the

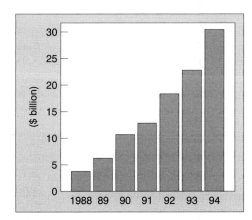

Figure 13.3.10
China's trade surplus with the USA, 1988–1994

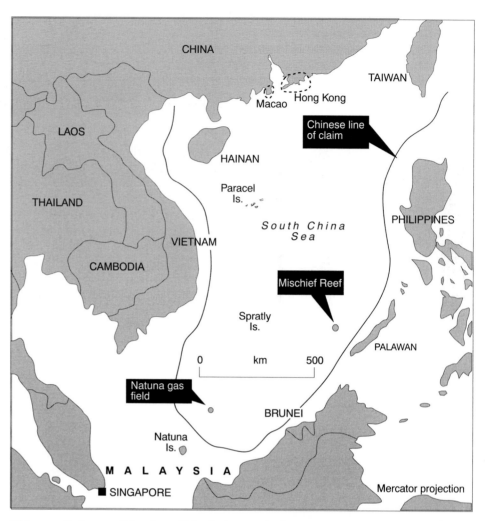

Figure 13.3.11
Chinese territorial claims in the South China Sea

Chinese. It is possible that China's increasingly aggressive stance may force ASEAN members into a collective security arrangement involving the United States, creating a kind of South-East Asian NATO. Already China's southern neighbours have sought assurances that the United States will keep at least 100 000 troops in Asia, compared with the Cold War peak of 135 000. This the Americans have agreed to do, despite being evicted from their military bases in Thailand in the 1970s, and the Philippines in the 1990s.

A volatile region

Even after the collapse of the Soviet Union, and even though most Asian countries nowadays seem more interested in making money than in making war, East Asia remains one of the world's most dangerous regions. Apart from China's territorial ambitions, other potential flashpoints include:

- Korea – North Korea, the last of the hard-line communist states, threatens to become a nuclear power and to invade capitalist South Korea, as it did in 1950, triggering the Korean War. Some 35 000 American troops are still stationed in the South as a token of US determination to resist further aggression. Even so, in 1996 a sizeable North Korean armed incursion had to be repulsed by South Korean forces.
- ASEAN – Unresolved border disputes and ethnic tensions among the organisation's present members have created the world's fastest-growing arms market, funded by their increasing prosperity. This situation may be exacerbated as ASEAN accepts new members, some of whom have been recent enemies.

The United States' principal strategic commitment in the region – to defend Japan – reduces some of the regional tension, since it obviates the need for the Japanese to over-arm themselves and for their neighbours to follow suit. From the military point of view, it is vital that US–Japanese relations remain harmonious; the repercussions of a rift between the two countries could be much more horrendous than the present trade deficit.

QUESTIONS

1 (a) Referring to Figure 13.3.9, how many times wealthier than the Chinese themselves are the inhabitants of Hong Kong and Taiwan?

   (b) Describe and account for the general reaction in these 'lost provinces' to the prospect of reunification with China.

2 Using the information on Figures 13.3.8 and 13.3.10, for the years 1990–1994 draw a composite graph to illustrate the trade surpluses of both China and Japan with the United States. Comment on the pattern revealed.

3 As a specialist in East Asian affairs, you have been commissioned (i) to brief US military chiefs on the present strategic situation in the region; (ii) to advise on how the United States might best maintain the region's peace and stability. Referring to Figure 13.3.11 and the account, outline the report that you would provide, making special reference to Japan and China.

Index

DATE DUE FOR RETURN

This book may be recalled before the above date.